Manufacturing and Production Engineering: Planning and Control

Manufacturing and Production Engineering: Planning and Control

Edited by
Jeff Hansen

Willford Press
www.willfordpress.com

Published by Willford Press,
118-35 Queens Blvd., Suite 400,
Forest Hills, NY 11375, USA

ISBN: 978-1-68285-417-4

Cataloging-in-Publication Data

Manufacturing and production engineering : planning and control / edited by Jeff Hansen.
p. cm.
Includes bibliographical references and index.
ISBN 978-1-68285-417-4
1. Production engineering. 2. Industrial engineering. 3. Production planning.
4. Production control. 5. Production management. I. Hansen, Jeff.
TS176 .M36 2018
658.5--dc23

For information on all Willford Press publications
visit our website at www.willfordpress.com

Willford Press

Printed in China.

Contents

Preface VII

Chapter 1 **Accelerated life tests as an integrated methodology for product design, qualification and production control: A case study in household appliances** 1
M. Tucci, F. De Carlo, O. Borgia and N. Fanciullacci

Chapter 2 **Numerical modeling and optimization of machining duplex stainless steels** 17
Rastee D. Koyee, Siegfried Schmauder, Uwe Heisel and Rocco Eisseler

Chapter 3 **Developing location indicators for Agricultural Service Center: A Delphi-TOPSIS-FAHP approach** 65
Morteza Zangeneh, Asadolah Akram, Peter Nielsen and Alireza Keyhani

Chapter 4 **Development of a predictive model for determining mechanical properties of AA 6061 using regression analysis** 90
Joseph Achebo

Chapter 5 **Inventory classification based on decoupling points** 106
Joakim Wikner and Eva Johansson

Chapter 6 **An extension for dynamic lot-sizing heuristics** 124
Fabian G. Beck, Eric H. Grosse and Ruben Teßmann

Chapter 7 **On decoupling points and decoupling zones** 140
Joakim Wikner

Chapter 8 **Inclusive design drivers and barriers - a manufacturing perspective from Pakistan** 189
Amjad Hussain, Aftab Ahmad and Keith Case

Chapter 9 **Validating quality productivity improvement framework for sponge iron industry in India** 210
Vinod S. Gorantiwar and R. L. Shrivastava

Permissions

List of Contributors

Index

Preface

Manufacturing engineering is a subdiscipline of industrial engineering and intersects with mechanical engineering. It mainly studies the processes and practices involved in manufacturing and production of products. It encapsulates numerous subdisciplines such as mechanics, drafting, computer integrated manufacturing, textile engineering, etc. This book discusses theories and concepts of manufacturing engineering and technology. The extensive content of this book provides the readers with a thorough understanding of the subject. It aims to equip students and experts with the advanced topics and upcoming concepts in this area.

This book is a comprehensive compilation of works of different researchers from varied parts of the world. It includes valuable experiences of the researchers with the sole objective of providing the readers (learners) with a proper knowledge of the concerned field. This book will be beneficial in evoking inspiration and enhancing the knowledge of the interested readers.

In the end, I would like to extend my heartiest thanks to the authors who worked with great determination on their chapters. I also appreciate the publisher's support in the course of the book. I would also like to deeply acknowledge my family who stood by me as a source of inspiration during the project.

Editor

1

Accelerated life tests as an integrated methodology for product design, qualification and production control: a case study in household appliances

M. Tucci*, F. De Carlo, O. Borgia and N. Fanciullacci

IBIS Lab, Department of Industrial Engineering, University of Florence, Viale Morgagni 40, Florence 50134, Italy

Cost reduction and product quality are key factors in the present competitive market. Product reliability is strongly correlated with customer satisfaction. Accelerated life tests (ALTs) represent a methodology able to investigate product reliability performance in a short time with respect to the classical testing methods in the design. The aim of this paper is the proposal and development of an integrated procedure based on ALTs in order to evaluate the reliability performance of a new product; the use of such testing methods is useful to systematically support the design and qualification phases and to help the service demand forecasting before the product launch. The smart choice of the acceleration parameter and the knowledge developed during the upstream phase in the product lifecycle can be useful and cost effective for its utilization in the downstream lifecycle, as an acceptable substitute of specific advanced final inspection methodologies. The methodology has been tested during the development of a new model of washing machine. The application of ALTs was able to stress one of the most important failure modes of the product, returning important validation data for the design phase and qualification phase and giving good expectation for its fruitful utilization for final inspection in production.

Keywords: accelerated life test; reliability performance; product development; final inspection

1. Introduction and literature review

The household appliances market has reached a very high level of competitiveness. The main reasons can be identified as the increase in the number of companies operating in this business area and in the expansion of emerging countries. The concentration of all these companies in the same market sector leads them to invest a lot of resources and time in product research and development (R&D) in order to persuade customers to choose their products. Continuous technological development and increase of capital investments are key requirements to follow customer needs and to keep pace with competitors (Minderhoud, 1999). Because of the competition, companies try to develop products with increasing performance and new features. Moreover, they try to reduce the time to market, to introduce innovation and to avoid losing their competitive advantage over competitors. As customer satisfaction remains the key point for a successful product, one of the main issues of the R&D activities is to design products with a

*Corresponding author. Email: mario.tucci@unifi.it

longer life far beyond the warranty expiration, ensuring long availability and proper functioning of the goods. Reliability and durability (Silvestri, Falcone, De Felice, & Di Bona, 2010) are important quality features of a new product which are, unlike static attributes, measurable on the bench with immediate results; they are dynamic and they must be estimated in two ways. The first approach is the reliability data analysis of similar products, applicable when there are similarities among them. For example, it is possible to estimate the assembly reliability using a reliability block diagram combined with a Weibull analysis (Kim, 2011; Zhai, Lu, Liu, Li, & Vachtsevanos, 2013).

The second way to evaluate reliability is through very long tests. New and innovative product reliability performances are commonly estimated through experimental tests, to simulate environmental life stress. Usually, such tests are time consuming and need dedicated resources. In the industrial setting, furthermore, the approach to these tests is not rigorous and scientifically robust. As a consequence, most of them don't provide the expected results and are not cost effective (O'Connor, 2001). Overlooked or poorly designed testing phases can lead to a product of poor quality, because some unapparent failure modes could go undetected (Yadav, Singh, & Goel, 2006).

In the household appliances field, and especially for washing machines, R&D achieved a very high level of reliability as the standard. Therefore, testing procedures of products have become very lengthy, since these are not prone to fail, while manufacturing companies need to estimate reliability performance more rapidly under the pressure of shorter time to market. Traditional reliability performance tests are no longer efficient enough and solutions are needed to accelerate them while not losing significance and correspondence with user experience. Already in the 70s, Nelson (1974) introduced the theory of accelerated life tests (ALTs) and showed its validity. ALTs are a category of test methodology able to assess reliability performance of a product or component in a short time, through the use of stress condition higher than the that in real time (Nelson, 2004). These tests are based on the hypothesis that there must be a relationship between the stress level and the product life. Hence, starting from a parametric form of this relation, a series of tests are performed to experimentally estimate it. Escobar and Meeker (2006) presented a review of ALTs and of their various applications. They are grouped based on: methods of acceleration, type and number of overstresses and life–stress relations. Several studies regarding different issues of ALTs are available in literature: Meeker (1984) compares compromise test plan and optimum test plan, Nelson (1980) used a step-stress procedure to further accelerate tests, while Fard and Li (2009) determined the optimum duration for each step. Yang (1994) deals with test durations using different censoring times for each level of overstress while Zhu and Elsayed (2013) propose an approach to ALT design with multiple stresses depending on different objectives and constraints. Yang and Zaghati (2006) increase the usage rate of a compact relay to accelerate its failure mechanism. ALTs can be used also to predict product field reliability using a relationship between ALT results and field returns as shown by Meeker, Escobar, and Hong (2008). There are also some ALTs' pitfalls, investigated by Meeker and Escobar (1998).

According to the presented studies, ALTs are typically implemented in the product lifecycle after the design phase and they integrate with the other production stages, as shown in Figure 1.

The diagram reported in Figure 1 shows the main stages that a new product goes through before it is put into production. ALTs are usually carried out after the final design step. The results of the experiments are mainly used for the finalization of the

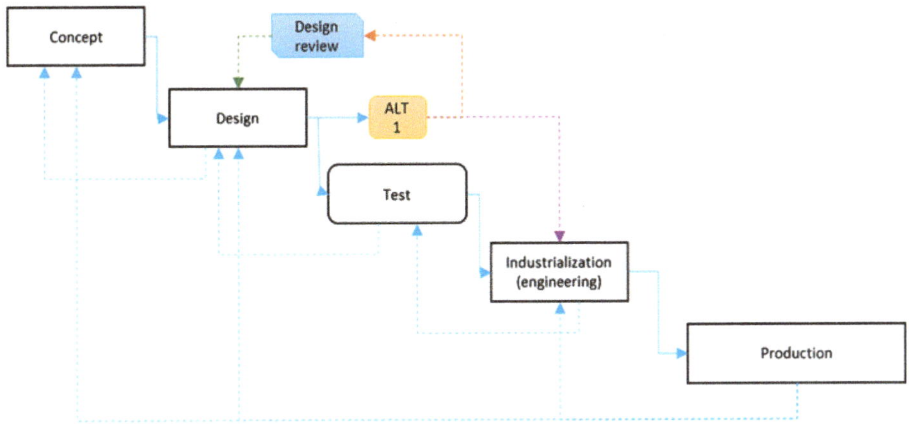

Figure 1. Integration of ALTs in the product design and development stages. The solid lines indicate a material flow, while the dashed lines represent the flow of information. ALT1 indicates the traditional application of accelerated testing.

design, highlighting the critical issues that the normal tests fail to reveal. The experimental evidence can also be useful in the production stage, going hand in hand with the usual tests on prototypes.

Although a lot has been written in literature on ALTs, at the moment they do not represent the standard test methodology when long-term product reliability must be investigated. Even though ALTs are more efficient than the traditional tests, they involve risks and technical difficulties. In fact, they need a lot of resources for design and implementation; this often worries the producer who sees in ALTs a risky methodology, also because of the high cost and for the chance of getting wrong results. Actually, to obtain correct and accurate results, the analyst must thoroughly investigate each interesting failure mode and pay attention on the sample size and on the number and type of overstresses applied. The bigger the sample size, the more accurate the results are. On the other hand, wider sample sizes would increase the costs, since more items and operators should be involved and the total test time will be greater.

The definition of the reliability performance must not be limited only to the design phase, but it should be updated even during production. In fact, during its life cycle, a product might be subjected to productive modifications (suppliers change, production layout modification, assembly sequence change, etc.). All these specific factors could bring the process out of control affecting product reliability. Thus they must be discovered, investigated and revised in order to elaborate corrective actions and to bring the process back under control. Some authors faced these problems: Meeker and Hamada (1995) presented in a tutorial the most suitable statistical tools for each phase of product lifecycle, pointing out how ALT is mainly suitable for the upstream phase (namely, concept, design, test and industrialization) and only for unavoidable degradation failures and known infant mortality, while other methodologies are able to capture unexpected infant mortality, known accidental failures or unexpected accidental failures. For the latter, the early field data and the warranty return data are obviously important, but if the company waits until the availability of these data, that means the economical and image damage was already done. Hobbs (2000) proposes an ad hoc methodology, the highly accelerated stress screens (HASS) just for such kind of failures both in the design and production phases.

Recently, Isaic-Maniu and VodĂ (2009) reviewed the reliability sampling plans to be used as final inspection of the production, developing some specific applications of ISO 2859 sampling plans to survival performance on items produced and inspected by means of accelerated tests. Hence, as far as we know, we can say that in the normal practice white goods industry does not employ such advanced methodologies.

With so much effort devoted to the development and tuning of the ALTs for a new product, it appears convenient to try to extend their deployment also to the control of the production process. Limiting their use only to the reliability performance assessment could be a mistake, while they could be suitable to make other kind of defects emerge, such as production problems, defective quality controls and so on (O'Connor & Kleyner, 2011; Raheja & Gullo, 2012). To this extent, design ALTs are not the most appropriate tool in themselves, but they can become useful if the choice of the accelerated parameters is modified according to those employed in HASS (Hobbs, 2000): vibrations and temperature variations.

The aim of this study is to increase the efficiency of ALTs to support the production stage, choosing conveniently the accelerated parameters, so to produce vibrations and temperature increase. In this paper a new methodological approach to ALTs is proposed, which essentially consists of applying them to three steps of the product life: testing, engineering and production. In particular, ALTs will be used not only as presented in Figure 1 (where these tests are useful for a limited period of the product lifecycle) but also during production and even in the wider after-sale phase (service, warranty, warehouse management, etc.).

As a case study, the reliability performances of a domestic washing machine were investigated using ALTs during the production phase.

The remainder of the present paper is as follows: after a description of the methodology in the methods section, the following results section presents a brief introduction on the case study, the experimental procedure adopted and the experimental outcomes. The discussion section follows, analysing the main achievements and showing a further possible area for ALTs deployment. Finally, some conclusions and interesting future developments are shown with a suggestion of a new possible way to use ALTs together with final inspections for statistical quality control (SQC).

2. Methods

ALT is a methodology based on the assumption that there is a relationship between the stress level and product life. The modelling function depends on two main factors: the component type under analysis and the stress type. Typically, the lifetime of a mechanical component is affected by operating parameters such as temperature, vibrations, environmental conditions and load. Literature review suggests the 'Inverse Power Law' model as the most suitable and flexible model for mechanical components if the stress is non-thermal. Escobar and Meeker (2006), Yang (2007) indicate the Inverse Power Law as the most suitable relationship that describes the dependence of the washing machine life L on mechanical stress.

$$L = \frac{A}{S^B} \tag{1}$$

where L is the nominal life, S is the stress (in this case a mechanical load) while A and B are constants, dependent on material properties, product design, etc.

The independent variable chosen for the reliability model was the number of standard washing cycles, even if in the following formulas it is referred to as t, recalling its link to the lifetime. As it will be explained in the next section, the tests were 'Type I censoring' with a censoring time fixed at 500 cycles. The estimation of the reliability function at the user level, obtained from the overstress values, was carried out with the software ALTA© (ReliaSoft Corporation, Tucson, Arizona, USA). To assess the reliability of the machines at the user level (normal operating conditions), it was necessary to find the probability function that best represents the data collected at each level of overstress; in this case, a two-parameters Weibull function was chosen. The probability density function (p.d.f.) is shown in Equation (2):

$$f(t) = \frac{\beta}{\alpha^{\beta}} t^{\beta-1} e^{\left[\left(\frac{t}{\alpha}\right)^{\beta}\right]}, \quad t > 0 \tag{2}$$

where α is the scale parameter, also called characteristic life (it is the life at which 63.2% of the population will be failed) and β is the shape parameter. When $\beta > 1$ ($\beta < 1$), the distribution has an increasing (decreasing) hazard rate, while for $\beta = 1$ it has a constant hazard rate.

The probability that a washing machine will fail during an interval [0, t] is defined by the cumulative distribution function $F(t)$. The latter is obtained by integrating the p.d.f. within the interval [0, t]:

$$F(t) = \int_0^t f(t)dt = 1 - e^{-\left[\left(\frac{t}{\alpha}\right)^{\beta}\right]} \tag{3}$$

As a consequence, the reliability function $R(t)$, derived by the cumulative function, is given by:

$$R(t) = 1 - F(t) = e^{-\left[\left(\frac{t}{\alpha}\right)^{\beta}\right]} \tag{4}$$

In some cases, when performing ALTs, we can define a useful parameter called accelerating factor. It is defined as (5):

$$A_f = \frac{t_p}{t'_p} \tag{5}$$

where t_p is the 100th percentile of the life distribution at user stress level and t'_p is the 100th percentile of the life distribution at the overstress level. Using A_f ,it is possible to calculate user-level reliability performance like test hours (cycles), life percentiles, etc. from those obtained with higher stress levels. For example, known the test hours with high stress level, A_f can be used to obtain the equivalent test hours at the user stress level. The choice of an acceleration factor is based on the hypothesis that the reliability function for all stress levels has the same shape parameter; because we are using a Weibull function, it means that β must be the same for all the stress levels. As a consequence, once that β is calculated for a certain level of stress, the scale parameter of the Weibull function for each stress level is the only missing parameter to be estimated.

The two Weibull parameters (α and β) were calculated using the maximum likelihood estimation (MLE) as it is the most suitable method with several censored data (Nelson & Meeker, 1978). The chosen unilateral lower confidence interval is 97%.

The reliability function at user stress level was obtained using data from tests implemented with three different mechanical overstress levels. As previously stated, to find the relationship between life and stress, the Inverse Power Law correlation function (1) was used.

In this case, the acceleration factor between two stress levels is,

$$A_f = \left(\frac{S'}{S}\right)^B \tag{6}$$

where S' is the highest stress level.

3. Results

As already mentioned in the introduction, the aim of the present study is to show how ALTs can support the industrialization phase, providing useful information about the long-term product reliability. These results are critical, for example, for a more correct initial sizing of inventories, for the prediction of the extents of the after-sales service, for understanding the issue of service calls management under warranty. The possible representation of the interaction of ALTs with the life cycle of the product is shown in Figure 2.

Figure 2 shows how this new ALTs application allows to interact directly on the production process, as a feedback. The comparison between the results of ALT1 and ALT2 allows you to assess the long-term performance of production with those foreseen in the design. For any discrepancy, you will be able to undertake a control and corrective action.

In addition, the predictions of the ALT2 can be compared with the actual performance of the product, rising from the field. So you will be able to promptly detect abnormalities and you might intervene with measures at different levels: from design to engineering, even up to the redesign of ALTs, if problems arise in the experimental phase of the accelerated testing.

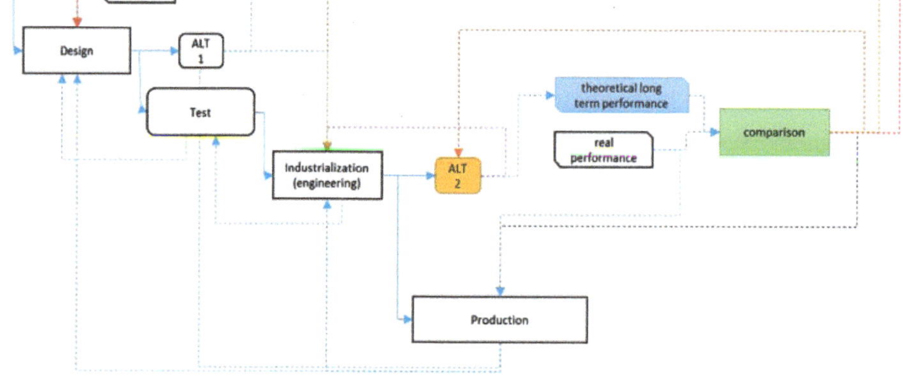

Figure 2. Possible interactions between industrialization ALTs and the other phases of product development. The second application of ALTs (ALT2) can have many interactions (dashed lines) with the other stages of development, making it possible to improve the process as a whole.

3.1. Case study

To verify the goodness of fit of ALTs as a greater support to the design, engineering and production steps, we applied them to a real case, to sustain the process of developing a new model of washing machine. The manufacturer decided to launch onto the market a new model developed on the basis of an existing top product with important innovations regarding the enhancement of the maximum drum capacity without altering the case size. This new specification had some relevant consequences since it involved the redesign of one of the most important components of the product: the oscillating system.

Reliability performance of the unchanged components was previously studied by the manufacturer, and a lot of information coming from warranty data analysis was available. So the real challenge was to assess the reliability performance of the new oscillating system during the industrialization phase, but the classical tests were unable to highlight any sources of criticality.

The manufacturer was interested in this issue for several reasons. First of all there was the need to validate the design and industrialization steps through an experimental approach in order to discover the real limit of the new solution and, if necessary, redesign the new component.

Another important goal was to adopt a methodology in order to keep the quality of the manufacturing process under control. Line set-up, suppliers changes, process capability demonstrated in the past to be important issues that could influence the company's product reliability. Therefore, some deviations or drifts should be evaluated as quickly as possible in order to avoid consequences on customer satisfaction due to any increase in faults. Moreover, an early and fast reliability performance assessment can be always very useful in the viewpoint of warranty analysis and service demand forecasting and planning.

The application of ALT to this case study is not trivial because, according to international scientific publications, no study was published regarding the specific application of ALTs to the oscillating system of a washing machine. Only a paper was presented by Park, Park, Kim, and Cho (2006) regarding the implementation of ALTs on the drain pump of a washing machine. Nevertheless, the research expertise of the authors in the context of accelerated degradation and testing as in De Carlo, Borgia, and Tucci (2014) and the comparison between in-house and field reliability have allowed us to design and fulfil the tests, see Borgia, De Carlo, Tucci, and Fanciullacci (2013). Also the ability to extend the applications of known methods and to apply new approaches to existing methods falls within the competence of the authors, as shown in De Carlo, Borgia, and Tucci (2011).

3.2. Experimental procedure

As said, the most important role for the identification of the product reliability performance using ALT is played by the testing phase. Before performing tests, it is fundamental to plan and determine the characteristic of the sample and the stressscheme. Planning activities were performed according to the specifications, the budget and the time constraints imposed by the manufacturer. These led to: (1) the choice of the component to be tested; (2) the failure mode to be stressed and the different levels of overstress; and (3) the test length and the number of samples tested at each level.

ALTs focused on the new mechanical oscillating group (formed by the tub and the drum) being the main innovation introduced in the new model, with an unknown behaviour. The new mechanical group was bigger than the previous one, in order to contain a larger quantity of clothes. The improvements were: a new material, a decreased thickness of the drum and of the tub and several new reinforcement ribs. The design changes implied also a lower gap between the tub and the drum both in static and dynamic conditions. It is important to analyse the latter parameter because it influences the probability that the two components collide, causing a failure: the wider the distance between the two components, the lower the probability that the drum comes into attrition with the tub. According to the root cause analysis of the field repairs, the contact between these two components is one of the most common failure modes. Hence, we decided to perform tests with only one overstressing parameter. The mechanical overstress chosen was a load imbalance placed inside the drum, since it intensifies the drum deformation and its attitude to get in contact with the tub. It also overloads the bearing group, producing heat and possibly deforming its polymeric housing.

The drum imbalance is normally regulated by the electronic control unit that stops the washing cycle if it exceeds a threshold value decided by the manufacturer. The maximum user-allowable imbalance was the baseline. Three equally spaced levels of overstress for testing the washing machines were elected: 165, 200 and 237% of the baseline. These values came from the results of a previous FEM analysis and were confirmed by the designer of the assembly.

The three overstress levels were applied during a programmed washing cycle that, after a marketing survey, was identified to be the most common washing cycle performed by the standard user. It consists of two phases: 90 min of low speed cycles and 20 min of spin cycle. The experimental measures acquisition was carried out only during the spin phase, when the group reaches its most severe dynamic conditions (maximum rotating speed).

It should be added that, according to the results of some preliminary tests, it was found that the usage rate does not affect the failure mode considered in this study. Hence, cycles carried out at low speed can be neglected without altering the results.

For each level of overstress, eight washing machines were used. The sample sizes were chosen accordingly to the planning sets proposed by Yang (1994, 2007), and Yang and Jin (1994) with the 'compromise test plans' and to the requirement of the company in terms of resources and time dedicated, while overstress levels were imposed by designer knowledge. With the 'compromise test plan' methodology, test variables were chosen to minimize the asymptotic variance of the estimate of a life percentile at the user stress level. Given as an input the higher stress level, the censoring time (500 cycles), the total number of machine to be tested (24) and the shape (β) and scale (α) parameter at the user and at the highest stress level (derived from historical data), we could calculate the asymptotic variance of the MLE of the mean life of the Weibull distribution at the user stress:

$$\mathrm{Var} = \frac{\sigma^2}{n} V \tag{7}$$

where $\sigma = 1/\beta$, n is the total number of units tested and V is the standardized variance defined by Yang and Jin (1994). Hence, because σ and n are predetermined, we can calculate the lower and the intermediate stress levels and the number of units to be allocated at each stress level by minimizing V. Yang (2007), using a numerical method, presents a table where, for various sets of input parameters, the optimum value of V is calculated.

The sample sizes obtained mathematically were also dynamically compared with the experimental evidences obtained with previous tests.

Tests were designed as right censored by imposing 500 cycles as censor 'time', which corresponds to an average of 2 years of use. For each test, every 50 cycles, a cycle called 'characterization' was performed. During this cycle, the load was set at the user level; afterwards, the machine was loaded again with the imbalance already established. The aim of the characterization cycle will be explained in the next paragraph.

3.3. Experimental set-up

A laboratory with three positions for the simultaneous testing of washing machines was set up in order to implement ALTs. In each test station, the washing machines were put into metal cages for safety reasons. In fact, in case of contact between the drum and the tub, especially with high imbalances, the drum coupling could break and could be thrown outside the washing machine by its inertia, with severe risks for the safety of workers.

During the tests, three different types of sensors were set on each washing machine in order to monitor those parameters useful for a following degradation analysis. This would be performed using also the data acquired during the 'characterization' cycle.

The measurement system consisted of three types of sensors, each one with its hardware and software acquisition equipment. To measure the drum–tub gap, four sensors were used: three longitudinally placed on the top of the tub, and the fourth at the back side.

Vibrations were measured with three sensors which monitored the acceleration along the axes of a Cartesian tern with origin on the top of the tank, z-axis vertically upwards, x-axis parallel to the axis of the drum and directed anteriorly. The vibration acquisition system consisted of a data logger with signal-processing capabilities connected to the three accelerometers.

The bearing temperature was measured by two thermocouples in contact with the bearing. The two values were then averaged to obtain a more reliable value.

Test activities were preceded by a calibration phase in which the measuring instruments and their software were set to ensure a proper operation. Finally, the data obtained from experimental tests were appropriately post-processed in order to make them suited for an eventual degradation analysis.

3.4. Experimental outcomes

ALTs were performed on 24 machines equally distributed over the three overstress levels according to the standard methodology. The first important outcome was the identification of a new (hidden) failure mode (in addition to those already known by the not accelerated tests and experience) that never occurred before. It involved the new oscillating system and it occurred seven times out of a total of eight failures. Although this fact is not a proof of the equivalence of ALT and HASS, it is an empirical evidence of some similarities, at least for industrial product with significant mechanical components, where the vibration and heat produced by the functioning of the product itself can be used as stress factors for other components, even revealing hidden failure modes.

The results obtained from the three overstress levels are shown in Table 1; next to the number of cycles performed by the machine, in brackets, it is indicated if the test finished without any failure (C – censored) or if a failure occurred (F – failed).

Table 1. Outcomes of the 24 tests (censored C, failure F). The test is right censored at 500 cycles (corresponding to 2 years of use).

Machine	Load imbalance 65%	Load imbalance 80%	Load imbalance 95%
1	500 (C)	500 (C)	50 (F)
2	500 (C)	500 (C)	350 (F)
3	500 (C)	500 (C)	250 (F)
4	500 (C)	500 (C)	300 (F)
5	500 (C)	350 (F)	300 (F)
6	500 (C)	500 (C)	350 (F)
7	500 (C)	500 (C)	500 (C)
8	500 (C)	500 (C)	400 (F)

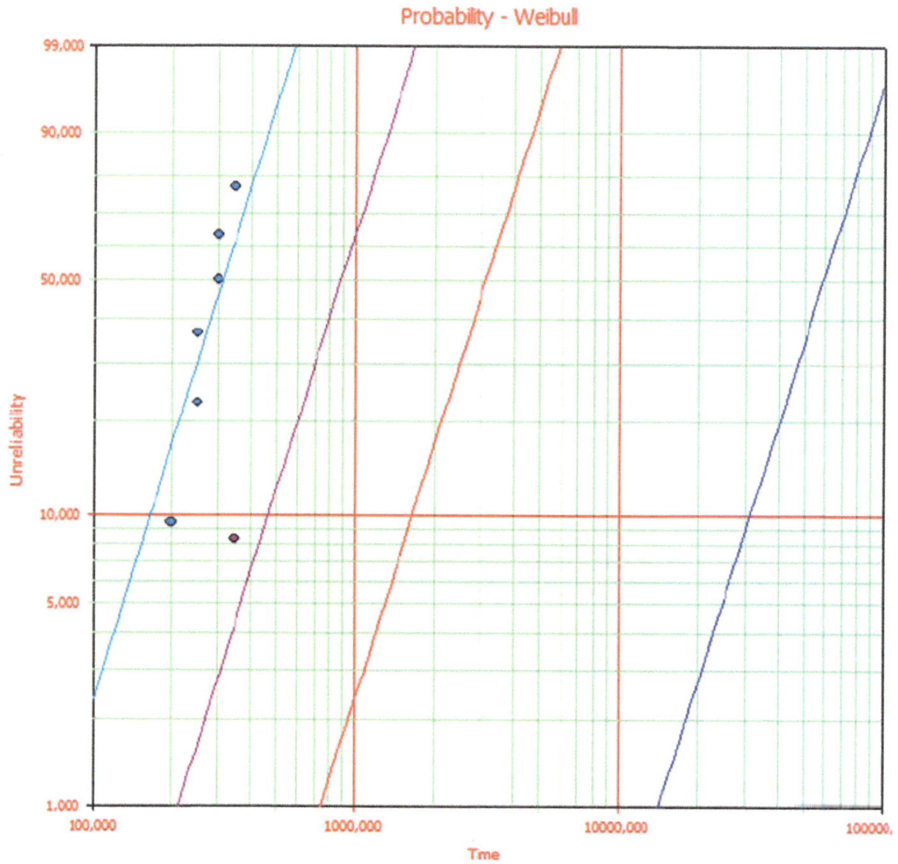

Figure 3. Weibull chart of the four life data distribution. From the left to the right you can see 237, 200, 165 and 100% distributions. Parallel lines indicate that the distributions have the same shape factor.

Data represented in Table 1 are used to estimate the reliability distribution at the user stress level. As already mentioned, a two-parameters Weibull distribution was used to fit the life data obtained during testing activities at the three overstress levels.

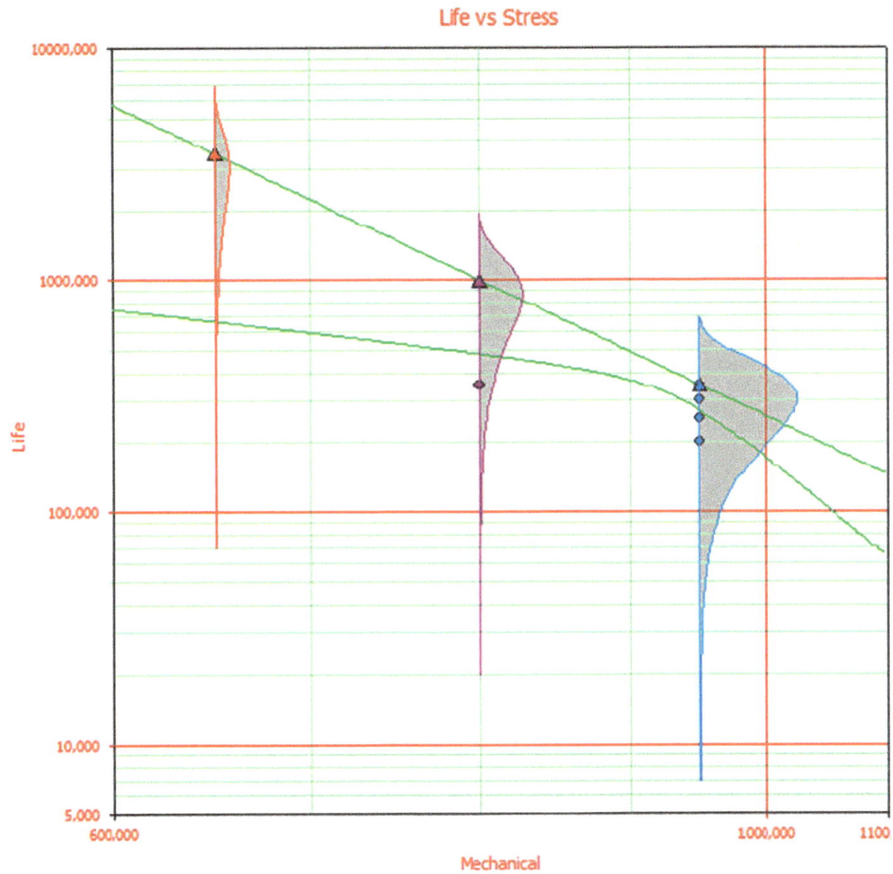

Figure 4. Correlation between the three different levels of overstress (abscissa) and the component life (ordinate) with the 97% lower confidence interval of the frequency distributions.

Based on the maximum likelihood method, the shape parameter β was calculated. Its value is 2.97 and it is equal in every distribution at all levels of overstress (165, 200 and 237%). The assumption of constant shape parameter is evident from the graphical representation of the four straight parallel lines on the Weibull chart (Figure 3).

Hence, the reliability at the user stress level follows a Weibull distribution with $\beta = 2.97$ and $\alpha = 53{,}166$ cycles. This distribution is derived from the correlation between life and mechanical stress obtained by the inverse power law (Figure 4).

Fundamental for service planning were the reliability performances at 500 and 1250 cycles; in fact, 500 cycles correspond to two years of standard usage – the normal warranty period – while 1250 cycles represent an extended warranty (5 years of usage). As regards these values, the estimated reliability performance was very close to one. It is important to underline that the values found were the average ones; as a precautionary measure, we have adopted the lower limit of the confidence interval, because it indicates the pessimistic reliability values for each number of cycle. Hence with the unilateral confidence level chosen (97%) there was a lower reliability of 99.2% for 500 cycles and 89.78% for 1250 cycles (Figure 5). Furthermore the user level was set to maximum

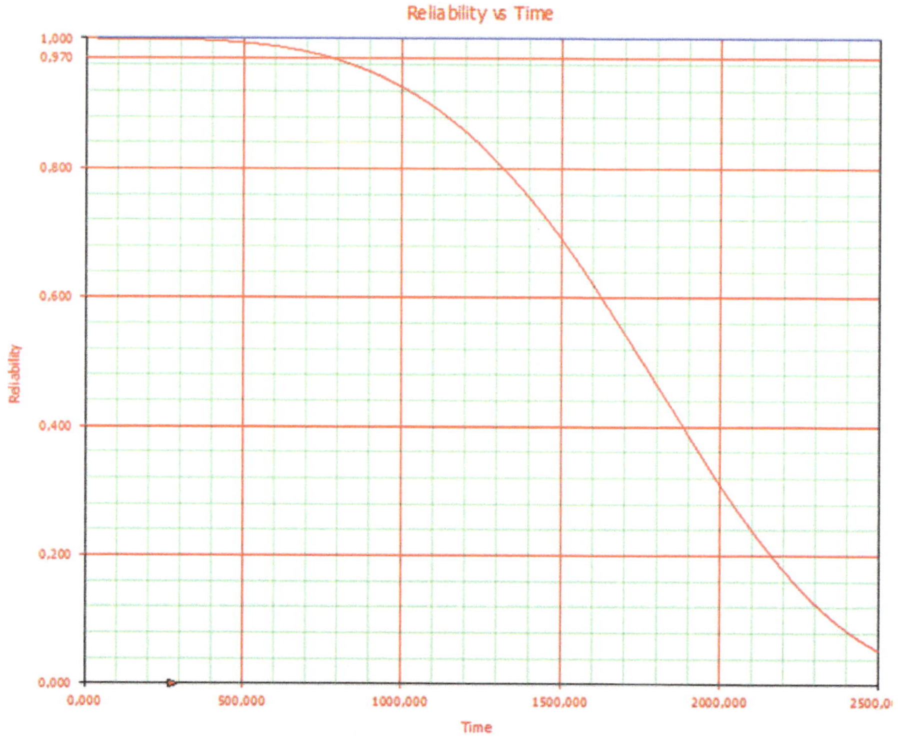

Figure 5. Lower 97% confidence interval limit of the estimated reliability function at the user stress level.

unbalance allowed by the control unit, which is not the average unbalance that loads the product during its life.

The value of 97% was chosen because it is the best reliability performance level to obtain a good trade-off between production costs and service costs. In accordance with the manufacturer, reliability performance specification was then used to calculate the number of cycles that ensures a reliability performance of 97%: 800 cycles that correspond to 3 years and 2 months. The mean life was equal to 1248 cycles (about 5 years).

As a further elaboration of the test results, it was possible to obtain the reliability surface as a function of the stress level and of the number of the performed cycles (Figure 6).

4. Discussion

The aim of the study was to develop a test methodology able to simultaneously identify the reliability performances of a domestic washing machine and to use it to sustain the industrialization phase. This procedure was carried out fulfilling deadlines and tight budget constraints.

By applying ALTs in an innovative manner (after the engineering phase and in parallel to the production), the experimental results identified some critical events related to the production phase of the new mechanical assembly. Satisfying the test duration requirements, ALTs gave results in less than 30% of the traditional test time. During the

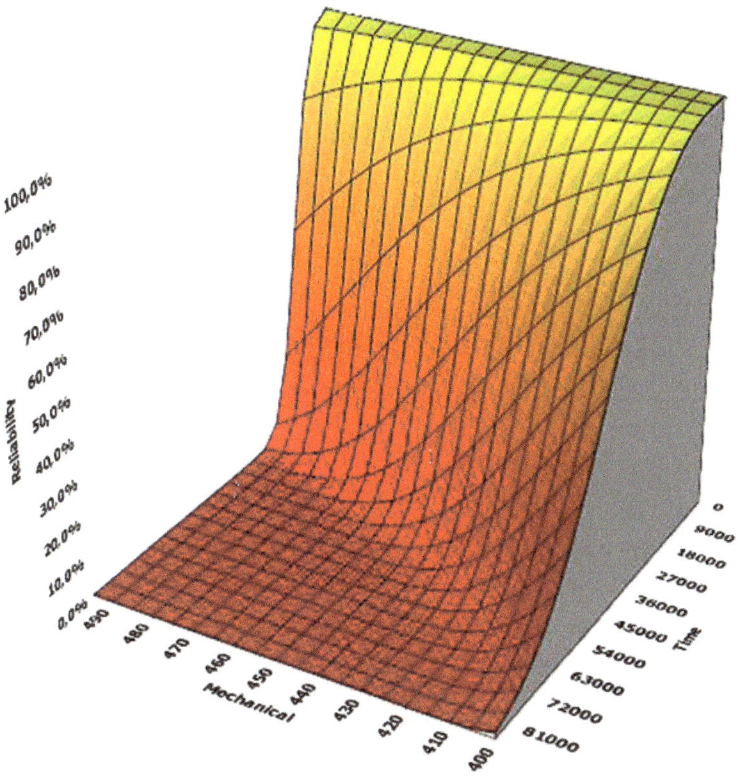

Figure 6. Reliability surface. In a three-dimensional space, the reliability values depend on the number of cycles performed and on the level of stress applied. By this graphical representation, it was simple to assess the reliability performance of the washing machine defining the stress level and the number of cycles.

FMEA analysis, particular attention was dedicated to determinate a realistic overstress that could lead to a truthful failure mode (Lu, Loh, Brombacher, & Ouden, 2000). In fact, although imbalance values that cause failures were higher than the maximum value that was normally allowed, their systematic occurrence should led the manufacturer to well investigate the causes of such phenomena. The failure mode found was catastrophic because it could stop the functioning of the washing machine and cause severe damages. Although reliability performance was not alarming, they were useful to give indications to production managers and designers about the root cause of such a problem which was identified in a supplier change.

ALTs have, therefore, demonstrated their potential in identifying production problems, but might they also be integrated with field data? As a matter of fact, it may happen that the reliability values obtained through the life test performed in the laboratories inside the company (in-house) are different from those that come from field (estimated using the service data). This difference can be due to the fact that ALTs were designed incorrectly because they do not exactly represent the actual product use conditions. Comparing the in-field and the in-house reliability data could give a quick validation about the goodness of the test design that took place in the early product stages, giving the opportunity to the company for continual improvement of these tools along the lifecycle of the product.

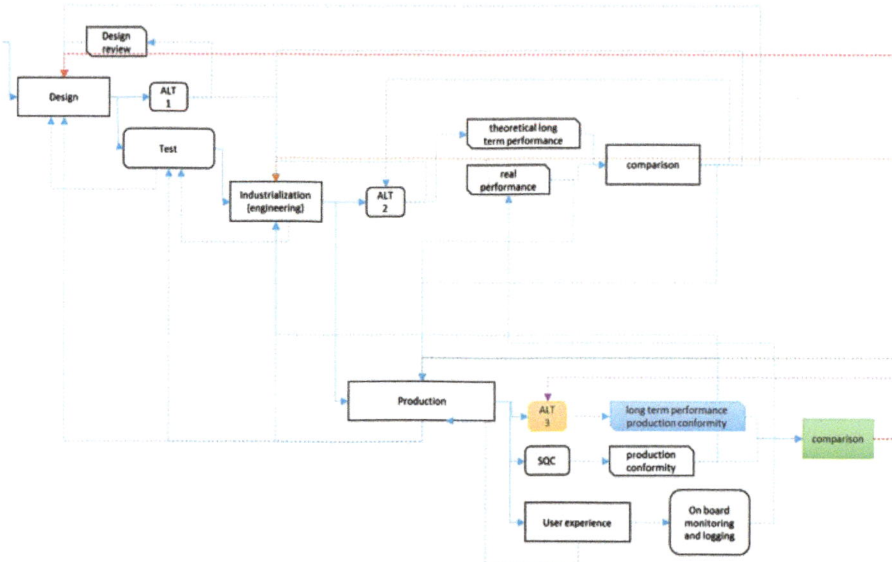

Figure 7. A possible third application of ALTs (ALT3). In the figure is shown how the experimental results of accelerated tests can be integrated in to the SQC to predict the long-term performance of the goods produced.

The favourable results and the demonstrated ability of these tests to reveal hidden failure modes as well, prompted us to think that ALTs could be a tool to be used even after the industrialization phase. Their results could be fundamental for the ultimate estimation of the optimal warranty period and for the full sizing of the supply system involved with the service. If the warranty period is known, the manufacturer could plan and manage all the activities necessary for the service, reducing costs while ensuring a high quality of the service offered. Consider, for example, the ability to plan the production and distribution of all the spare parts and their stocks or the possibility to extend the warranty period.

To sum up, in the previous paragraphs we introduced the typical use of ALTs, which reflects the ideas on which they were originally developed (Figure 1) and the use of ALTs as they have been implemented in the present work by the authors (Figure 2). The results of the study are very encouraging and have also suggested to the authors to propose an additional application area of ALTs to support industrial production. This third mode includes the steps of production and of use by end-users, as shown in Figure 7.

The idea behind this application is that the accelerated testing provides an insight into the long-term performance of the goods under investigation. This investigation, usually made on prototypes or first series' items, can be also extended to the production stage, in which it could stand beside the widespread techniques of SQC. This deployment is feasible because of the small sample size, compatible with the nature of a destructive test performed on production batches; it is also effective because they are able to reveal hidden failure modes, differently from traditional SQC in final testing. Finally, differently from the field data returned analysis, the production ALTs (ALT3 in Figure 7) make it possible to intervene in the production more promptly. In this way the manufacturer is able to proactively develop and maintain the production under control, contrasting negative production trends. As a further implementation, if the product is equipped with a system of on-board data logging and monitoring, the information downloaded from the assets during technical assistance may provide a benchmark for comparing the ALTs results. Ultimately, the possibilities of integration of ALTs during the phases of design, engineering and production are many and allow you to fully support each of these phases.

The use of ALTs as a tool to be applied in parallel with SQC is still in progress in the case study, as the production of the new model of washing machine has just started yet. The experience done in the design phase and the hidden defects discovered and corrected, albeit not related to the failure mode studied, gives an excellent outlook on its effectiveness for the planned downstream utilization.

ALTs showed their potential but also highlighted some critical issues and limitations. For example it is essential to have a dedicated laboratory for the experimental measurements and to have scrupulously investigated the aspects of all the possible failure modes (e.g. by means of an FMEA). After finding all possible failure modes, before running ALTs, you need to identify the most frequent or catastrophic ones through a deep Pareto analysis. It should analyse and classify all the data coming from the service reports. It is therefore clear which enormous effort must be taken, even before the execution of the tests. A great work across all the departments involved in the product development (design, materials, structural, etc.) is needed. Consequently, a considerable amount of resources and time is necessary to develop the preparation phase of the tests, in addition to their execution. Companies that have invested in this direction will be able to easily overcome the problems related to implementation and evaluation of reliability performance such as determination of threshold values and the choice of parameters. The results obtained have wide confidence intervals; it would be interesting to increase sample sizes for the three overstress levels in order to obtain more confident outcomes. In particular, it would be interesting to increase the number of machines tested at the lowest level of stress (165%). Another alternative could be the implementation of unequally spaced overstress levels, increasing only the lower stress level; in this way we could obtain more failure data without incrementing the number of washing machines tested.

A future development of the case study will be to characterize the standard functioning of the washing machine using the data acquired with the sensors and logged in the control unit; these can be also used to compare, at any time, the standard performance with those of machines just produced. Any difference found will be the issue of investigations aimed at finding non-compliances of production process.

References

Borgia, O., De Carlo, F., Tucci, M., & Fanciullacci, N. (2013). Service demand forecasting through the systemability model: A case study. *International Journal of Engineering & Technology, 5*, 4312–4319. ISSN:0975-4024.

De Carlo, F., Borgia, O., & Tucci, M. (2011). Risk-based inspections enhanced with Bayesian networks. *Proceedings of the Institution of Mechanical Engineers, Part O: Journal of Risk and Reliability, 225*, 375–386.

De Carlo, F., Borgia, O., & Tucci, M. (2014). Accelerated degradation tests for reliability estimation of a new product: A case study for washing machines. *Proceedings of the Institution of Mechanical Engineers, Part O: Journal of Risk and Reliability*, in press.

Escobar, L. A., & Meeker, W. Q. (2006). A review of accelerated test models. *Statistical Science, 21*, 552–577.

Fard, N., & Li, C. (2009). Optimal simple step stress accelerated life test design for reliability prediction. *Journal of Statistical Planning and Inference, 139*, 1799–1808.

Hobbs, G. K. (2000). *Accelerated reliability engineering: HALT and HASS*. Chichester: Wiley.

Isaic-Maniu, A., & VodĂ, V. G. (2009). Reliability sampling plans: A review and some new results. *Journal of Applied Quantitative Methods, 4*, 190–201.

Kim, M. C. (2011). Reliability block diagram with general gates and its application to system reliability analysis. *Annals of Nuclear Energy, 38*, 2456–2461.

Lu, Y., Loh, H. T., Brombacher, A. C., & Ouden, E. (2000). Accelerated stress testing in a time-driven product development process. *International Journal of Production Economics, 67*, 17–26.

Meeker, W. Q. (1984). A comparison of accelerated life test plans for Weibull and lognormal distributions and type I censoring. *Technometrics, 26*, 157–171.

Meeker, W. Q., & Escobar, L. A. (1998). Pitfalls of accelerated testing. *IEEE Transactions on Reliability, 47*, 114–118.

Meeker, W. Q., Escobar, L. A., & Hong, Y. (2008). Using accelerated life tests results to predict product field reliability. *Technometrics, 51*, 146–161.

Meeker, W. Q., & Hamada, M. (1995). Statistical tools for the rapid development and evaluation of high-reliability products. *IEEE Transactions on Reliability*, 44, 187–198.

Minderhoud, S. (1999). Quality and reliability in product creation – Extending the traditional approach. *Quality and Reliability Engineering International, 15*, 417–425.

Nelson, W. B. (1974). A survey of methods for planning and analyzing accelerated tests. *IEEE Transactions on Electrical Insulation, EI-9*, 12–18.

Nelson, W. B. (1980). Accelerated life testing – Step-stress models and data analyses. *IEEE Transactions on Reliability, R-29*, 103–108.

Nelson, W. B. (2004). *Accelerated testing: Statistical models, test plans, and data analyses*. Hoboken, NJ: Wiley-Interscience. Retrieved from www.wiley.com

Nelson, W. B., & Meeker, W. Q. (1978). Theory for optimum accelerated censored life tests for Weibull and extreme value distributions. *Technometrics, 20*, 171–177.

O'Connor, P. D. T. (2001). *Test engineering: A concise guide to cost-effective design, development, and manufacture*. Chichester: Wiley.

O'Connor, P., & Kleyner, A. (2011). *Practical reliability engineering*. Chichester: Wiley.

Park, S. J., Park, S. D., Kim, K. S., & Cho, J. H. (2006). Reliability evaluation for the pump assembly using an accelerated test. *International Journal of Pressure Vessels and Piping, 83*, 283–286.

Raheja, D. G., & Gullo, L. J. (2012). *Design for reliability*. Hoboken, NJ: Wiley.

Silvestri, A., Falcone, D., De Felice, F., & Di Bona, G. (2010). New reliability allocation methodology: The integrated factors method. *International Journal of Operations and Quantitative Management, 16*, 67–85. ISSN:1082-1910.

Yadav, O. P., Singh, N., & Goel, P. S. (2006). Reliability demonstration test planning: A three dimensional consideration. *Reliability Engineering & System Safety, 91*, 882–893.

Yang, G. (1994). Optimum constant stress accelerated life test plans. *IEEE Transactions on Reliability, 43*, 575–581.

Yang, G. (2007). *Life cycle reliability engineering*. Hoboken, NJ: Wiley.

Yang, G., & Jin, L. (1994). Best compromise test plans for Weibull distributions with different censoring times. *Quality and Reliability Engineering International, 10*, 411–415.

Yang, G., & Zaghati, Z. (2006). *Accelerated life tests at higher usage rates: A case study*. In *Proceedings of the Reliability and Maintainability Symposium* (pp. 313–317). Philadelphia, PA: IEEE.

Zhai, L.-Y., Lu, W.-F., Liu, Y., Li, X., & Vachtsevanos, G. (2013). Analysis of time-to-failure data with Weibull model in product life cycle management. In A. Y. C. Nee, B. Song, & S.-K. Ong (Eds.), *Re-engineering manufacturing for sustainability* (pp. 699–703). Singapore: Springer.

Zhu, Y., & Elsayed, E. A. (2013). Design of accelerated life testing plans under multiple stresses. *Naval Research Logistics (NRL), 60*, 468–478.

2

Numerical modeling and optimization of machining duplex stainless steels

Rastee D. Koyee[a]*, Siegfried Schmauder[b], Uwe Heisel[a] and Rocco Eisseler[a]

[a]*Institute for Machine Tools, University of Stuttgart, Holzgartenstr. 17, D-70174 Stuttgart, Germany;* [b]*IMWF, University of Stuttgart, Pfaffenwaldring 32, D-70569 Stuttgart, Germany*

The shortcomings of the machining analytical and empirical models in combination with the industry demands have to be fulfilled. A three-dimensional finite element modeling (FEM) introduces an attractive alternative to bridge the gap between pure empirical and fundamental scientific quantities, and fulfill the industry needs. However, the challenging aspects which hinder the successful adoption of FEM in the machining sector of manufacturing industry have to be solved first. One of the greatest challenges is the identification of the correct set of machining simulation input parameters. This study presents a new methodology to inversely calculate the input parameters when simulating the machining of standard duplex EN 1.4462 and super duplex EN 1.4410 stainless steels. JMatPro software is first used to model elastic–viscoplastic and physical work material behavior. In order to effectively obtain an optimum set of inversely identified friction coefficients, thermal contact conductance, Cockcroft–Latham critical damage value, percentage reduction in flow stress, and Taylor–Quinney coefficient, Taguchi-VIKOR coupled with Firefly Algorithm Neural Network System is applied. The optimization procedure effectively minimizes the overall differences between the experimentally measured performances such as cutting forces, tool nose temperature and chip thickness, and the numerically obtained ones at any specified cutting condition. The optimum set of input parameter is verified and used for the next step of 3D-FEM application. In the next stage of the study, design of experiments, numerical simulations, and fuzzy rule modeling approaches are employed to optimize types of chip breaker, insert shapes, process conditions, cutting parameters, and tool orientation angles based on many important performances. Through this study, not only a new methodology in defining the optimal set of controllable parameters for turning simulations is introduced, but also the optimum set of process input variables for turning duplex stainless steels is defined.

Keywords: 3D-FEM; duplex stainless steel; JMatPro; VIKOR method; FANNS; machining performance index

1. Introduction

Empirical machining models are of limited use: They are generally only calibrated to be valid for a limited process range. Meanwhile, due to the simplification, analytical machining models are only partially suitable for describing complex processes such as can be described by finite element modeling (FEM). FEM can provide a comprehensive and in some cases complementary approach to empirical, mechanistic, or analytical approaches to study machining process. It makes possible to illustrate complex tools in

*Corresponding author. Email: rastee.ali@ifw.uni-stuttgart.de

2D or 3D configurations while simultaneously taking plasto-mechanical and thermal processes into consideration. However, it should be clear that in 2D-FEM of many machining operations: turning, drilling, milling, etc. (except broaching, sawing, etc.) the obtained final surface doesn't correspond to the final one that is obtained in 3D-FEM. Thus, some prediction issues cannot have any realistic meaning (Klocke, 2011; Quiza & Davim, 2011).

The success and reliability of any FEM depends upon accurate mechanical (elastic constants, flow stress, friction, fracture stress/strain, etc.) and thermo-physical (density, thermal conductivity, heat capacity, etc.) data. Characterization of material flow stress behavior is needed for the extreme conditions of machining which involve plastic strains of 5–20, strain rates up to 10^6 s^{-1}, temperatures between 500 and 1400 °C, temperature gradient (1000 °C/mm), high heating rates close to 10^6 °C s^{-1}, and high cutting pressures of 1–3 GPa (Arrazola, Özel, Umbrello, Davies, & Jawahir, 2013). For this purpose, researchers recently applied modified constitutive equations, a great variety of which are exist in the literature (He, Xie, Zhang, & Wang, 2014; Hou & Wang, 2010; Li, Li, Wang, Liu, & Wu, 2013; Liu et al., 2013; Samantaray, Mandal, & Bhaduri, 2009; Song, Ning, Mao, & Tang, 2013; Wang et al., 2013). Some of the researchers have utilized Java-based Materials Properties (JMatPro) software to model the material properties and behavior of steels and Ni-based superalloys (Guo, Saunders, Miodownik, & Schille, 2007; Olovsjö, Hammersberg, Avdovic, Ståhl, & Nyborg, 2012; Saunders, Guo, Li, Miodownik, & Schille, 2004). Inverse identification of the constitutive model parameters has been proposed as an alternative method of defining the model coefficients, so that the models are valid over large ranges of conditions during machining (Klocke, Lung, & Buchkremer, 2013; Pujana, Arrazola, M'Saoubi, & Chandrasekaran, 2007; Shrot & Bäker, 2012; Umbrello, M'Saoubi, & Outeiro, 2007). Recent researches proposed methods for characterization friction and heat partition coefficients at the tool-work material interface in metal cutting operations (Arrazola & Özel, 2010; Attanasio & Umbrello, 2009; Bonnet, Valiorgue, Rech, & Hamdi, 2008; Heisel, Storchak, & Krivoruchko, 2013; Iraola, Rech, Valiorgue, & Arrazola, 2012; Puls, Klocke, & Lung, 2012; Rech et al., 2013; Smolenicki, Boos, Kuster, Roelofs, & Wyen, 2014; Ulutan & Özel, 2013). Several contributions have improved the FEM of serrated chip formation, chip flow, and chip breaking (Aurich & Bil, 2006; Buchkremer et al., 2014; Calamaz, Coupard, & Girot, 2008; Chagas, Barbosa, Barbosa, & Machado, 2013; Ducobu, Rivière-Lorphèvre, & Filippi, 2014; Guo & Yen, 2004; Lorentzon & Järvstråt, 2008; Rhim & Oh, 2006; Sima & Özel, 2010; Vaziri, Salimi, & Mashayekhi, 2011). Other works have focused on FEM of machining stainless steel materials (Koné, Czarnota, Haddag, & Nouari, 2011, 2013; Maranhão & Paulo, 2010; Outeiro, Umbrello, & M'Saoubi, 2006). Finally, recent works have also investigated the machining of duplex stainless steels experimentally (De Oliveira Junior, Diniz, & Bertazzoli, 2014; Koyee, Schmauder, & Eisseler, 2013; Koyee, Heisel, Eisseler, & Schmauder, 2014; Krolczyk, Legutko, & Gajek, 2013; Nomani, Pramanik, Hilditch, & Littlefair, 2013; Paro, Hänninen, & Kauppinen, 2001).

A vividly testimony to the potential interest on duplex stainless steels as a key research topic by various researches in the world is obvious when a search with the keywords 'duplex stainless steel' in popular database such as Science Citation Index-Expand or Scorpus returns tens of recent publication. However, one can hardly find a research that addresses the FEM simulation of machining duplex grades. Furthermore, despite significant recent advances, FEM itself remains a 'plug and play' technique for predicting some process output, depending on the assumed boundary conditions, including the friction. For example, it has been reported that FEM simulation of the same processes can produce different results (Arrazola & Özel, 2009).

To address the FEM simulation of machining DSS correctly, the plug and play technique must be limited first. For this purpose, this study introduces a new method of inverse identification through converting the overall differences between simulation results and experimentation ones into a single measure using Taguchi-VIKOR method. FEM modeling under mixed Taguchi design $L_{18}(2^1 \times 3^7)$ is performed to tune input parameters. Thermal contact conductance h_{tc}, cutting speed v_c, feed rate f_r, cutting tool–workpiece interface hybrid Coulomb μ_c and shear μ_s friction coefficients, Taylor–Quinney coefficient κ_t, percentage reduction of original flow stress $\%p$, and Cockcroft–Latham critical damage criterion $D_{\text{crit.}}$ are considered as controllable input parameters. On the other hand, cutting experimentations are conducted and different performances are measured, recorded, and analyzed. The percentage difference between numerically and experimentally obtained performances such as thrust cutting force $\%E_t$, feed cutting force $\%E_f$, main cutting force $\%E_c$, chip thickness E_h, and tool nose temperature $\%E_T$ are considered as performance characteristics and are unified into a single index called VIKOR method. The derived indices are then optimized globally to determine the optimum set of input parameter using an effective neural network-based nature-inspired meta-heuristic algorithm known as Firefly Algorithm Neural Network System (FANNS). The optimum sets are next validated through experimentations. In the next stage of the research work, Taguchi optimization procedure is employed to numerically optimize the chip breaker types *CB*, insert geometries *Geo.*, cooling medium *CM* such as still air, water-based and cryogenic coolants, cutting conditions such as cutting speed v_c and feed rate f_r, and tool orientation angles such as normal rake α_n and inclination angle λ. Resultant cutting forces (*R*), effective plastic stresses, chip-tool interface cutting temperatures, and tool wear rate are designated as performance characteristics. An expert system based on fuzzy rule modeling approach is adopted to derive a new index called numerical machining performance measure (NMPM). Finally, analysis of means (ANOM) is applied on the computed NMPMs to define the optimum levels of control factors. A schematic diagram summarizing the methodological framework developed in this study is shown in Figure 1.

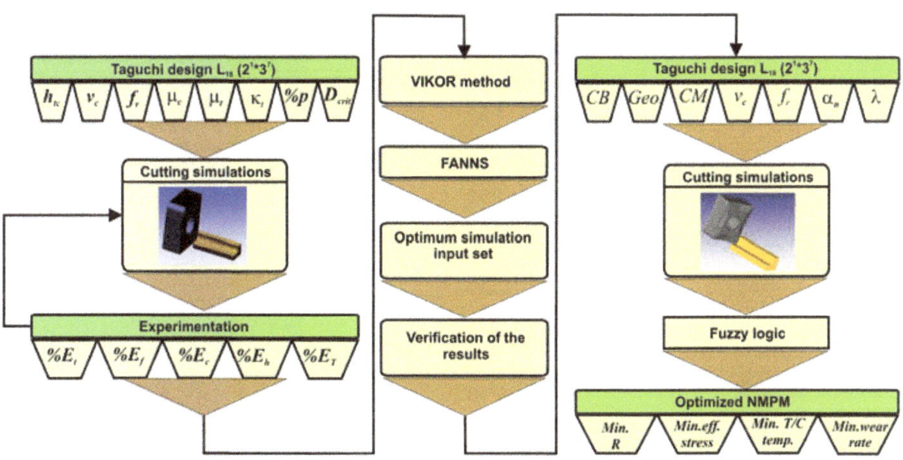

Figure 1. Framework of the research.

2. Experimentations

2.1. Longitudinal turning operation

EN 1.4462 standard duplex and EN 1.4410 super duplex rods of circular cross section (diameter 55 mm, length 200 mm) were machined on a CNC lathe (Gildemeister CTX 420 Linear V5) having maximum drive power of 25 kW and speed range of 35–7000 revolutions/min. The tool holder was a Sandvik Coromat referenced PCLNL 2525 M 12, and the inserts were uncoated rhombic cemented carbide of ISO designation; CNMA 120412-IC20. They had neither chip breaker nor coating. The reason behind the selection of these basic-shape inserts was mainly to facilitate the optimization of the chip breaker geometry in the final stage of the study. The chemical composition and mechanical properties of the involved work materials are tabulated in Table 1.

In the literature, depth of cut has shown very little influence on chip thickness, contact stress at the tool–chip interface, tool wear rate, and average temperature (Astakhov, 2006). Therefore, in this study, the cutting tests are conducted at constant cutting depth of 1.5 mm with the following cutting data: cutting speed v_c = 80,160 and 240 m/min, and feed rate f_r = 0.15, 0.225 and 0.3 mm/rev. Finally, to avoid complications in cutting temperature measurements, the turning tests were carried out dry.

2.1.1. Cutting forces

The cutting force components F_c, F_f, and F_t are measured using piezoelectric dynamometer from Kistler. To ensure a steady state force signal and reduce the tool wear effect on cutting forces and temperatures, a new cutting edge per each experimental trial was employed. The effect of feed rate and cutting speed on the components of cutting force is shown in Figure 2. For instance, increasing the cutting speed from v_c = 80–240 m/min at constant f_r = 0.225 mm/rev while turning EN 1.4462 had reduced F_c, F_f, and F_t by 6.594, 40.613, and 19.8%, respectively. On the other hand, increasing feed rate from f_r = 0.15 to 0.3 mm/rev at constant v_c = 160 m/min while turning EN 1.4410 had increased F_c, F_f, and F_t by 47.39, 22.076, and 52.82%, respectively.

Table 1. Properties of the work materials.

Chemical composition (%Weight)	EN 1.4410	EN 1.4462
Carbon	.015	.018
Chromium	24.92	22.42
Nickel	6.91	5.44
Molybdenum	4.06	3.12
Manganese	.75	.84
Silicon	.25	.37
Nitrogen	.3	.18
Phosphor	.021	.025
Sulfur	.0007	.003
Copper	.1	–
Mechanical properties		
Yield strength (MPa)	579	514
Tensile strength (MPa)	826	737
Hardness (BHN)	236	212
Elongation (%)	40	41

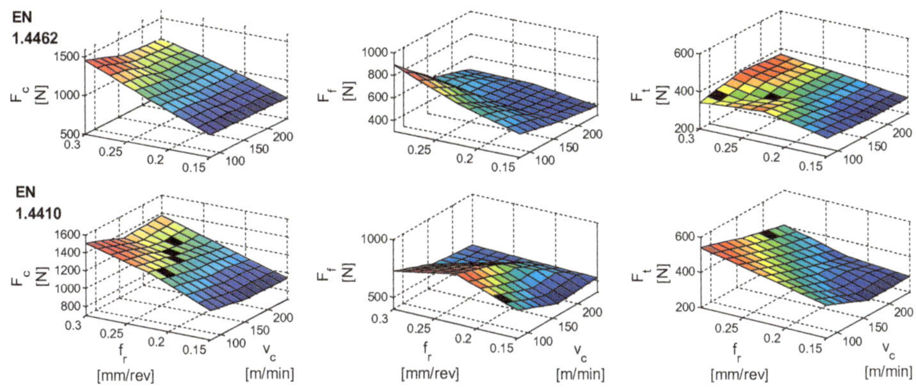

Figure 2. Dependency of the components of cutting force on the cutting conditions.

2.1.2. Cutting temperatures

Although infrared (IR) cameras are quite expensive, they remain one of the most promising solutions for temperature measurement since they allow for a non-contact, extensive measurement of temperature. Thus, the problem of perturbation of the heat flow in the tool and changes the results is avoided. The fast response of the IR camera lets high, cutting speeds to be used in machining experiments, since it can capture the transient changes and the stages of chip formation and entanglement around the workpiece, the tool holder, and the tool post. In this study, high-resolution thermo-graphic camera Image IR8300 by InfraTec was used to film the chip formation and measures the chip temperature. Each film made for an experimental trial is composed of 800 images. The IR camera is placed straight above the rake face of the tool. The lens of the camera is protected against possible impacts of the chips flying about. The images in the films are examined, and the mean of maximum temperature in the middle of the chips is recorded. Figure 3 shows the setup of IR camera.

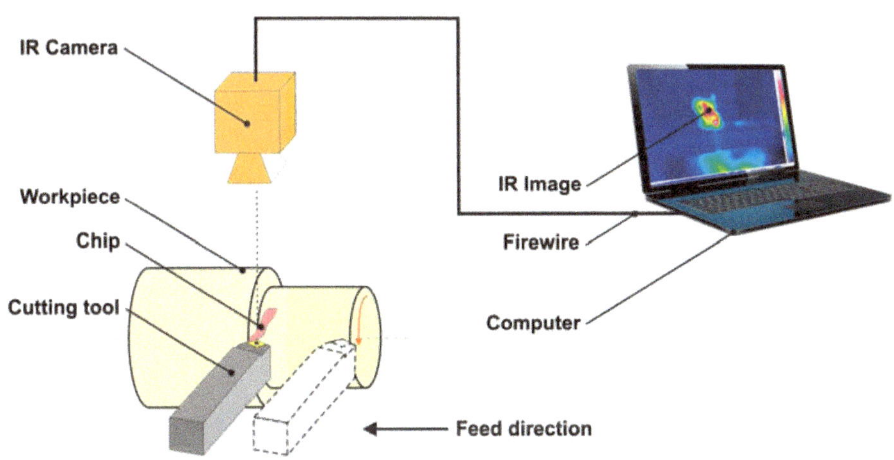

Figure 3. Experimental setup of IR camera.

The dependency of maximum chip temperature on the cutting condition and the machined material can be seen in Figure 4. Examining the mean maximum chip temperature values, the average maximum chip temperature when machining EN 1.4410 was higher than EN 1.4462 by 3–5%. Experimental results have shown that high cutting speed and feed rates will not always lead to high chip temperatures. This seems in contradiction with the known proportional relations between cutting conditions and cutting temperature. The most likely explanation to this is that when the cutting speed and feed rate are increased, higher rate of removed volume is expected. In this case, the higher heat flux entering the chip resulting from the higher interface temperature is divided over a larger volume, hence lower intensity.

Due to the continuous entanglement of the DSS chips around the cutting tool and workpiece surface, the chip surface was not always available for direct observation by the IR camera. Furthermore, due to the concave surface of the scanned workpiece, highly reflective nature of DSS, surface and difficulties in accurately measuring surface emissivity, the erroneous temperature reading of tool–chip interface was inevitable (see Figure 5). During the early efforts to compare the maximum chip temperatures, similar

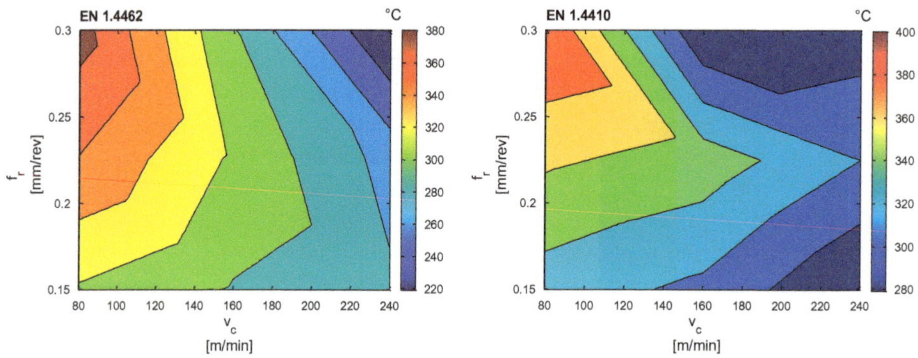

Figure 4. Dependency of maximum chip face temperature on the cutting conditions.

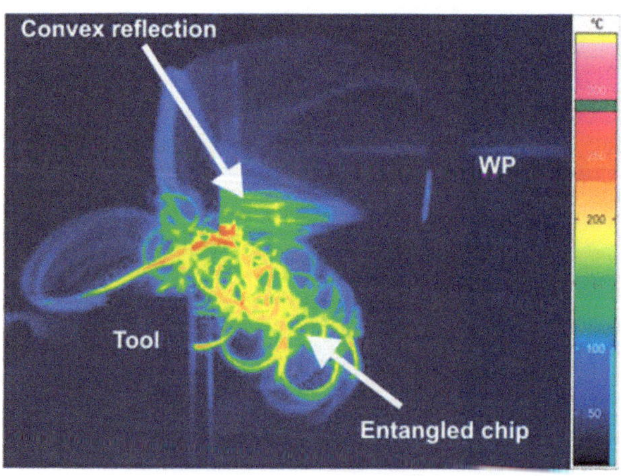

Figure 5. Problems associated with IR camera temperature measurement during machining DSS.

problems of measuring maximum chip temperature due to the obstructed chip's free face have also been encountered. Therefore, the application of the obtained maximum chip temperature results is restricted only to the experimentation phase. Instead, the calibrated IR images of the black tool tip with emissivity value of 0.93 were obtained very shortly (0.25 sec) after the feed was halted and the temperature of the cutter at the end of steady cutting is recorded. The experimentally measured temperature data are then utilized to validate the results obtained in the simulation phase.

Figure 6 maps the experimentally measured maximum tool surface temperatures at location of 0.5 mm < 45° from the origin of the nose curvature. It can be seen that the maximum temperature on the tool face increases with cutting speed. With the increase in feed rate, the cross section of chip and tool–chip contact length increases and consequently friction rises. This is also involves the increase of temperature.

2.1.3. Micrographs of the chips

The serrated chip geometries, which were detected in some machining occasions, have to be correctly configured and measured. Thus, the chips are carefully collected and prepared for detailed analyses. Among many chip-related measurements, maximum and minimum chip thickness and degree of serration are recorded using two different techniques. In the first, the collected chips are embedded in transparent epoxy, polished using increasingly finer diamond grit, and etched in chemical solution. Despite the process time-consuming nature and reduction of the 3D structures to 2D planes, this method enabled the ease tracings of the streamline of material flow and paths of the grains. The variation of plastic deformation during a chip formation cycle is clearly shown in Figure 7(a). Higher plastic deformation during the loading stage and lower plastic deformation during the unloading stage of the chip formation cycle have influenced the grain shapes, so that in the zone of high plastic deformation, the grains are severely deformed and in the zone of low plastic deformation, the grains are moderately deformed. The tips of the cracks can be clearly observed at the boundary between the zones. The chip structure shown in Figure 7(b) is an example of saw-tooth continuous transitional chips obtained when DSSs are machined.

In the second measurement technique, the chips are photographed and measured using a high magnification optical microscope. The average values of five consecutive chip thickness ratios (h_{max}/h_{min}) are defined as the degree of serration:

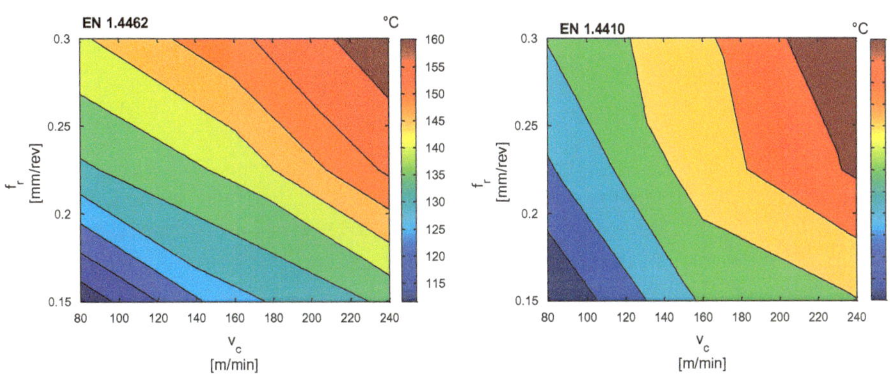

Figure 6. Dependency of tool tip temperature on the cutting conditions.

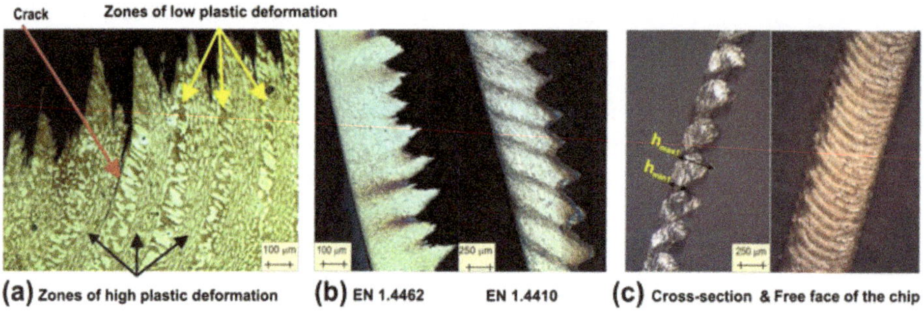

Figure 7. Examples of formed chips when machining DSS. (a) EN 1.4462 at $v_c = 80$ m/min and $f_r = 0.225$ mm/rev, (b) Left, EN 1.4462 $v_c = 80$ m/min and $f_r = 0.15$ mm/rev, right, EN 1.4410 $v_c = 80$ m/min and $f_r = 0.3$ mm/rev, and (c) EN 1.4410 $v_c = 160$ m/min & $f_r = 0.3$ mm/rev.

$$H = \frac{1}{5}\sum_{i=1}^{5}\frac{h_{\max_i}}{h_{\min_i}} \tag{1}$$

where $h_{\max}$ and $h_{\min}$ are the perpendicular distances measured from the bottom of the chip to the peaks and valleys, respectively (see Figure 7(c)). The presence of fragments is also revealed by showing the image of free side of the chip. Only in the cases when $H \approx 1$, a smooth continuous chip is expected. Otherwise, the higher the ratio H, the more serrated the type of the chip is. Comparing the cutting of DSS, EN 1.4462 has generally led to produce friendlier to the machine, thinner and more segmented types of chips. This is possibly due to the higher content of chip breaking elements, such as phosphor and sulfur (see Table 1). In general, as cutting speed and feed rate are increased, there are increases in the degree of serration (H) as observed in the Figure 8. It is also observed that cutting speed, which adversely affected the $h_{\max}$, has contributed positively in the chip thinning process through reducing the consecutive distance between the peeks.

3. Prerequisite conditions of machining numerical simulations

3.1. *Control factors*

Since the success and reliability of machining modeling depends upon accurate mechanical and thermo-physical data, a brief description of the involved FEM control factors is considered important here, so that deeper understanding of the simulation of cutting DSS is gained.

3.1.1. Damage

One of the reasons behind the poor machinability of standard and super DSS is their high tensile strength. Therefore, when machining of DSS, tensile stress is expected to play an important role. In this research, the effect of tensile strength is included in the material response employing Cockcroft–Latham's damage criterion, which is expressed mathematically as follows:

$$D = \int_0^{\varepsilon_f} \sigma_1 d\varepsilon \tag{2}$$

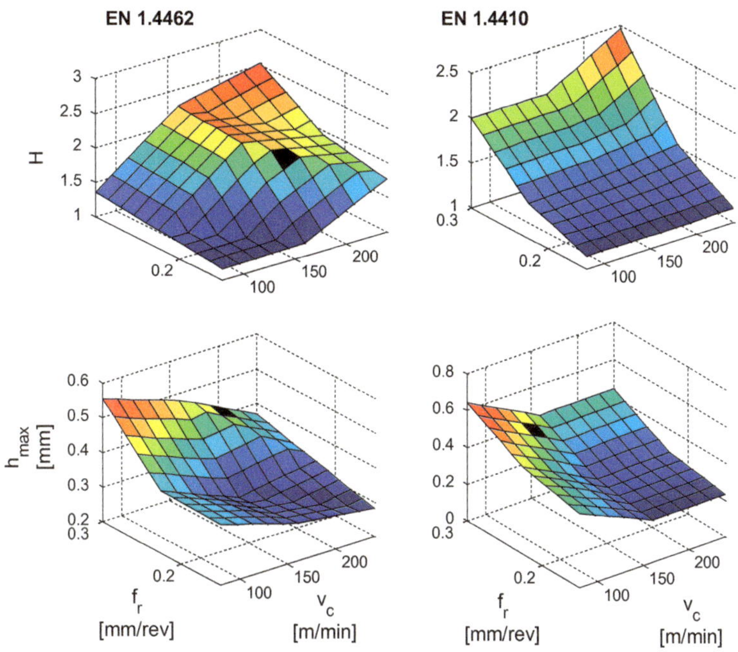

Figure 8. Dependency of the chip characteristics on the cutting conditions.

where ε_f is the effective strain, σ_1 is the principal stress, and D is a material constant. Cockcroft and Latham's criterion states that when the integral of the largest tensile principal stress component over the plastic strain path in Equation (2).

Reaches the critical value $D_{crit.}$, usually called damage value, fracture occurs or chip segmentation starts, the flow stress is reduced to a lower value %p, which is expressed as percentage of the original flow stress (Umbrello et al., 2007).

3.1.2. Friction

In this study, Zorev's friction model is adopted to overcome the shortcomings of the Coulomb and the shear friction law simultaneously. So that in the range of low normal pressure, the Coulomb law is applied while for high normal pressures, the shear friction law is used. This can be expressed by means of the following formulation:

$$\tau_f(x) = \begin{cases} \tau_p : \mu\sigma_n \geq \tau_p, & (l_p < x < l_c) \text{ Sticking region} \\ \mu\sigma_n : \mu\sigma_n < \tau_p, & (0 < x < l_p) \text{ Sliding region} \end{cases} \tag{3}$$

where the τ_f, σ_n, and μ are the frictional stress, normal stress, and coefficient of friction, respectively (Zorev, 1963).

3.1.3. Thermal contact conductance h_{tc}

The thermal contact conductance h_{tc}, also known as the heat transfer coefficient, describes the heat flux of two solids in contact and is defined as follows:

$$h_{tc} = \frac{q}{\Delta T} \tag{4}$$

in which ΔT is the temperature difference at the contacting surfaces and q, the heat flux, defined as:

$$h_{tc} = \frac{d}{dA}\left(\frac{dQ}{dt}\right) \tag{5}$$

where A is the area of contact surface and Q is the conducted heat in (J). It is recognized that thermal contact conductance is a function of several parameters, the dominant ones being the type of contacting materials, the macro- and micro-geometry of the contacting surfaces, the temperature, the interfacial pressure, the type of lubricant or contaminant, and its thickness (Rosochowska, Balendra, & Chodnikiewicz, 2003).

3.1.4. Taylor–Quinney coefficient

As the mechanical behavior is affected by temperature (softening effect), the plastic deformation is accompanied by heat generation which results in temperature rise. The heat generation due to this phenomenon is described by the following relationship:

$$\dot{q}_p = \kappa_t \bar{\sigma}\,\dot{\bar{\varepsilon}}_p \tag{6}$$

where $\dot{q}_p$ is the volumetric heat generation due to plastic work in (W/m^3), $\bar{\sigma}$ is the von Mises equivalent stress in (MPa), $\dot{\bar{\varepsilon}}_p$ is the von Mises equivalent plastic strain, and κ_t is the Taylor–Quinney coefficient which represents the proportion of plastic works converted into heat. (Haddag & Nouari, 2013) have shown strong dependency of κ on both strain and strain rate for various engineering materials. Within the confines of their constitutive framework, the assumption that κ is constant is inconsistent with the rate independence of the stored energy of cold work, which is a fundamental consequence of thermodynamics. It would seem that the only justification for a priori assumptions on κ is a lack of information on the stored energy of cold work. Therefore, this coefficient is considered as an unknown input value that has to be inversely determined in the next cutting simulations.

3.2. Prediction of chip serration

Serrated chips (also called segmented or non-homogeneous chips, see Figure 7(c)) are semi-continuous with large zones of low shear strain and small zones of high shear strain; hence, the latter zone is called shear localization. Metals with low thermal conductivity and strength that decreases sharply with temperature (thermal softening) exhibit this behavior (Kalpakjian & Schmit, 2006). Adiabatic shear bands have been observed in the serrated chip during high strain rate metal cutting process of titanium (Sima & Özel, 2010), carbon steel (Rhim & Oh, 2006), nickel-based alloys, and stainless steels (Lorentzon, Järvstråt, & Josefson, 2009). Adiabatic shear bands are narrow zones with thickness of the order of few micro-meters where shear deformation is highly localized. Each material has a different susceptibility to adiabatic shear because it depends on properties such as heat capacity, heat conductivity, strength level, microstructure, geometry, defects, and strain rates. It is also known that adiabatic shear banding precedes material failures at high strain rates. Adiabatic shear banding is usually accompanied by a loss in stress capacity owing to intense thermal softening in the shear bands and, in many cases, shear bands serve as sites for crack initiation and growth during subsequent dynamic fracture (Odeshi, Al-ameeri, & Bassim, 2005). Localized adiabatic shearing can be considered a unique consequence of severe plastic deformation at high strain rates. As both thermal and strain softening lead to rapid deformation localization, a shear band forms via a nearly adiabatic process. Also of note is that grain refinement can occur within shear bands and severe plastic strain (which can reach 5–20) can also appear within these shear bands (Xue, Liao, Zhu, & Gray, 2005).

In order to capture the formation of chip serrations numerically, a material model which accounts for thermal and strain softening should be employed. In last several years, numerous attempts have been made to predict serrated chip formation by finite element method. However, it was found to be difficult to predict the shear band and the serrated chip formation by the FEM technique, especially in three-dimensional configuration. In addition to the improper meshing size and strategy, the application of conventional flow stress models is one of the main reasons.

Unfortunately, a universal material model suitable for all cutting simulations remains one of the main unaccomplished tasks. Due to the typical machining high strain, strain rate temperature, and temperature gradient, it is not always easy to determine the flow stress curves experimentally. Additionally, experimental methods such as split Hopkinsonbartest are quite expensive and require a large number of experiments to conduct. On the other hand, empirical material laws which describe the flow stress as a function of strain, strain rate, and temperature contain specific material constants that have to be determined by regression analyses or by the least-squares method and verified experimentally. During the last three decades, many of such models are suggested. The most notably are the engineering-based and physically based models. Results of the first evaluative simulations trials carried out in this study have shown that chip morphologies predicted by engineering-based model such as Johnson–Cook (JC) model and physically based model such as Zerilli–Armstrong lacked the formation of the shear bands and strain localizations which are often denoted as characteristics of chips formed during machining materials that possess high ultimate and yield strength and low thermal conductivity such as duplex stainless steels. For example, the following conditions were considered in a 3D-FEM and extruded 2D-FEM modeling:

- Tool: normal rake angle = −5°, relief angle = 7°, inclination angle = 0°, radius of the cutting edge = 0.03 mm, and tool material is tungsten carbide (WC).
- Work material: EN 1.4462
- Work material models: (a) JMatPro and (b) JC model. The JC constitutive model is given by equation:

$$\sigma = (A + B\varepsilon^n)\left(1 + C\ln\left(\frac{\dot{\varepsilon}}{\dot{\varepsilon}_0}\right)\right)\left(1 - \left(\frac{T - T_r}{T_m - T_r}\right)^m\right) \quad (7)$$

where σ is current von Mises flow stress; A is initial yield strength; B is strain hardening coefficient; ε is equivalent plastic strain; n is strain hardening exponent; C is strain rate coefficient; $\dot{\varepsilon}$ is equivalent plastic strain rate; $\dot{\varepsilon}_0$ is reference plastic strain rate; m is thermal softening exponent; T is current temperature; T_m is melting temperature; T_r is reference temperature at which material constants are determined (Nasr, Ng, & Elbestawi, 2007).

The elastic-plastic, viscosity, and thermal softening terms are represented by the first, second, and third brackets of the above equation, respectively. The procedure of computing the constants of Johnson–Cook model from JMatPro stress flow curves involved the following steps:

- The value of A is calculated from the yield stress of the metal given in Table 1, that is, $A = 514$ MPa.
- At $T = T_r$ and $\dot{\varepsilon} = \dot{\varepsilon}_0$, the curve $\ln(\sigma - A)$ vs. $\ln(\varepsilon)$ is plotted. The values of B and n are extracted from the intercept and slope of this plot, respectively.
- Substituting the values of A, B, and n in Equation (7) and assuming $T = T_r$ which eliminates the thermal softening term, the strain rate coefficient (C) is obtained from the slope of the graph $\{\sigma/(A + B\varepsilon^n)\}$ vs. $\ln(\dot{\varepsilon})$.
    - At $\dot{\varepsilon} = \dot{\varepsilon}_0$, the viscosity term is eliminated. The exponent (m) is obtained from the slope of the graph $\ln[1 - \{\sigma/(A + B\varepsilon^n)\}]$ vs. $\ln\{(T - T_r)/(T_m - T_r)\}$. Finally, the overall correlation coefficient was found satisfactory at $R^2 = 0.85567$. The JC parameters of the EN 1.4462 DSSs are given in Table 2.
- Cutting regime: cutting speed $v_c = 100$ m/min and uncut chip thickness $f_r = 0.25$ mm.
- Damage criteria: $D_{\text{crit.}} = 50$ MPa and $\%p = 20$.
- Shear coefficient $\mu_t = 0.4$ and $h_{tc} = 45$ N/(mm s K).
- Mesh size = 20 μm and aspect ratio = 7.
- Stress flow curves of the work material are illustrated in Figure 9.

The left and right sides of the Figure 10 show the states of the deformation zones for JMatPro and JC models, respectively. For instance, Figure 10(b) shows the plastic strain distribution in the deformation zone, while Figure 10(c) presents the temperature distribution in the deformation zones. As clearly seen from the figures, the formation of transitional or saw-tooth chip types is hardly predicted in the chips correspond to the JC

Table 2. Parameters of the Johnson–Cook model for EN 1.4462.

Parameter	A (MPa)	B (MPa)	n	C	m	$\dot{\varepsilon}_0$ (1/s)
Value	514	612.96	.1801	.01194	.9765	1

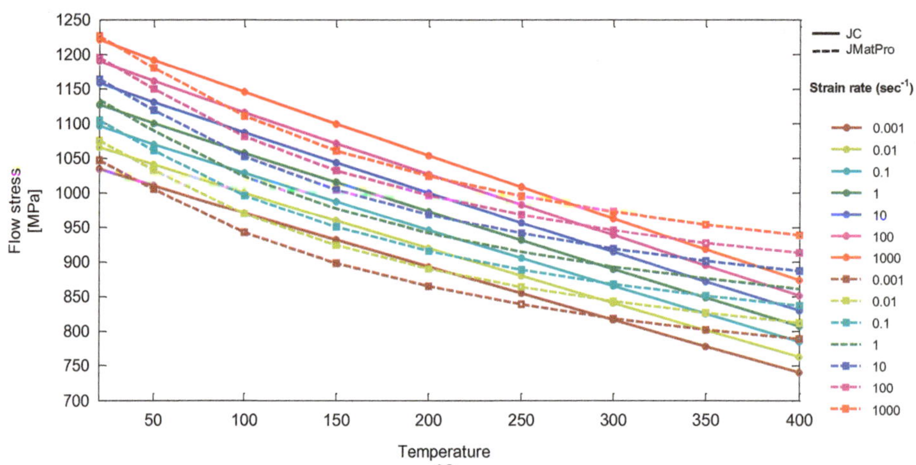

Figure 9. EN 1.4462 JC and JMatPro flow stress curves at $\varepsilon = 1$.

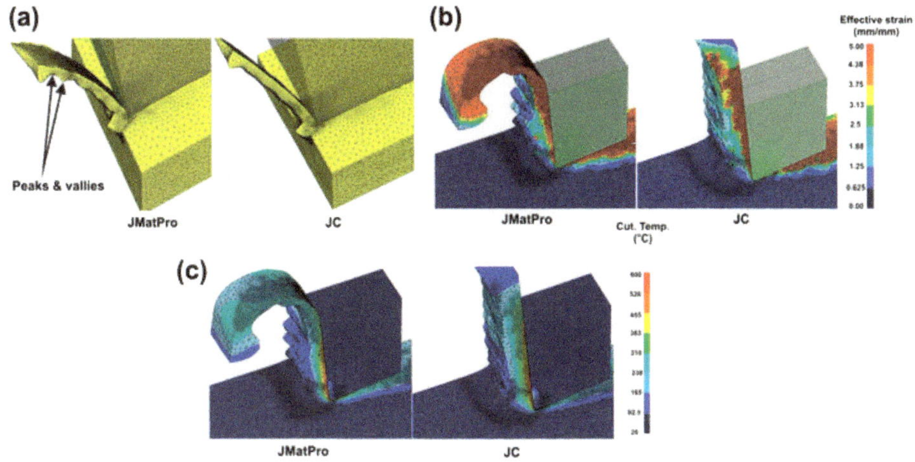

Figure 10. JMatPro and JC-based cutting simulation results.

model. Furthermore, the morphologies of the chips when cutting stainless steel materials are often related with chip up-curling and side-curling are also hardly noticed in JC material model cases. On the other hand, FEM results employed JMatPro flow stress curves seem to perform better in terms of the prediction of chip curling and formation adiabatic shear bands phenomenon. Therefore, the option of implementing flow stress data generated by JMatPro software is adopted. A text file was first generated for each workpiece material with chemical composition and mechanical properties shown in Table 1 and copied into the keyword file of the actual simulation. The software is also used to model the elastic and thermo-physical behavior of the work material.

3.3. Finite element simulations

The classical representation of cutting process by 2D models of chip formation suitable for orthogonal cutting has a limited assistance to tool and process designers. Moreover, the obtained final surface does not correspond to the final one that is obtained in 3D. Consequently, some prediction performances like residual stresses cannot have any realistic meaning. For these reasons, the oblique turning models are designed in 3D-FEM environment. The predicted thermo-mechanical properties of the work materials are shown in Figure 11.

The simulations are carried out with commercial software DEFORM-3D v10.1. The software is based on the implicit updated Lagrangian formulation. The workpiece, each 10 mm size, was considered as a plastic object and meshed with approximately 140,000 elements. The tool considered a rigid object meshed with more than 100,000 elements. It is oriented according to the cutting angles set in experimental test and moves along a straight path. The cutting tool material was uncoated tungsten carbide (WC) and assigned directly from the available material database of the software. Thermal and mechanical properties of WC tools are tabulated in Table 3.

To improve the gradients of temperature, stress, strain, and strain rate distributions, the mesh is refined in the vicinity of areas of cutting tool and the workpiece where the primary and secondary shear zone will be located. However, fixing the minimum element size at a very fine one could make the computation very expensive in terms of

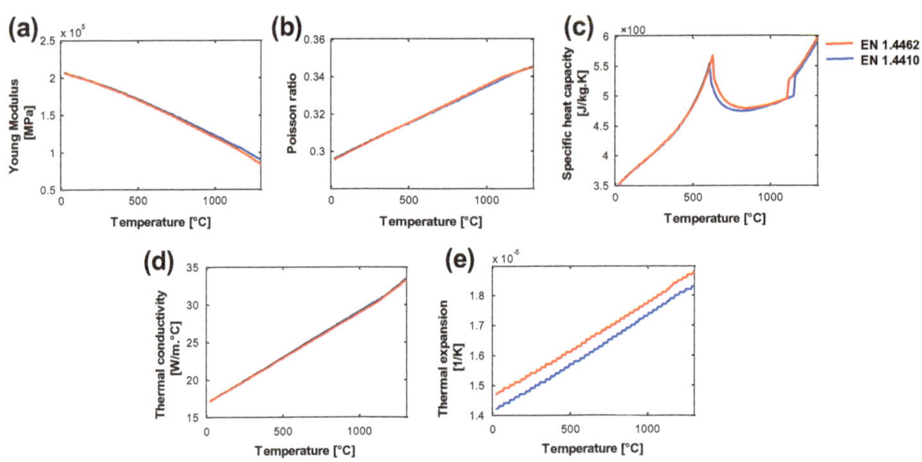

Figure 11. Temperature-dependent mechanical and physical properties of the work materials.

Table 3. Thermal and mechanical properties of WC.

Elastic modulus	650 GPa
Poisson's ratio	.25
Thermal conductivity	59 W/m K
Specific heat capacity	15 J/kg K
Thermal expansion	5×10^{-6} 1/K

running time and size of the final database. Therefore, the minimum mesh element size of the workpiece and aspect ratio was fixed at ¼ of the minimum uncut chip thickness and 7, respectively. The top sides of the workpiece as well as all sides of the cutting tool were allowed to exchange heat with the environment; the convection coefficient is considered constant at 0.02 N/(s mm K), which is the default value for dry cutting in DEFORM 3D. On the other hand, meshing of the cutting tool model expected to be a decisive factor for the simulation of the heat flux, temperature, and stresses in the material decohesion/deformation zone (Niesłony, Grzesik, Chudy, & Habrat, 2014). Therefore, the meshing density on the tool is increased in the potential chip-tool contact surface, while the rest of the tool is meshed with relatively coarser mesh. The mesh and boundary conditions for the finite element model are shown in Figure 12.

3.4. Design of the FEM experiments

When the experimental design is a full factorial one, the final number of experiments is usually large, because all the possible combinations should be introduced in the design. In order to allow improvement in the methodology of simulation tests and study the entire control factors space with the smallest possible number of experiments, the concept of design of experiment (DOE) is applied. DOE is defined as process of planning of an experiment so that appropriate data will be collected, which are suitable for further statistical analyses resulting in valid and objective conclusions. One of the robust methods of experimental design is Taguchi method, which provides a simple, efficient, and systematic approach for the optimization of experimental designs for performance

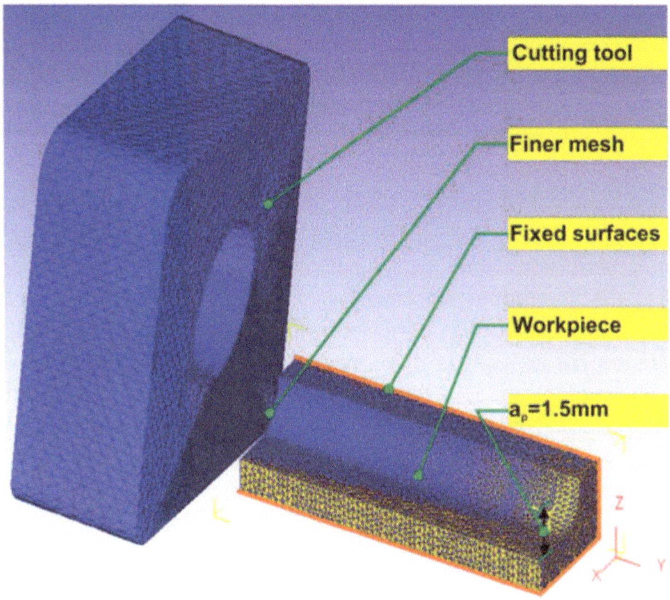

Figure 12. Objects of the FEM model.

quality and cost. To evaluate the effect of each control factor or its interaction effects on the process outputs, Taguchi employs standard orthogonal arrays. After conducting the experiments and collecting the desired outputs according to the Taguchi arrays, the values of outputs have to be transferred into signal-to-noise (S/N) ratios. Usually, there are three S/N ratios available, (η), depending on the type of characteristic; the lower-the better (LB), the higher-the better (HB), and the nominal-the better (NB). Owing to the non-beneficiary nature of the considered performances (outputs) here, the LB is adopted as:

$$\eta_{ij} = -10\log_{10}\left(\frac{1}{n}\sum_{i=1}^{n} Y_{ijk}^2\right) \quad (8)$$

where η_{ij} is the S/N ratio for the response j of experimental number i, and Y_{ijk} is the experiment result for the response j of the experiment i, in the kth replication; n is the total number of replications, and Y_0 is the nominal response. Thereafter, analysis of means (ANOM) and analysis of variance are performed on the S/N ratio of orthogonal array to identify the optimal factor level combination and estimate the error variance for the effects and variance of the prediction error, respectively (Keleştemur, Arıcı, Yıldız, & Gökçer, 2014).

In this study, the eight considered control factors are as follows: (1) thermal contact conductance h_{tc}, (2) cutting speed v_c, (3) feed rate f, (4) Coulomb friction coefficient μ_c, (5) shear friction coefficient μ_t, (6) Taylor–Quinney coefficient κ_t, (7) percentage decrease in flow stress $\%p$, and (8) Cockcroft–Latham critical damage value $D_{\text{crit.}}$. The third and fourth control factors are referred to the cutting conditions and are included in the design to give the design more flexibility in dealing with other sets of cutting conditions that are planned in the second stage of the study. Taguchi mixed design L_{18} ($2^1 \times 3^7$) is adopted so that the underestimation and overestimation problems are avoided by keeping the most parameter levels at three. Additional advantage of this factorial design is the reduction of experimentation trials from 18 to 9, because cutting speed and feed rate are the only input variables that can be designated as controlled factors during experimentation phase of the study. The control factors and their levels and the orthogonal array are shown in Tables 4 and 5, respectively.

4. Results of the DOE

The first stage of performing numerical simulations has confirmed the strong influence of control factors on the cutting performances. Among the performances that have been strongly affected were the cutting temperatures, cutting forces, strain rates, and chip morphologies. Figure 13 depicts the impact of sets of control factors on the cutting temperature and chip morphology after cutting 8 mm of the workpiece length. It is difficult to draw clear conclusion points on the parametric effect of each control factor based on

Table 4. Control factors and levels for the design of simulation experiments.

Factors	Sym.	Unit	Levels 1	2	3
Thermal contact conductance	h_{tc}	N/(s mm K)	100	1000	
Cutting speed	v_c	m/min	80	160	240
Feed rate	f_r	mm/rev	.15	.225	.3
Coulomb friction coefficient	μ_c	–	.5	.75	1
Shear friction coefficient	μ_t	–	.6	.9	1.2
Taylor–Quinney coefficient	κ_t	–	.8	.9	1
Reduction in flow stress	$\%p$	–	10	30	50
Critical damage value	$D_{crit.}$	MPa	50	100	150

Table 5. L_{18} $(2^{1\times}3^7)$ orthogonal array.

Exp. No.	Control factors: h_{tc} (N/mm s K)	v_c (m/min)	f_r (mm/rev)	μ_c	μ_t	κ_t	$\%p$	$D_{crit.}$ (MPa)
1	100	80	.15	.5	.6	.8	10	50
2	100	80	.225	.75	.9	.9	30	100
3	100	80	.3	1	1.2	1	50	150
4	100	160	.15	.5	.9	.9	50	150
5	100	160	.225	.75	1.2	1	10	50
6	100	160	.3	1	.6	.8	30	100
7	100	240	.15	.75	.6	1	30	150
8	100	240	.225	1	.9	.8	50	50
9	100	240	.3	.5	1.2	.9	10	100
10	1000	80	.15	1	1.2	.9	30	50
11	1000	80	.225	.5	.6	1	50	100
12	1000	80	.3	.75	.9	.8	10	150
13	1000	160	.15	.75	1.2	.8	50	100
14	1000	160	.225	1	.6	.9	10	150
15	1000	160	.3	.5	.9	1	30	50
16	1000	240	.15	1	.9	1	10	100
17	1000	240	.225	.5	1.2	.8	30	150
18	1000	240	.3	.75	.6	.9	50	50

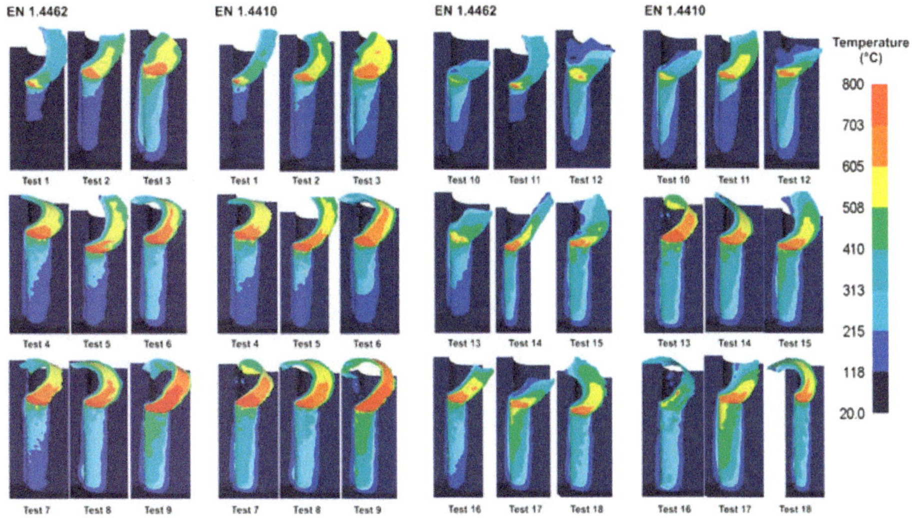

Figure 13. Simulated workpiece temperature distributions.

the images shown, since at each cutting step, the values of performances change considerably. Therefore, statistical tools have to be employed to analyze the results and help emphasizing the effect of control factors separately.

Based on the Taguchi optimization procedure, the next step after recording the results of experimentation is to calculated the S/N ration of obtained performance characteristics such as cutting forces in feed (F_f), cutting (F_c), and thrust (F_t) directions, strain, strain rate, principal stress, mean tool–chip interface temperature, chip temperature, and chip thickness for each simulation. Owing to the non-beneficial nature of the studied performances, the lowest value is always desirable. Therefore, Equation (8) is employed in this case. The simulation of randomly specified trials was repeated and no differences between replications were noticed (i.e. $k = 1$). The overall mean of (η_{ij}) associated with n trials is then determined using the principles of analysis of means (ANOM). The level which returns maximum S/N ratio is designated as the optimum, since it optimizes the response and minimizes the effect of the noise factors simultaneously.

Figure 14 shows the results of performing ANOM after normalization. The mean S/N ratios are normalized between the worst '0' and the best '1'. However, it should be noted here that explaining the main effect plots of each involved control factor on obtained performance characteristics is an enormous task and requires lots of paper space. For example, the optimum $factor^{level}$ combinations which minimize the main cutting force for machining EN 1.4462 and EN 1.4410 are $h_{tc}^1\ v_c^3\ f_r^2\ \mu_c^1\ \mu_t^1\ \kappa_t^3\ \%p^1\ D_{crit.}^1$ and $h_{tc}^1\ v_c^3\ f_r^1\ \mu_c^2\ \mu_t^1\ \kappa_t^3\ \%p^1\ D_{crit.}^1$, respectively. Another example, in the case of turning 25 EN 1.4410, simulation results showed that increasing the cutting speed value from $v_c = 80$ m/min to $v_c = 240$ m/min had reduced the main effects of F_c, F_f, F_t, strain, and chip thickness by 43.6%, 32.08%, 46.08%, 13.244%, and 39.44%, respectively. While the main effects of strain rate, normal stress, and chip/tool interface temperature had increased by 37.98%, 7.25%, and 19.045%, respectively.

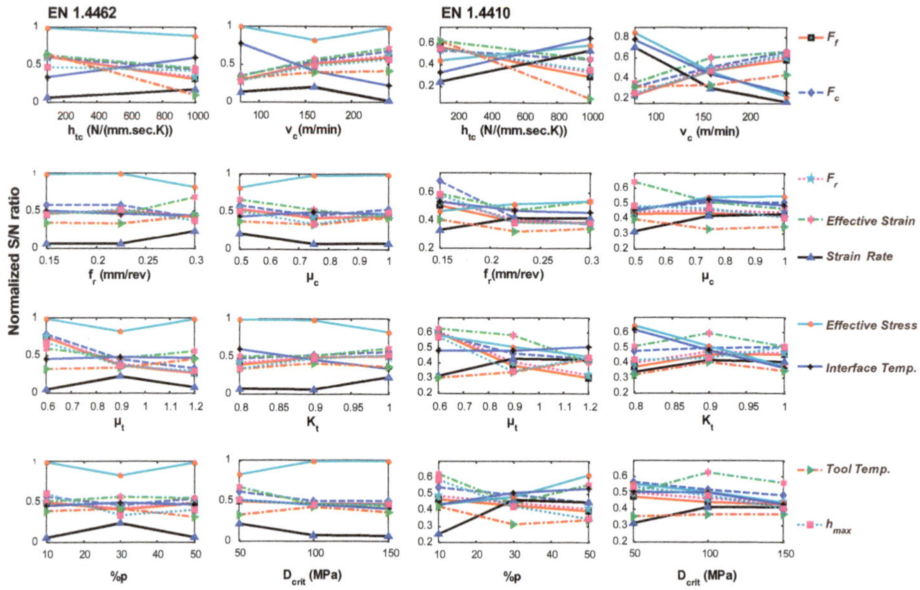

Figure 14. Main effect plots of the effect of control factors on the cutting performances during turning EN 1.4410 and EN 1.4462 DSSs.

5. Proposed methodology

The percentage of differences between the experimentally measured cutting forces, chip temperatures and thicknesses, and the FEM-predicted ones are computed using the percentage of difference expression (%E):

$$\%E = \frac{Y_{\mathrm{EXP}} - Y_{\mathrm{FEM}}}{Y_{\mathrm{FEM}}} \times 100 \tag{9}$$

where Y_{EXP} represents the experimental performance values and Y_{FEM} the simulated performance values. To estimate the effect of the control factors and important interactions on the $\%E$, percentage of contribution of each control factor on variance of the corresponding error percentage should be computed. The control factor that returns highest contribution percentage is designated as the prime factor. For the sake of more convenience, percentages are drawn in pie charts and shown in Figure 15. For instance, the control factors which are expected to play important roles in minimizing the percentage differences of feed force $\%E_f$, cutting force $\%E_c$, thrust force $\%E_t$, tool temperature $\%E_T$, and chip thickness $\%E_h$ during turning EN 1.4462 were μ_t, κ_t, μ_t and $h_{tc} \times v_c$, μ_t and h_{tc}, and μ_t, respectively.

Results of $\%E$ computations vs. the experimental order are illustrated in Figure 16. It can be seen that the predicted feed and thrust force components are generally underestimated except for some few cases at $D_{\mathrm{crit.}} = 150$ MPa. Fluctuations of $\%E_f$ and $\%E_t$ were generally in the range from −40 to 195%. On the other hand, $\%E_c$ has shown better agreements with the range of approximately −40 to 45%. Furthermore, comparing the T_{tool} and $h_{\max}$, the $\%E$ between experimental and FEM values also fluctuate between overestimation and underestimation, with very few cases where the absolute $\%E$ is less than 5%. These results show the strong impact of control factors on the outputs of the

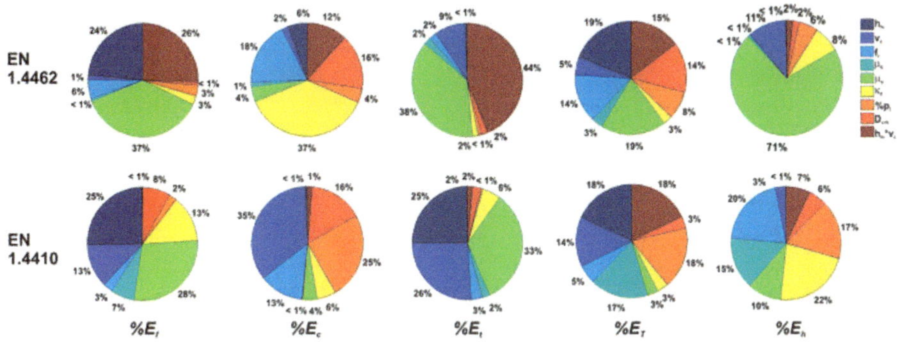

Figure 15. The percentage contribution of FEM control factors of the percentage error.

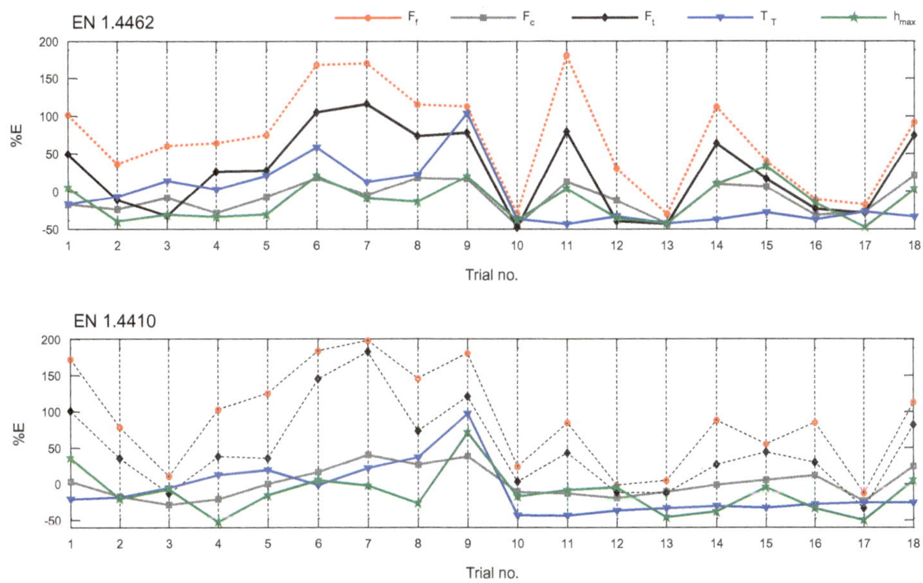

Figure 16. Percentage error between the experimental and predicted values for cutting performances.

FEM. The next sections describe the study approach intended to simultaneously minimize all the differences between measured and predicted performances.

5.1. VIKOR-based Taguchi

The VIKOR method was developed for multi-criteria optimization of complex systems. It determines the compromise ranking-list, the compromise solution, and the weight stability intervals for preference stability of the compromise solution obtained with the initial (given) weights. This method focuses on ranking and selecting from a set of alternatives in the presence of conflicting criteria. It introduces the multi-criteria ranking index based on the particular measure of 'closeness' to the 'ideal' solution. On the other hand, VIKOR-based Taguchi provides a systematic and efficient methodology for determining the optimum combination of the control factors such that the output is effective and has high performance and also is robust to the noise factors. The multiple-percentage of error problem at this stage is converted into a single-response problem as follow:

Step 1. Construction of the decision matrix: The decision matrix in this case is comprised of the values of S/N ratios for computed percentages of errors, that is, η_{ij} of $\%E_j$ where $i = 1, 2, \ldots,$ number of simulation experiments ($m = 18$), $j = 1, 2, \ldots,$ number of performances percentage of errors ($n = 5$).

$$D = \begin{bmatrix} \eta_{11} & \eta_{12} & \cdots & \eta_{1n} \\ \eta_{21} & \eta_{22} & \cdots & \eta_{2n} \\ . & . & \cdots & . \\ . & . & \cdots & . \\ \eta_{m1} & \eta_{m2} & \cdots & \eta_{mn} \end{bmatrix} \tag{10}$$

Step 2. Calculation of normalized ratings by the vector normalization:

$$r_{ij} = \left(\frac{\eta_{ij} - \min(\eta_j)}{\max(\eta_j) - \min(\eta_j)} \right) \tag{11}$$

Step 3. Calculate the relative weight of performances using the standard deviation method:

$$w_j = \frac{\mathrm{STDV}_j}{\sum_{j=1}^{n} \mathrm{STDV}_j} \tag{12}$$

where $STDV_j$ is the standard deviation of performance j. The summation of the individual weights should always yield one.

Step 4. Calculate weighted normalized ratings (ϑ_{ij}):

$$\vartheta_{ij} = w_j r_{ij} \tag{13}$$

Step 5. Identification of positive ideal and negative ideal solutions (A^* and A^-): The positive ideal solution, A^* (A_i^*; $i = 1, 2, \ldots, m$), is made of all the best values (maximum S/N ratio) and the negative ideal solution, A^- (A_i^-; $i = 1, 2, \ldots, m$), is made of all the worst values(minimum S/N ratio) at the responses in the weighted normalized decision matrix. They are calculated using Equations (14) and (15). where J is the set of cost type criteria and J' is the set of benefit type criteria.

$$A^* = \{\vartheta_1^*, \ldots, \vartheta_i^*\} = \{(\max \vartheta_{ij} | j \in J), (\min \vartheta_{ij} | j \in J')\} \tag{14}$$

$$A^- = \{\vartheta_1^-, \ldots, \vartheta_i^-\} = \{(\min \vartheta_{ij} | j \in J), (\max \vartheta_{ij} | j \in J')\} \tag{15}$$

where $J = \{j = 1, 2, \ldots, n | \vartheta_{ij},$ if desired response is large$\}$; $J' = \{j = 1, 2, \ldots, n | \vartheta_{ij},$ if desired response is small$\}$.

Step 6. Compute the values of utility measure S_j and regret measure R_j by respective relations:

$$S_i = \sum_{j=1}^{n} w_j \frac{\left(\vartheta_j^* - \vartheta_{ij}\right)}{\left(\vartheta_j^* - \vartheta_j^-\right)} \tag{16}$$

$$R_i = \max_j \left[w_j \frac{\left(\vartheta_j^* - \vartheta_{ij}\right)}{\left(\vartheta_j^* - \vartheta_j^-\right)} \right] \tag{17}$$

Step 7. Compute the VIKOR index (Q_V) values for attribute $j = 1, 2, \ldots, n$, by the relation

$$Q_V = \xi \frac{(S_i - S^*)}{(S^- - S^*)} + (1 - \xi) \frac{(R_i - R^*)}{(R^- - R^*)} \tag{18}$$

where $S^* = \min_j S_j$, $S^- = \max_j S_j$; $R^* = \min_j R_j$, $R^- = \max_j R_j$.

ξ is introduced as weight of the strategy of the maximum group utility, usually $\xi = 0.5$.

Step 8. Rank the alternatives, sorting by the values S, R, and Q_V. The results are three ranking lists.

Step 9. For given attribute weights, propose a compromise solution, alternative A_{b1}, which is the best ranked by Q_V, if the following two conditions are satisfied:

Condition 1: "Acceptable advantage": $Q_V(A_{b2}) - Q_V(A_{b1}) \geq (1/(m-1))$, where A_{b1} and A_{b2} are the first and second-best alternatives in the ranking order by Q_V.

Condition 2: "Acceptable stability in decision making": alternative A_{b1} must also be the best ranked by S and/or R. This compromise solution is stable within a decision-making process, which could be 'voting by majority rule' (when $\xi > 0.5$ is needed) or 'by consensus' (when $\xi \approx 0.5$) or 'with veto' (when $\xi < 0.5$). If one of the conditions is not satisfied, then a set of compromise solutions is proposed, which consists of:

(a) Alternatives A_{b1} and A_{b2} if only condition 2 is not satisfied.
(b) Alternatives A_{b1}, A_{b2}, …, A_{bi} if condition 1 is not satisfied; A_p is determined by the relation $Q_V(A_{b2}) - Q_V(A_{b1}) \approx (1/(m-1))$ (Rao, 2013).

The procedure of calculating VIKOR indices for each work material case is started with the representation of decision matrix. S/N values are normalized using Equation (11). The criteria weights are then specified using the standard deviation method as per Equation (12). Thereafter, normalized decision matrices are calculated using Equation (13). The best and the worst values of all the criteria in weighted normalized decision matrices are identified using Equations (14) and (15). The values of utility (S) and regret (R) measures are calculated using Equations (16) and (17), respectively. The VIKOR index values (Q_V) are calculated using Equation (18). For both work material cases, the values of the first and second ranked alternatives are in acceptable advantage range, that is, $Q_V(A_{b2}) - Q_V(A_{b1})$ is always greater than $1/(18–1) = 0.05882$. For example, the Q_V (A_{b2}) and Q_V (A_{b1}) of EN 1.4462 are 0.106 and 0.012, respectively, which satisfies condition 1, that is, $0.106–0.012 = 0.094$ is greater than 0.05882. Both regret and utility measure values by consensus prove the stability in decision-making. Therefore, the values of Q_V are directly used in the next analyses. Based on the outcomes of the

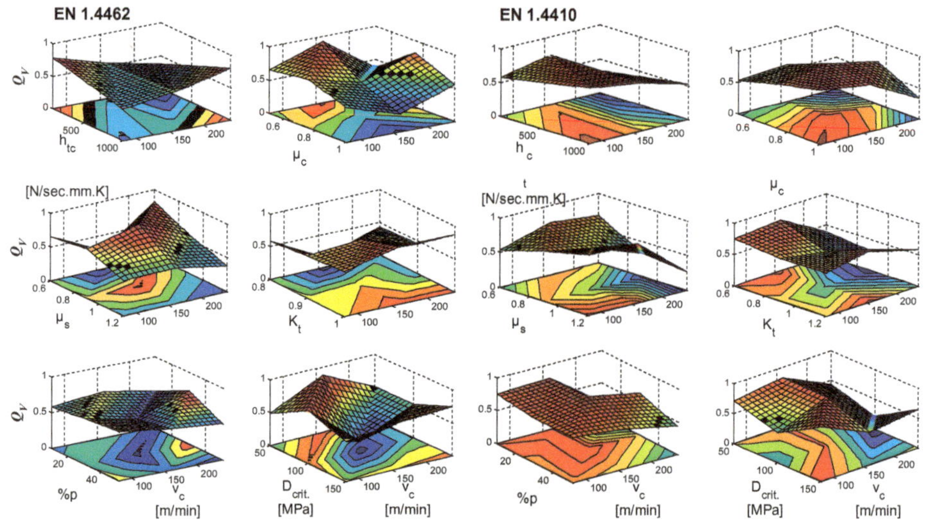

Figure 17. Surface and contour plots of the VIKOR indices (Q_V).

VIKOR method, the best alternatives when cutting EN 1.4410 and EN 1.4462 were alternatives No. 15 and No. 6, respectively. Performing the ANOM, it can be proved that the optimum *factor*^{level} sets when machining EN 1.4462 and EN 1.4410 are h_{tc}^2 v_c^1 f_r^1 μ_c^2 μ_t^2 κ_t^3 $\%p^2$ $D_{\text{crit.}}^3$ and h_{tc}^2 v_c^2 f_r^3 μ_c^2 μ_t^3 κ_t^3 $\%p^2 D_{\text{crit.}}^3$, respectively. The combination of these control factors should minimize the difference between experimental and numerical results. Figure 17 maps the effect of cutting speed interactions with the rest of control factors on the Q_V values. The dark blue areas represent the favored regions where the difference between experimental and numerical performances is minimum.

5.2. Firefly algorithm neural network system

Unfortunately, owing to the missing impact of interactions in Table 5, the choice of the selecting the optimal set is not always straightforward. Additionally, the above optimum sets are case specific to the given cutting conditions. To comprehend the Q_V, so that it covers all the domain of the control factors and their corresponding interactions, an effective modeling technique is required. Unfortunately, the traditional modeling techniques such as response surface methodology were seen not quite efficient to accurately predict all values of Q_V. Furthermore, to account for all interactions, adopting quadratic model for example, comprise 44 terms which makes the problem even more complex. Therefore, a multi-layer perceptron (MLP) artificial neural network (ANN) is proposed to accomplish the task. Once, this step is accomplished, a recently developed stochastic optimization algorithm has to be integrated with the proposed ANN to minimize Q_V and determine the optimum set at any required cutting condition.

5.2.1. Artificial neural network

Neural network is a logical structure with multi-processing elements, which are connected through interconnection weights. The knowledge is presented by the interconnection weights, which are adjusted during the training phase. The process of training a neural network can be broadly classified into two typical categories: supervised learning and unsupervised learning. The first requires using both the input and the target values for each sample in the training set. The most common algorithm in this group is the back propagation, used in the Multi-Layer Perceptron (MLP), but it also includes most of the training methods for recurrent neural networks, time delay neural networks, and thrust basis networks. Unsupervised learning is used when the target pattern is not completely known (Quiza & Davim, 2011).

To establish a useful relationship between independent variables (h_{tc}, v_c, f_r, μ_c, μ_t, κ_t, %p, and $D_{crit.}$) and dependent variable (Q_V), supervised feed-forward MLP-neural networks with back propagation (BP) as learning algorithm were adopted. The back propagation algorithm consists of two phases: the forward phase, where the activations are propagated from the input to the output layer, and the backward phase, where the error between the observed actual and the requested nominal value in the output layer is propagated backwards in order to modify the weights and bias values. This procedure is repeated many times until the sum of squared error term reaches an acceptable level.

The MLP-neural networks used in this study have two layers: one hidden layer and one output layer (see Figure 18). The hidden layer uses a sigmoid-type transference function:

$$F(\chi) = \frac{1}{1 + \exp(-b - \sum \omega_i \chi_i)} \tag{19}$$

while the output layer uses a linear function:

$$F(\chi) = b + \sum \omega_i \chi_i \tag{20}$$

where ω and b are the weights and biases of the network, respectively. To carry out the training process, all the inputs were normalized using the following equation:

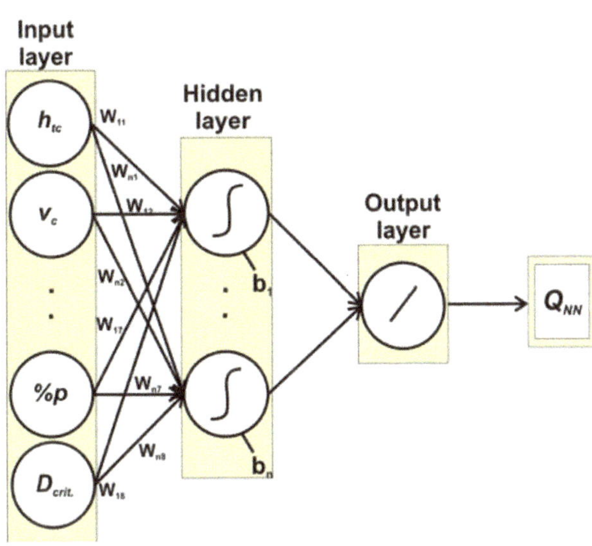

Figure 18. The neural network architecture used in the FANNS.

$$\chi_i = \frac{2(\chi - \chi_{\min})}{(\chi_{\max} - \chi_{\min})} - 1 \tag{21}$$

The final output (Q_{NN}) of input vector (8*1)-number of hidden neurons (N)-output feed-forward neural network structure can be mathematically expressed as:

$$Q_{NN} = \log\text{sig}\left(\begin{bmatrix} \omega_{11} & . & . & \omega_{18} \\ . & . & . & . \\ . & . & . & . \\ \omega_{81} & . & . & \omega_{88} \end{bmatrix} \times \begin{bmatrix} h_{tc} \\ . \\ . \\ D_{\text{crit.}} \end{bmatrix} + \begin{bmatrix} b_1 \\ . \\ . \\ b_N \end{bmatrix}\right) \times [\omega_1 \quad . \quad . \quad \omega_N] + b_{N+1} \tag{22}$$

In this study, the MLP uses sigmoid transfer functions in the hidden layers. These functions are often called 'squashing' functions, because they compress an infinite input range into a finite output range. Sigmoid functions are characterized by the fact that their slopes must approach zero as the input gets large. This causes a problem when you use steepest descent to train a multi-layer network with sigmoid functions, because the gradient can have a very small magnitude and, therefore, causes small changes in the weights and biases, even though the weights and biases are far from their optimal values. The purpose of the resilient back propagation training algorithm (RPROP) is to eliminate these harmful effects of the magnitudes of the partial derivatives. Only the sign of the derivative can determine the direction of the weight update; the magnitude of the derivative has no effect on the weight update. The size of the weight change is determined by a separate update value. The number of learning steps is significantly reduced in comparison to the original gradient-descent procedure as well as to other adaptive procedures, whereas the expense of computation of the RPROP adaptation process is held considerably small. Another important feature, especially relevant in practical application, is the robustness of the new algorithm against the choice of its initial parameter (Riedmiller & Braun, 1993).

The scaling or normalization ensures that the ANN will be trained effectively without any particular variable skewing the results significantly. The weights and biases of the network are initialized to small random values to avoid immediate saturation in the activation functions. The neural networks trained using the gradient descendent with adaptive velocity and momentum back propagation algorithm to model Q_V; where it is not easy to obtain analytical and good empirical relations. The optimum architecture was found out by varying network characteristics in MATLAB using trial and error technique. It was found that when the training function, RPROP; the number of hidden neurons, 13; maximum number of epochs to train, 100; the learning rate, 0.01; the increment to weight change, 1.2; decrement to weight change, 0.5; initial weight change, 0.07; and maximum weight change, 50, the root mean square of errors were the minimum.

Because neural networks are purely empirical models, validation is critical to operational success. Figure 19 shows the comparison of experimental results and modeling in verifying the network generalization capabilities. The results are almost identical.

5.2.2. Firefly algorithm (FA)

FA is one of the recent swarm intelligence methods developed by Yang (2010) in 2008 and is a kind of stochastic, nature-inspired, meta-heuristic algorithm that can be applied for solving the hardest optimization. The algorithm is inspired by the flashing lights of

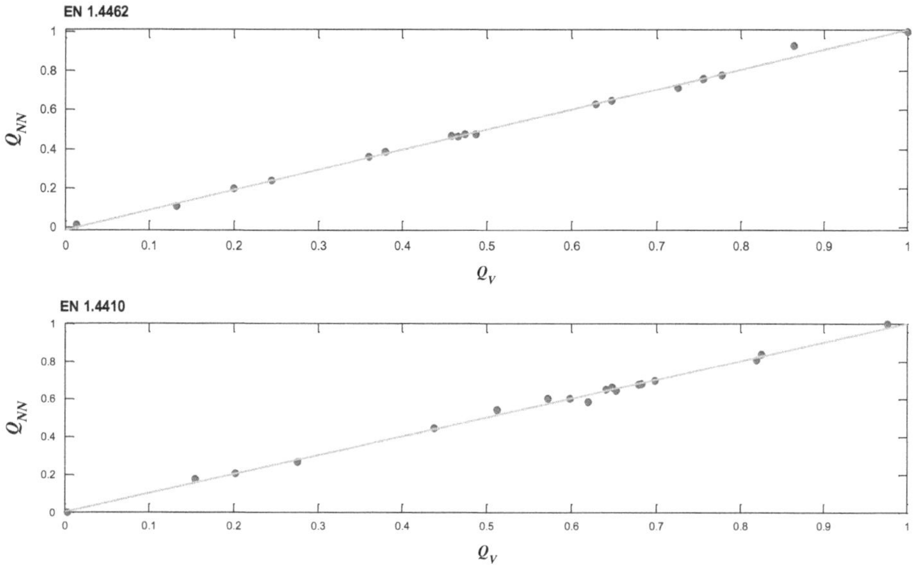

Figure 19. Scatter plot of actual vs. ANN-predicted VIKOR indices.

fireflies in nature. Such flashing light may serve as the primary courtship signals for mating. Besides attracting mating partners, the flashing light may also serve to warn off potential predators. The algorithm has been formulated by assuming:

(1) All fireflies are unisex so that one firefly will be attracted to other fireflies regardless of their sex.
(2) Attractiveness is proportional to their light intensities. The less bright will be moving toward the brighter one. It will move randomly if there is no brighter one.
(3) The brightness of a firefly is affected or determined by the landscape of the objective function.

The light intensity and attractiveness are in some way synonymous. While the intensity is referred to as an absolute measure of emitted light by the firefly, the attractiveness is a relative measure of the light that should be seen in the eyes of the beholders and judged by other fireflies. The attractiveness *β(δ)* and light intensities *I(δ)* of each firefly are considered to decrease monotonically depending on the distance δ as:

$$\beta(s) = \beta_0 e^{-\gamma_a \delta_r^2} \tag{23}$$

$$I(s) = I_0 e^{-\gamma_a \delta_r^2} \tag{24}$$

where β_0 and I_0 denote the maximum attractiveness and light intensities, respectively (i.e. at $\delta_r = 0$), and γ_a is the light absorption coefficient, which controls the decrease of the light intensity. The light intensity I of a firefly representing the solution s_i and is proportional to the value of fitness function $I(s_i) \propto f(s_i)$. The distance between any two fireflies s_i and s_j is expressed as the Euclidean distance by the base firefly algorithm, as follows:

$$\delta_{ij} = \|s_i - s_j\| = \sqrt{\sum_{k=1}^{k=d} (s_{ik} - s_{jk})^2} \tag{25}$$

where d denotes the dimensionality of the problem. The movement of the ith firefly is attracted to another more attractive firefly j. In this manner, the following equation is applied:

$$s_i = s_i + \beta_0 e^{-\gamma_a s_{ij}^2}(s_j - s_i) + \alpha\psi_i \tag{26}$$

where ψ_i is a random number drawn from Gaussian distribution and α is the randomization parameter. In summary, FA is controlled by three parameters: the attractiveness β, the absorption coefficient γ_a, and the randomization parameter α. Finally, the basic steps of the Firefly Algorithm (FA) can be summarized as the pseudo code and shown in Figure 20.

In order to obtain solutions that will provide useful information to the user during the phase of inverse identification of input parameters, neural network models should be integrated with the FA. The neural network models integrated with the FA optimizer were named FANNS and its architecture is shown in Figure 21.

The target of the optimization process in this study is to determine the optimal values of the input parameters that lead to the minimum value of ANN-predicted Q_{NN} at any given cutting condition. The optimization problem can be defined as:

$$\text{Objective} = \text{Max}\ (1 - Q_{NN}) \tag{27}$$

The decision variables are h_{tc}, v_c, f_r, μ_c, μ_t, κ_t, $\%p$, and D_{crit}. They are constrained within the range of simulation experiments (refer to Table 4):

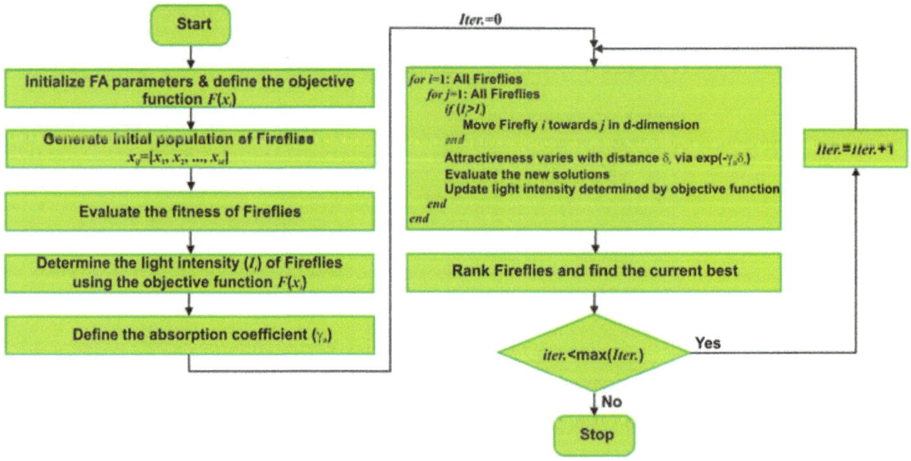

Figure 20. Flow chart of the Firefly Algorithm (FA).

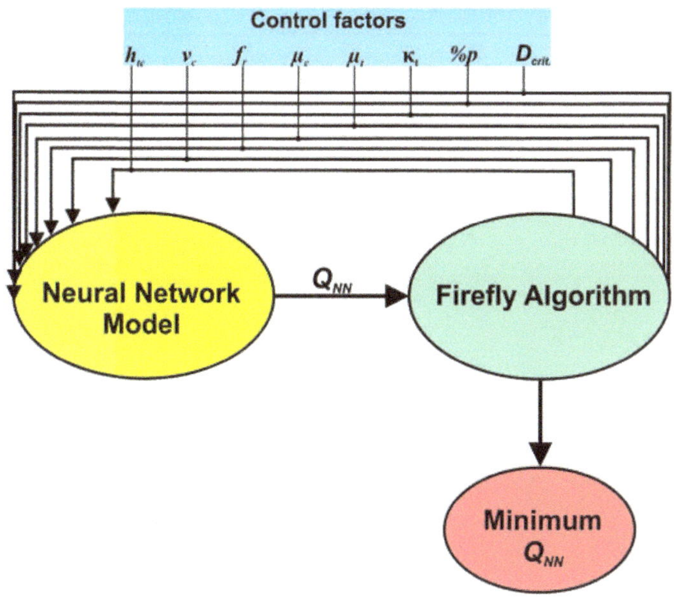

Figure 21. Firefly algorithm neural network system (FANNS).

$$\begin{gathered} 100 \le h_{tc} \le 1000 \text{ N/(s mm K)} \\ 80 \le v_c \le 240 \text{ m/ min} \\ 0.15 \le f_r \le 0.3 \text{ mm/rev} \\ 0.5 \le \mu_c \le 1 \\ 0.6 \le \mu_t \le 1.2 \\ 0.8 \le \kappa_t \le 1 \\ 10 \le \%p \le 50 \\ 50 \le D_{crit.} \le 150 \text{ MPa} \end{gathered} \tag{28}$$

5.2.3. Application of FANNS to the global optimization of VIKOR indices and validation of the results

The proposed FANNS approach has been applied to effectively model and optimize the VIKOR indices. The FA initializing optimization parameters were as follows: population size = 20, number of iterations = 2000, randomization (α) = 0.5, attractiveness (β_0) = 0.2, and attractiveness variation (γ_a) = 1. The optimal sets of control factors that lead to the optimum Q_V values are tabulated in Table 6, and the results of the FANNS are shown in Figure 22. The obtained optimization results showed that FANNS is highly reliable, converge consistently and quickly (the computation time was less than 3 min) to the optimum solution.

To validate the global FANNS results, the experimental cutting conditions should be identical to the numerical ones. Therefore, three longitudinal turning experiments per each work material were carried out at the exact optimum cutting conditions as listed in Table 6. The averages of the measured performances are computed and used to calculate the percentage difference between the numerical and experimental ones using Equation (9).

Table 6. Optimum sets of control factors.

Optimum control factor	Unit	EN 1.4462	EN 1.4410
Thermal contact conductance (h_{tc})	N/(s mm K)	652.306	647.945
Cutting speed (v_c)	m/min	181.175	130.601
Feed rate (f_r)	mm/rev	.21321	.27441
Coulomb friction coefficient (μ_c)	–	.63151	.91255
Shear friction coefficient (μ_t)	–	.74952	1.01455
Taylor–Quinney coefficient (κ_t)	–	.95352	.96525
Reduction in flow stress (%p)	–	19.3549	28.8121
Critical damage value ($D_{crit.}$)	MPa	69.2885	89.1997
Maximum VIKOR index (Q_{NN})	–	.92211	.95785

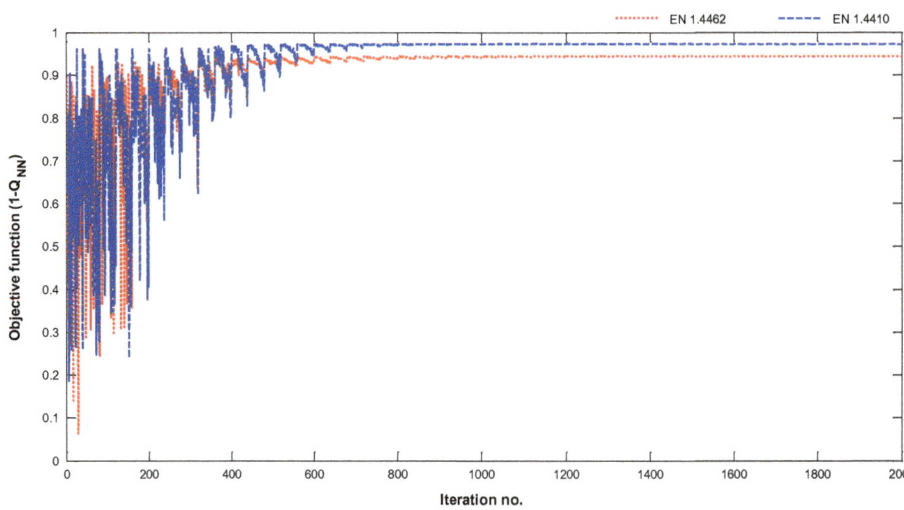

Figure 22. Minimization of the VIKOR indices using FANNS.

The percentage difference between numerically obtained cutting forces, chip temperature, and maximum chip thicknesses, and the experimental ones for each work material are tabulated in Table 7. It can be seen that the calculated performances are in close agreement to the experimental results. The global optimum difference percentages are also compared with difference percentages of the rank No. 1. The lower percentages of difference confirm the advantage of FANNS application in identifying the input parameters while simulating the machining of duplex stainless steels.

5.3. Extension of FANNS application: a case study

One of the aims of this study was to define a robust approach, so that the optimum set of control factors at any given cutting speed and feed rate is accurately determined. In this section, the adaptability of the described approach to the changing cutting speeds and feed rates is examined. Three arbitrary experimental cutting tests per each work material were conducted. The optimum sets of control factors at the exact cutting speeds

Table 7. Validation numerical and experimental results.

		F_f (N)		F_c (N)		F_t (N)		T_{tool} (°C)		h_{max} (mm)	
		Exp.	Num.	Exp.	Num.	Exp.	Num.	Exp.	Num.	Exp.	Num.
EN 1.4462		385.167	369.941	935.654	9,720,614	298.646	290.139	134.635	140.519	.3346	.3516
	%E	−4.116	3.745	−2.139	4.187	5.125					
EN 1.4410		761.152	723.686	1485.63	1569.54	517.254	491.827	142.714	148.709	.5535	.5843
	%E	−5.177	5.346	−5.169	4.031	5.254					

Table 8. Optimum sets of control factors at fixed cutting speed and feed rates.

	Parameter	Unit	A.1.	A.2.	A.3.
	v_c	m/min	75	150	225
	f_r	mm/rev	.325	.175	.2
EN 1.4462	h_{tc}	N/(s mm K)	665.558	771.698	820.049
	μ_c	–	.53225	.66125	.69922
	μ_t	–	.69525	.70525	.84552
	κ_t	–	.98308	.90325	.97825
	$\%p$	–	24.7349	21.1528	19.8571
	$D_{\text{crit.}}$	MPa	65.7898	77.6984	71.2524
EN 1.4410	h_{tc}	N/(s mm K)	795.523	826.375	862.815
	μ_c	–	.99125	.84258	.87134
	μ_t	–	1.04321	1.06925	1.08125
	κ_t	–	.98995	.94582	.92545
	$\%p$	–	30.7425	22.2481	26.4358
	$D_{\text{crit.}}$	MPa	92.506	100.678	111.452

and feed rates were determined using the proposed FANNS (see Table 8); meanwhile numerical models, each at the exact optimum set of optimized control factors, were prepared and executed. Figure 23 shows the measured and predicted cutting performances for EN 1.4410 and EN 1.4462. The average of absolute percentage differences between the experimental and numerical results for each material case is 3.8652 and 4.9956%, respectively. Considering the wide range of applied cutting speeds and feed rates, the proposed methodology of inversely identifying the simulation input factors shows excellent results with respect to the examined performances.

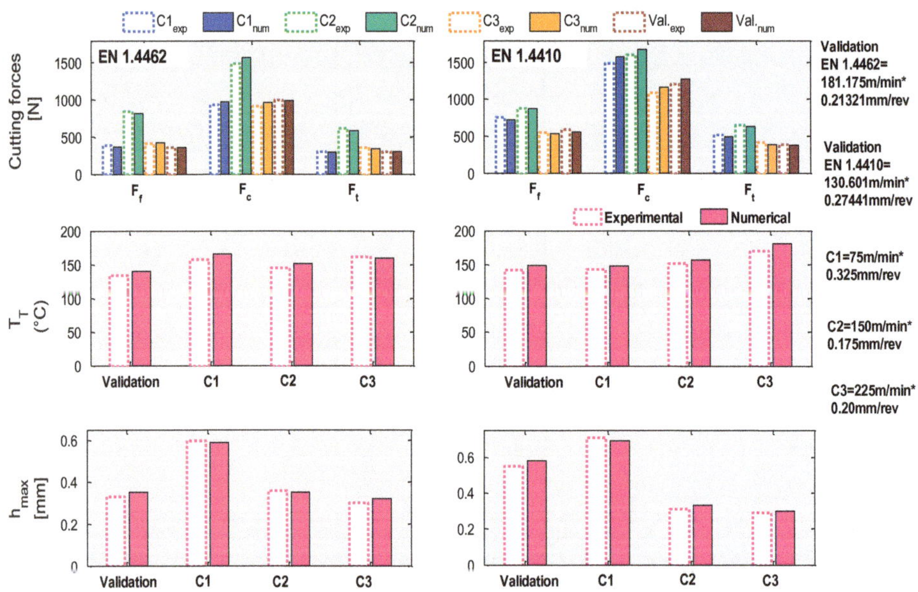

Figure 23. Experimental validation with respect to the performances for EN 1.4462 and EN 1.4410 DSS.

6. Inverse identification of the Usui's wear law constants

Tool wear is a major consideration in all machining operations. It adversely affects tool life, quality of machining surface, and its dimensional accuracy, and consequently, the economics of cutting operations. Tool wear and the changes in tool geometry during cutting manifest themselves in different ways, generally classified as flank wear, crater wear, nose wear, notching, plastic deformation of the tool tip, chipping and gross fracture (Kalpakjian & Schmit, 2006). The allowable average wear land for various machining operations is usually determined by the width of flank wear (*VB*) which occurs on the relief (flank) face of the tool.

To characterize tool wear for a cutting operation, there exist two main approaches: empirical tool life models and tool wear rate models. In order to derive a reliable empirical tool life model, a large number of experimental tests are essential which is usually time-consuming, cost-intensive, restricted to the investigated tool–workpiece combinations and cannot predict the influence of work material or tool materials on the values of constants in the models. On the other hand, tool wear rate models involve process variables that are not directly measurable or very difficult to measure during a cutting operation, such as normal stress and temperature on the tool face, chip temperature, and chip sliding velocity along the tool rake face. However, in the last several years, the FEM has been successfully applied to estimate such variables (Attanasio, Ceretti, Fiorentino, Cappellini, & Giardini, 2010). Consequently, a better understanding of the fundamentals of cutting mechanics, engineering analyses of tool wear, and systematic approach for the process optimization is possible.

Considering the Show's equation of adhesive wear, tool wear rate model derived by Usui and co-workers involves variables such as temperature T, normal stress σ_n, sliding velocity v_s at the contact surface, and two constants A and B. It is expressed as:

$$\frac{\mathrm{d}W}{\mathrm{d}t} = A\sigma_n v_s \exp\left(-\frac{B}{T}\right) \tag{29}$$

Their results have shown that both the flank and crater wear rates have the same functional form. The variables in Usui's wear rate model can be predicted by FEM simulation of cutting process or combining analytical method and FEM (Yang & Liu, 2002). In this study, the procedure of calculating the Usui's wear model constants can be simply summarized as follows:

(1) Conduct tool life experimentations under dry condition using a typically recommended insert grade to machine stainless steels to determine the wear rate (*dW/dt*).
(2) To estimate the variables in Equation (29), prepare identical to the experimentations, fully coupled thermo-mechanical, and FANNS-optimized 3D-FE models, and run the simulations.
(3) Collect the numerically estimated σ_n, T, and v_s.
(4) Fit the nonlinear Usui's model to determine the constants A and B.

Seven arbitrary multi-pass dry turning tests of EN 1.4410 and EN 1.4462 DSS bars were performed on a variable spindle speed CNC lathe using rhombic 80° CNMA 120412-IC20 uncoated cemented carbide inserts. The solid bars had outside diameter of 55 mm. Constant material volumes of 29,405 mm^3 (i.e. 6 mm in thrust and 60 mm in axial directions) were removed per each experimental trial. An optical microscope with magnification 200 times was then used to measure the maximum wear on the flank surface after each cutting pass. Thereafter, logarithmic plots of tool wear vs. cutting time were drawn, and the slopes (*dW/dt*) were determined. Meanwhile, the 3D-FE cutting simulations were carried out based on the proposed procedure. The tool is considered rigid with no wearing possibility. Values of T, σ_n, and v_s were directly extracted from simulation results. Matlab function nonlinear model fit was then utilized to fit the model and test its adequacy through ANOVA. The adjusted correlation factor (R_{adj}) was in very good agreement with predicted correlation factor ($R_{pre.}$) which supports the prediction power of the model and was generally above 0.94. The root mean square of errors (RMSE) for EN 1.4462 and EN 1.4410 Usui wear models were 0.000314 and 0.000469, respectively. The final forms of Usui's wear model for each workpiece material case are given as:

$$\left(\frac{\mathrm{d}W}{\mathrm{d}t}\right)_{EN\ 1.4462} = 6.0478 \times 10^{-9} \sigma_n v_s \exp\left(-\frac{1172.8}{T}\right) \tag{30}$$

$$\left(\frac{\mathrm{d}W}{\mathrm{d}t}\right)_{EN\ 1.4410} = 7.1704 \times 10^{-9} \sigma_n v_s \exp\left(-\frac{992.5203}{T}\right) \tag{31}$$

7. Optimization of Turning EN 1.4410 and EN 1.4462 DSS

In the second stage of this study, a hypothetical benchmark analysis based on the Taguchi optimization procedure is suggested. The objective was to select the best combination of criterions such as chip breakers type, tool geometry, process and cutting conditions, and cutting tool orientation angles for an effective machining of EN 1.4462 and EN 1.4410 DSS. For this purpose, Taguchi's mixed design $L_{18}(2^1 \times 3^7)$ is once again seen the most appropriate because of its attractive characteristics which combines the highest possible number of levels along with largest number of criterion and smallest number of experiments. Another attractive characteristic of the design is that the interaction between column 1 and 2 is orthogonal to all columns and hence can be estimated without sacrificing any column. The interaction can be estimated from the 2-way table of columns 1 and 2. The studied criterion and their corresponding levels are listed in Table 9. Columns 1 and 2 can be combined to form a 6-level column. Each level in column three represents an insert designation which is often commercially available. The next sub-sections briefly describe the studied criterion.

Table 9. Criterion and their corresponding levels.

			Levels		
Factors	Symbol	Unit	1	2	3
Chip breaker type	*CB*	–	M3 M	PP	
Insert shape	*Geo.*	–	DNMG	CNMG	WNMG
Coling medium	*CM*	–	Still air	Water	Cryogenic
Cutting speed	v_c	m/min	75	150	225
Feed rate	f_r	mm/rev	.1	.175	.25
Rake angle	α_n	degree	0	−6	−12
Inclination angle	λ	degree	0	−6	−12

7.1. *Control factors*

7.1.1. *Chip-breaker type*

In particular designed for machining stainless steels, chip breakers such as M3 M type are adapted with geometric features that improve the tool's life due to a reinforced cutting edge at the area where notch wear tends to occur when machining stainless steel, causing poor surface finish and risk of edge breakage (see Figure 24). On the other hand, PP-type chip breakers, which are also recommended for machining stainless steels, are characterized by having 3-step smart dot structures which provides smooth chip evacuation with a wide range of feed rates, and smooth taper cutting edge to reduce cutting forces.

7.1.2. *Cutting insert shapes*

In metal cutting, the primary goal was to achieve the most efficient separation of chips from the workpiece. One of the main factors that contribute in optimizing chip morphology is the insert shape. For this reason, the selection of the right cutting tool geometry is critical. The three basic insert shapes which have to be investigated in this study were diamond 55°, rhombic 80°, and trigon 80°. Other geometric features, such as clearance angle, tolerance, size, thickness, nose radius, and cutting edge preparation, were identical.

7.1.3. *Cooling medium*

Following the determination of the optimum thermal contact conductance of the tool–chip interface (h_{tc}) using FANNS approach, the influence of environmental heat convection in dry, wet, and cryogenic conditions on the cutting processes is investigated and optimized. A window for heat exchange was defined as shown in Figure 25. It was restricted to the secondary and tertiary deformation zones on the inserts using a cylindrical shape of 1.2 mm radius and 3 mm length and was not intended to change any other boundary conditions in the finite element model. Local convection coefficients in wet and cryogenic cutting were assigned in the areas covered by the window, while the rest of the cutting tool is still subjected to air convection. The convection coefficients of dry, wet, and cryogenic cutting mediums are taken from the literature as: 0.02, 10, and 5000 N/(s mm K), respectively (Pu et al., 2014; SFTC, 2010).

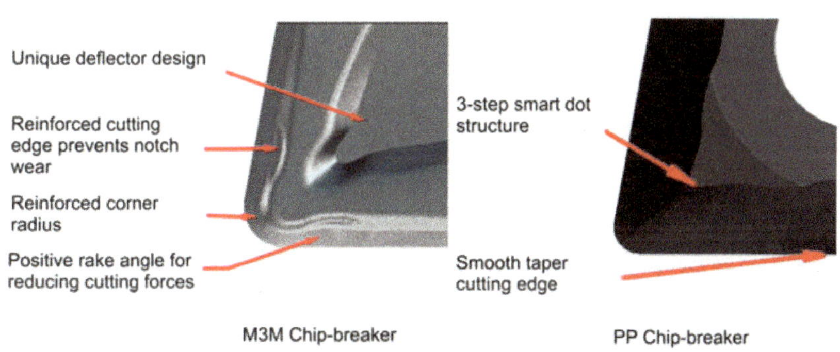

Figure 24. Investigated chip breakers characteristics.

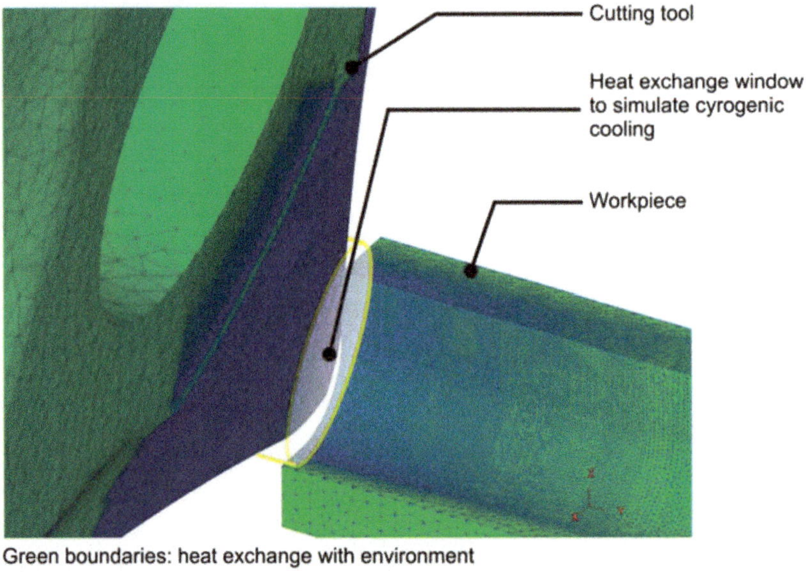

Figure 25. An Illustrative thermal boundary condition of test no. 13.

7.1.4. Cutting conditions

For all simulation trials, the finishing and semi-finishing set of cutting conditions are selected according to the maximum and minimum levels of cutting speed (v_c) and feed rate (f) as follows:

(a) Finising:

$$\begin{gathered} 150 \leq v_c \leq 225 \text{ m min} \\ 0.1 \leq f_r \leq 0.175 \text{ mm/rev} \end{gathered} \tag{32}$$

(b) Semi-finishing:

$$\begin{gathered} 75 < v_c < 150 \text{ m min} \\ 0.175 < f_r \leq 0.25 \text{ mm/rev} \end{gathered} \tag{33}$$

7.1.5. Tool orientation angles

Tool inclination angles affect many performances such as cutting forces, chip thickness, tool–chip contact length, chip flow angle, etc. When machining DSS, the formed chips are tough, abrasive to the tooling, and tend to entangle around cutting tool, workpiece, and tool post. A proper cutting insert inclination and rake angle for example may cause the chips to flow to a position at which they are less likely to become entangled and rub and scratch the finished surface of the workpiece. Geometry of oblique cutting process with straight cutting edge is shown in Figure 26. The process is performed with tools having the tool cutting edge angle $K_r \neq 90°$ and tool inclination angle $\lambda \neq 0°$. The rake angle may be measured in more than one plane, and hence, more than one rake angles can be defined for a given tool and angle of obliquity. The different rake angles in oblique cutting are named as: normal (α_n), velocity (α_v), and effective (α_e) rake angles. The flow of chip is at an angle to the normal to the cutting edge. The angle between normal to the cutting edge and chip velocity vector is called chip flow angle (Δ). The shear angles could also be measured in different planes, such as a plane normal to the cutting edge (φ_n) and plane of effective rake angle (φ_e). Since chip flow and shear angles are often designated as dependent cutting variables, and the rake and inclination angles as independent cutting variables, therefore, when constructing the 3D-FE models, cutting tools are only positioned according to the combination of inclination and rake angles.

7.2. 3D-FE models

Due to the incorporation of chip breaker geometry in the array of control factors, a great deal of geometrical resolution is essential. Therefore, the 3D-CAD models of the cutting inserts are created in Solidworks first and exported as Standard for the Exchange of Product model data (STEP) files into Transmagic software to increase the resolution. The models are saved in Standard Tessellation Language (STL) format and exported to the Deform-3D environment. Both workpiece and cutting tools are meshed with grids of varying tetrahedral mesh densities so that the highest densities of meshes are assigned to the locations where temperature, strain, strain rate, …, etc. have the highest gradient. The total number of mesh elements in workpiece and cutting tool were 80,000 and 250,000, respectively. In addition to the described geometrical parameters, the cutting tools had identical nose radius of 0.8 mm, cutting edge radius of 30 μm, and clearance angle of 5°. The *WC* cutting tools are considered rigid and are worn out under Usui's

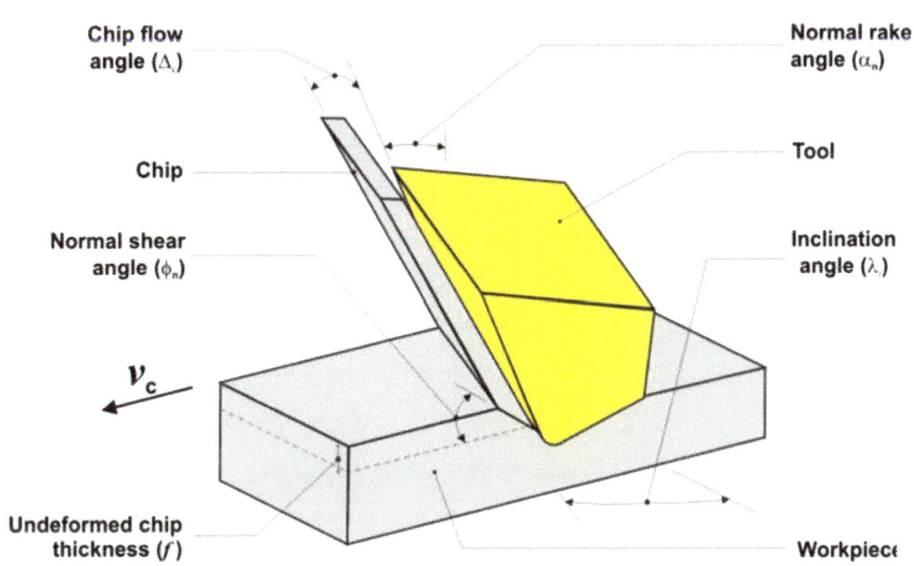

Figure 26. Oblique cutting process.

wear model. Other FEM settings such as elasto-visco plastic and thermo-mechanical properties of the workpiece, the boundary conditions, and process conditions are covered under previous sections of this study.

7.3. Simulation results and discussion

Once the turning simulations has been finished, cutting performances such as cutting forces, temperatures, …, etc. have been extracted. The S/N ratio of each performance is computed based on its beneficial or non-beneficial status. Almost all performances considered in this study were non-beneficial; therefore, *lower is better* concept applied and the values of S/N ratios are computed using Equation (8). To eliminate the complexity of problem, instead F_f, F_c, and Ft components, resultant cutting force (R) has been considered:

$$R = \sqrt{{F_f}^2 + {F_c}^2 + F_r^2} \tag{34}$$

The main effect of each control factor on each performance is determined by the ratio of sum of all S/N ratios corresponding to a factor at particular level to number of repetition of factor level. The optimum combinations corresponding to maximum average effect are directly depicted in the main effect plots as shown in Figure 27 below. Statistical analysis of variance (ANOVA) is performed to specify statistically significant control factors.

Figure 28 exhibits the percentage of contribution of each control factor on the variance of the corresponding cutting performance. In the following sub-sections, the effects of control factors on each single cutting performance are described.

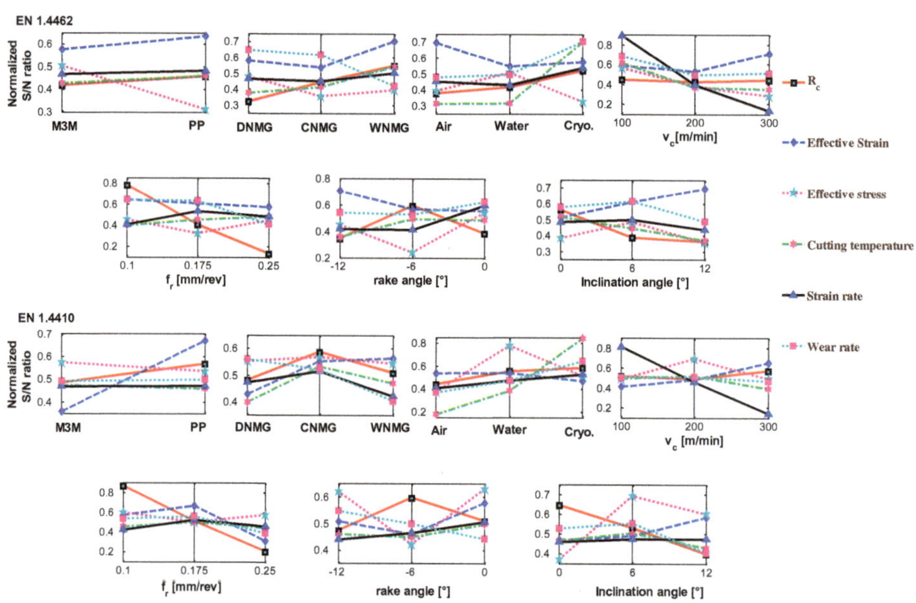

Figure 27. Main effect plots of the effect of control factors on the cutting performances during turning EN 1.4410 and EN 1.4462 DSSs.

7.3.1. Influence of control factors on the resultant cutting forces

Based on a quick review of the results, the following conclusion points can be drawn:

(1) The optimum insert designations which minimize *R* during turning EN 1.4462 and EN 1.4410 can be described in ISO norms as: WNMG 060408-PP and CNMG 120408-PP, respectively.

(2) In spite of the advantageous aspects of increasing cutting speed, applying coolant, and cryogenic conditions in increasing the productivity and reducing the overall *R* values, results have shown a limited influence of these control factors on the mean *R* values. This statement is supported by the slopes of the *R* curves in Figure 27.

(3) Among the considered control factors, feed rate had shown the strongest impact on the resultant cutting forces (see Figure 28).

(4) The optimum values of cutting tool orientation angles expressed by ranges of investigated rake and inclination angles are −6° and 0°, respectively.

7.3.2. Influence of control factors on the effective strain

Without relating the hardening parameters in the loading function to the experimental uniaxial stress-strain curve, the work-hardening theory of plasticity cannot be applied in practical terms. In order to correlate the test results obtained by different load programs, the introduction of any strain and stress variables, that are functions of plastic strain and plastic stress, and can be plotted against each other is considered useful. These variables are often called effective strain and effective stress.

DSSs are a group of stainless steel alloys that are characterized by high work-hardening rate. They are machined at very high strain rates which increase the work-hardening rate and causing higher resistance to plastic deformation. Generally, the following observation points can be made:

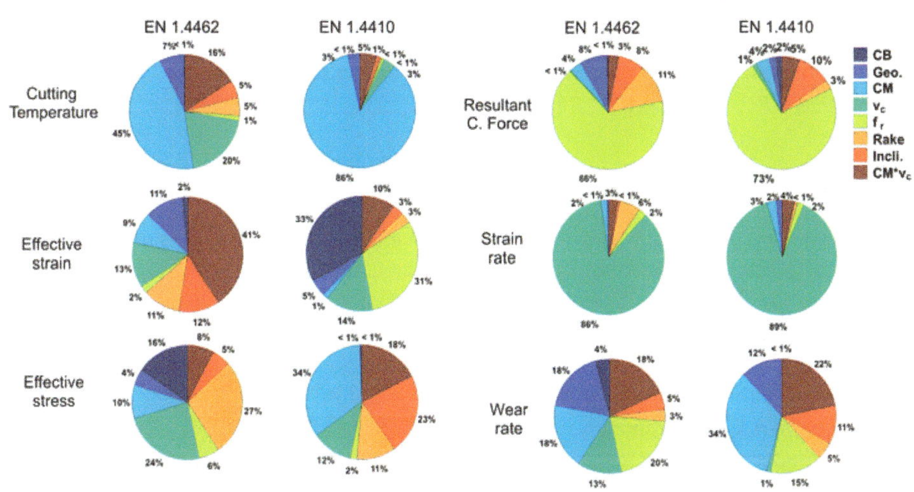

Figure 28. The percentage contribution of FEM control factors in the cutting performances variances.

(1) The average effective plastic strain when machining EN 1.4462 was 13.2% lower than the corresponding EN 1.4410.
(2) Employing chip breaker type PP has generally caused lower effective strain values.
(3) Detrimental control factors vary considerably per work material family.
(4) Higher values of effective strains are observed when the cooling medium on the tool is cryogenic.
(5) Main effects of effective strain decrease with increasing cutting speed.
(6) Mean effective strains are found to be minimum at the feed rate range of 0.1–0.175 mm/rev and inclination angle range of 6°–12°.
(7) Minimum effective strain value when cutting EN 1.4462 and EN 1.4410 at $\alpha_n = 0°$ and $\alpha_n = -12°$, respectively.

7.3.3. Influence of control factors on the effective stress

Effective plastic stress is considered a relevant characteristic of the cutting process which characterizes the level of resistance to cutting. The correlation between the state of this performance imposed by the cutting tool in the layer being removed and the fracture strain of the work material could be used to estimate the physical efficiency of the cutting process. In summary to the results, the following conclusions can be drawn:

(1) The effective stress encountered in cutting EN 1.4410 12% higher than EN 1.4462.
(2) The optimum control $factor^{level}$ combinations of cutting EN 1.4462 and EN 1.4410 were CB^1 $Geo.^1$ CM^2 v_c^1 f_r^1 α_n^3 λ^2 and CB^1 $Geo.^2$ CM^2 v_c^2 f_r^1 α_n^3 λ^2, respectively.
(3) Increasing the rake angle in negative direction and inclination angle in positive direction generally increases the effective stress.
(4) The minimum effective plastic stress of 1261.48 MPa and 1492.01 MPa when cutting EN 1.4462 and EN 1.4410 was recorded at experiment No. 9, respectively.
(5) When cutting EN 1.4462, process conditions such as v_c, f_r, CM, and $v_c \times CM$ interaction have contributed to the effective stress values by 48%. On the other hand, when cutting EN 1.4410, they had contributed by as much as 66%.

7.3.4. Influence of control factors on the tool–chip interface temperature

Mean contact temperature at the tool–chip interface (also is referred to as the cutting temperature) is the basic tribological characteristics of the tool–chip interface. It plays a major role in the formation of crater on the tool face and leads to failure of tool by softening and thermal stresses. This temperature is the most suitable parameter to correlate the tribological conditions with tool wear. Control factors that affect the tool–chip interface temperature are workpiece and tool material, tool geometry, cutting conditions, and cutting medium. Fortunately, most of these factors are included in the original design of simulation experimentations. Figure 29 depicts the state of cutting temperature distribution in the primary and secondary deformation zones of the workpiece. The results of the numerical studies on temperatures in cutting DSS can be summarized as follows:

(1) Mean cutting temperatures at dry cutting when machining EN 1.4462 and EN 1.4410 were 36.3 and 53.5% higher than at corresponding cryogenic process conditions.
(2) The contribution of water-based coolant in lowering the cutting temperature has not exceeded 7.8% in both material cases.
(3) Minimum mean cutting temperature when cutting EN 1.4462 and EN 1.4410 has been recorded at experimental run 16 and 15, respectively.
(4) Cooling medium, cutting condition, and their interaction accounts for most of contributions in cutting temperature variations.
(5) The optimum control $factor^{level}$ combinations of cutting EN 1.4462 and EN 1.4410 were CB^2 $Geo.^3$ CM^3 v_c^1 f^3 α_n^2 λ^1 and CB^1 $Geo.^2$ CM^3 v_c^1 f^2 α_n^3 λ^2, respectively.
(6) The average cutting temperature when machining EN 1.4462 was 9.85% lower than the corresponding EN 1.4410.

7.3.5. Influence of control factors on the strain rate

The flow stress curves shown in Figure 9 exhibit sensitivity to temperature and strain rate. This sensitivity is directly related with time and temperature dependency of the mechanisms that govern the deformation and the evolution of the deformation in the material. The main mechanism by which plastic strain takes place is thermally activated

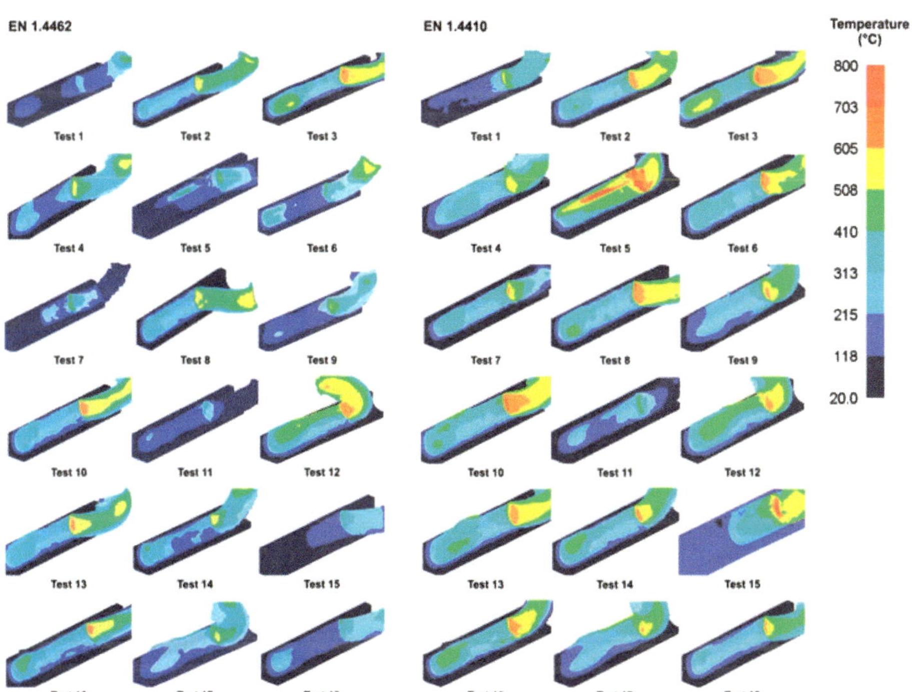

Figure 29. Contour of temperature distribution in three-dimensional models of cutting DSS.

motion of dislocations past obstacles that exist within the lattice over a wide range of strain rates and cutting temperatures. The material response is significantly affected by the nature and density of the obstacles (which may change as the deformation takes place). When dealing with metals, experimental results show that the stress required for plastic strain often reduces with the increase of temperature and with the decrease of plastic strain rate (Gilat & Wu, 1997). It can then be said that temperature and plastic strain rate greatly influence the material response. In general, the stress decreases with the increasing of temperature and decreasing the plastic strain rate. Actually, temperature and strain rate effects are coupled, since one influences the other. Temperature affects the rate of deformation, which is controlled mainly by a thermally activated mechanism. On the other hand, plastic strain at high rate generates significant heating and cause an increase in temperature which leads to mechanical instability and the localization of deformation into narrow sheets of material (the adiabatic shear bands), which act as pre-cursor for eventual material failure (Davim & Maranhão, 2009). As generalization to the given results in Figures 30 and 31, the following concluding points can be depicted:

(1) Average strain rate values under cutting EN 1.4462 and EN 1.4410 are of 2.067×10^5 1/sec and 3.23×10^5 1/sec, respectively.
(2) Increasing cutting speeds from 100 to 300 m/min at lower feed rates of 0.1 mm/rev had drastically increased the strain rate values.

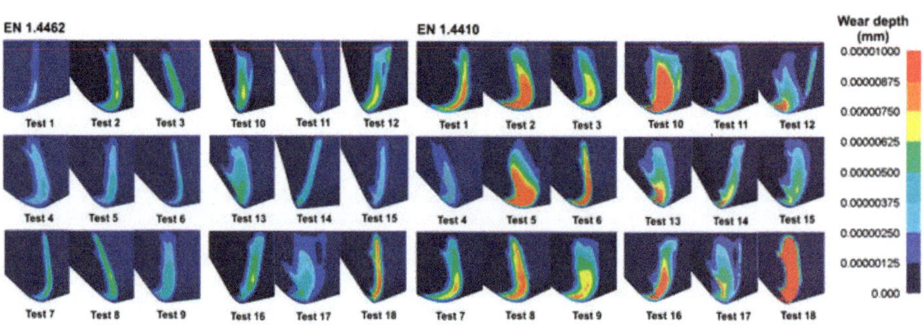

Figure 30. Contours of tool wear depth in three-dimensional models of cutting tools.

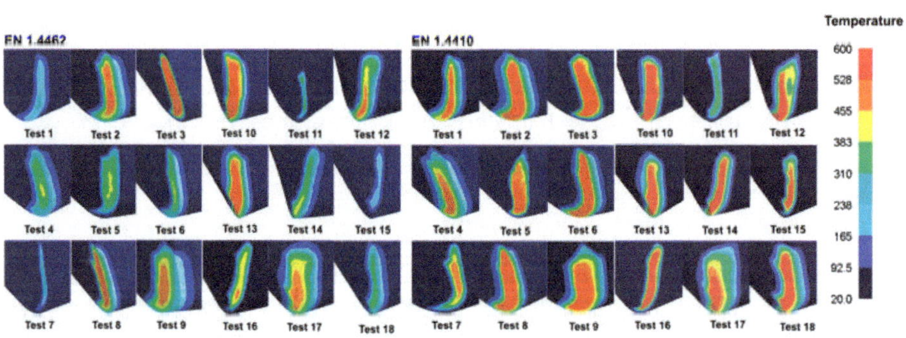

Figure 31. Contours of tool temperature distributions in three-dimensional models of cutting tools.

(3) Mean strain rate values in cryogenic cutting conditions during cutting EN 1.4462 and EN 1.4410 were, respectively, 9.158 and 11.863% lower than the corresponding still air conditions.
(4) $CB^2\ Geo.^3\ CM^3\ v_c^1\ f^2\ \alpha_n^3\ \lambda^2$ and $CB^2\ Geo.^2\ CM^3\ v_c^1\ f^2\ \alpha_n^{33}\ \lambda^3$ were the optimum control $factor^{\text{level}}$ combinations which have minimized the strain rate when cutting EN 1.4462 and EN 1.4410, respectively.

7.3.6. *Influence of control factors on the tools wear*

The Equation (34) has been used in conjunction with finite element simulation to model wear of WC tools in oblique cutting of DSS. The simulation results related to the tool wear were obtained in terms of tool wear rate, total wear depth, and tool temperature. Average values of tool wear rates are statistically analyzed and plotted as shown in Figures 27 and 28. Figure 30 maps the contours of tool wear depth in the nose area of the employed cutting tools after 7 mm longitudinal cutting. As it can be observed that the locations and intensity of tool wear depths are functions of work materials and control factor sets. Considering the effect of these factors, the summary of the findings is presented below.

(1) The optimum control $factor^{\text{level}}$ combinations which have minimized the tool wear rate during cutting EN 1.4462 and EN 1.4410 were $CB^2\ Geo.^1\ CM^3\ v_c^1\ f_r^1\ \alpha_n^3\ \lambda^2$ and $CB^2\ Geo.^1\ CM^3\ v_c^1\ f_r^1\ \alpha_n^1\ \lambda^2$, respectively.
(2) Major percentage of contributions of 69 and 72% in variance of tool wear rate were attributed to the process and cutting conditions and their interactions when cutting EN 1.4462 and EN 1.4410, respectively.
(3) The average tool wear rate when machining EN 1.4462 was 5.365% lower than the corresponding EN 1.4410.
(4) Contour profiles of tool wear depth shown in Figure 33 coincide with temperature distribution at rake and flank faces of the cutting tools in Figure 31. Maximum wear depths are occurred in positions at rake and flank surfaces where the tool temperature is the highest.

8. Numerical machining performance measure (NMPM)

The previous analyses of machining performances have showed different control $factor^{\text{level}}$ preferences. This will cause confusions to the process designer or the machinist who interested in finding the best compromise combination of control factors which simultaneously optimize the machining of DSS. Therefore, a new machining performance measure has to be defined. In this study, fuzzy logic system is employed to combine the normalized performances, such as the resultant cutting forces, effective stress, cutting temperature, and tool wear rate into a single characteristics index called machining performance index (NMPM).

Fuzzy logic unit is composed of four main sub units namely fuzzification, knowledge base, fuzzy inference engine, and defuzzification unit as shown in Figure 32. The fuzzifier converts the entered crisp value in to a suitable semantic value to be supplied to the fuzzy set for the first part of fuzzy rule. In the knowledge base units, information about of fuzzy rule base and membership functions is given. The fuzzy rule base is a database mainly for saving fuzzy rule, which is composed of if–then expression to indicate a conditional descriptive sentence. 'If' is called the first part, and 'then' is called the conclusion part. They are used to describe the relationship between input and output. Fuzzy values are determined by the membership functions (μ) that define the degree of membership of an object in a fuzzy set. So far there has been no standard method of choosing the proper shape of the membership functions for the fuzzy sets of the control variables; the present investigation adopts a triangular shape membership function (see Figure 33).

For a rule:

Rule$_j$: If x_1 is (A_{i1}) and x_2 is (A_{i2}), …, x_k is A_{ik} then $y_i = B_i$, $j = 1, 2, …, N$.

where N is the total number of fuzzy rules, x_i ($i = 1, 2, …, k$) are the input variables, y_i are the output variables, and A_{ik} and B_i are fuzzy sets modeled by membership functions μ_{Ai} (x_j) and μ_{Bi} (y_i), respectively. Based on the Mamdani implication method of inference, the membership function of the output of fuzzy reasoning can be expressed as:

$$\mu_{B_i}(y_i) = \max\{\min_i [\mu_{A_{i1}}(x_1),\ \mu_{A2}(x_2),\ \ldots,\ \mu_{A_{ik}}(x_k)]\} \tag{35}$$

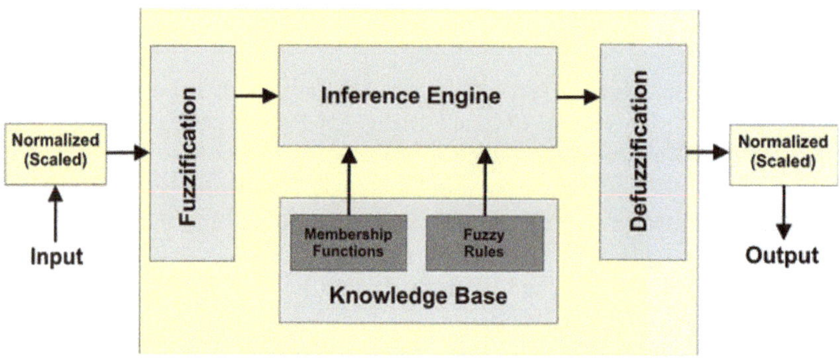

Figure 32. Fuzzy Logic Unit (FLU).

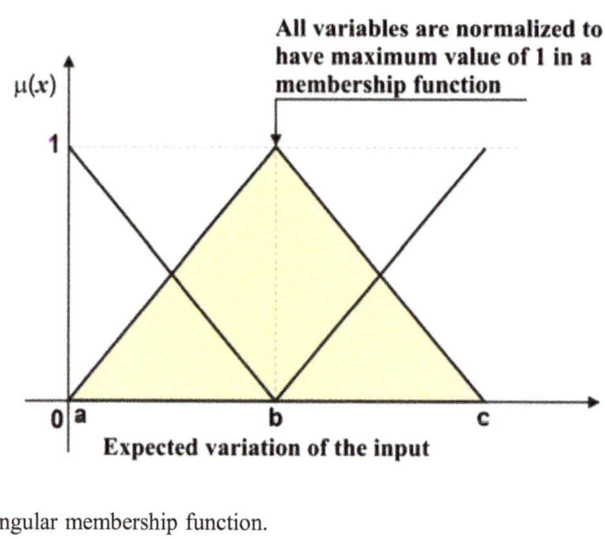

Figure 33. Triangular membership function.

Finally, a defuzzification method, called the center-of-gravity method, is adopted to transform the fuzzy inference output $\mu_{Bi}(y_i)$ into a non-fuzzy value y_0:

$$y_0 = \frac{\int y_i \mu_{B_i}(y_i)dy}{\int \mu_{B_i}(y_i)dy} \tag{36}$$

The yielded value is the final crisp output value obtained from the input variables. It is worth mentioning here that in most cases, the membership functions are defined within a normalized interval (universe of discourse). Therefore, before the fuzzification can be started, the input values have to be scaled (normalized) so that they fit into the universe of discourse (Gupta, Singh, & Aggarwal, 2011).

Matlab software was used to construct the inference model of the NMPM. The S/N ratios of performance values were first adjusted to a notionally common scale between null and one, so that the digit 'one' represents the most desirable and 'null' is the least desirable alternative. The four input variables are assigned with the following fuzzy sets: Small (S), Medium (M), and Large (Lg). The output variable has the following nine levels: Extremely Low (EL), Very Low (VL), Low (L), Lower Medium (LM), Medium (M), Upper Medium (UM), High (H), Very High (VH), and Extremely High (EH). Mamdani implication method is employed for the fuzzy inference reasoning. The relationship between system input and output is expressed by an 'if–then' type. Totally 3^4 fuzzy rules per material were formulated. Finally, the structural characteristics of the fuzzy inference system (FIS) in this study can be plotted as shown in Figure 34.

The main effect plots of the NMPM are presented in Figure 35. The figure indicates that the mean NMPM values of EN 1.4462 are generally higher than of EN 1.4410 and the final optimum $factor^{level}$ combinations for both materials are CB^1 $Geo.^3$ CM^3 v_c^1 f_r^1 α_n^3 λ^2 and CB^2 $Geo.^2$ CM^3 v_c^1 f_r^1 α_n^2 λ^2, respectively. Based on the NMPM values, it can be concluded that EN 1.4462 has better machinability in terms of the employed performances. Finally, performing ANOVA on the computed NMPMs has confirmed the major effects of process and cutting conditions and their interactions on the overall performance of cutting DSS (see Figure 36).

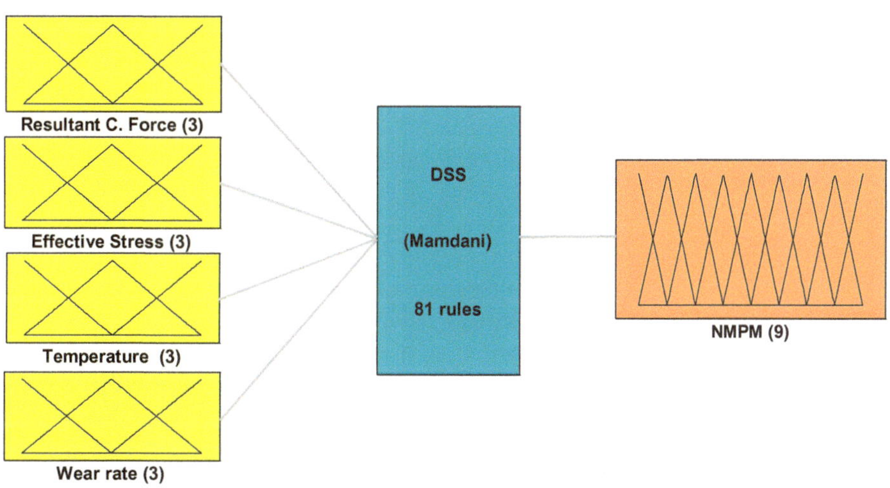

Figure 34. FIS-NMPM: 4 inputs, 1 output, 81 rules.

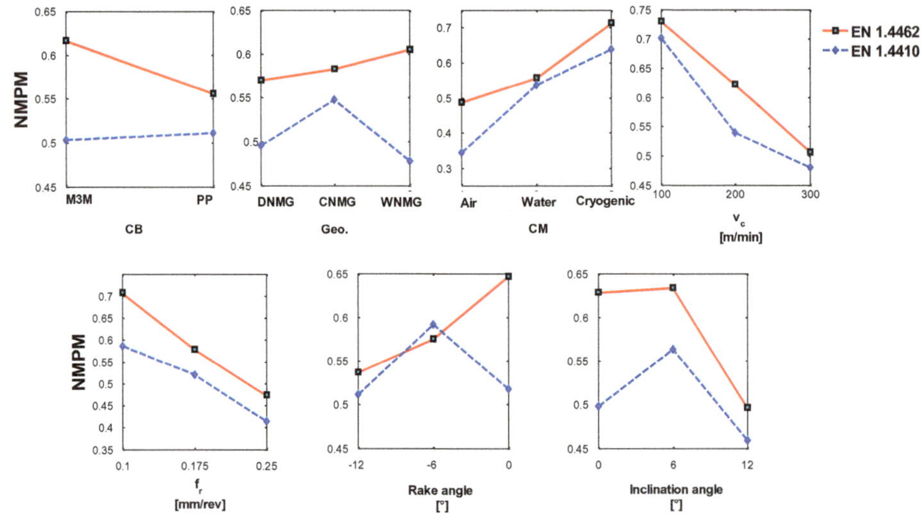

Figure 35. Main effect plots of the effect of control factors on the cutting performances during turning EN 1.4410 and EN 1.4462 DSS.

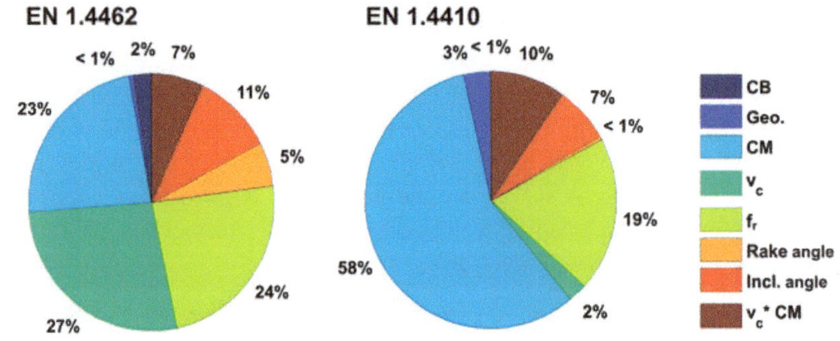

Figure 36. The percentage contribution of FEM control factors in NMPM variance.

9. Conclusions

In this study, three-dimensional numerical simulations of oblique cutting of standard EN 1.4462 and super EN 1.4410 DSS have been investigated. A new approach for an effective and accurate simulation of DSS using JMatSoft's-generated flow stress curves has been proposed. Taguchi design of experiments has been extensively used during the course of the study. Many FEM pre-processing variables are assigned as control factors for subsequent inverse identification process. The overall difference between the experimentally and numerically obtained results has been calculated and statistically analyzed based on the Taguchi optimization procedure. VIKOR method has been applied to combine the differences into one index. Thereafter, a new computational algorithm (FANNS) is suggested to effectively model and optimizes the indices in terms of the adopted control factors. Results of numerical simulation are validated and the approach prepared for the next stage of the investigation. In the second stage, a new methodology based on hypothetical numerical simulations is proposed to investigate the machining of DSS.

Parametric analysis of performances is undertaken, and many interim conclusion points are depicted. A new measure to evaluate machining performance of each of the employed DSS grades has also been suggested and the optimum machining setup is determined. Based on the experimental and numerical results and data analysis the following main conclusion points can be drawn:

(1) JMatPro-generated elasto-visco plastic and thermo-mechanical properties can be effectively used to numerically simulate the machining of DSS.
(2) Using time-dependent damage criteria such as Cockroft-Latham damage criteria, JMatPro has outperformed the JC material model in prediction of chip serrations under similar thermo-mechanical material properties.
(3) Pre-processing FEM control factors such as h_{tc}, v_c, f_r, μ_c, μ_t, κ_t, $\%p$, and $D_{\text{crit.}}$ have strong impacts on the percentage difference between experimental and numerical cutting performances.
(4) The lowest mean values of the overall error percentages for cutting EN 1.4462 and EN 1.4410 were, respectively, at the following pre-processing control *factor*^{level} combinations; $h_tc^2\ v_c^1\ f_r^1\ \mu_c^2\ \mu_t^2\ \kappa_t^3\ \%p^2\ D^3_{\text{circ.}}$ and $h_tc^2\ v_c^2\ f_r^3\ \mu_c^2\ \mu_t^3\ \kappa_t^3\ \%p^2\ D^3_{\text{circ.}}$.
(5) Validations of numerical results through experimentations have revealed that the proposed VIKOR-FANNS approach can efficiently minimize the overall percentage differences at any desired cutting conditions.
(6) Hypothetical benchmark analysis based on the Taguchi optimization procedure can be numerically employed to select the best combination of criterions such as: *CB*, *Geo.*, *CM*, v_c, f_r, α_n, and λ.
(7) Performing ANOM of the derived NMPM values has indicated that the final optimum *factor*^{level} combinations for cutting EN 1.4462 and EN 1.4410 are $CB^1\ Geo.^3\ CM^3\ v_c^1\ f_r^1\ \alpha_n^3\ \lambda^2$ and $CB^2\ Geo.^2\ CM^3\ v_c^1\ f_r^1\ \alpha_n^2\ \lambda^2$, respectively.

Acknowledgments

The authors wish to acknowledge the important logistic contributions from; Dr Uwe Diekmann, Herr. Dipl.-Ing. Michael Fiderer, Mrs M.A. Sibylle Krug, and Mr Bauer.

Disclosure statement

No potential conflict of interest was reported by the authors.

Funding

This work was supported by the Stuttgart University, Team 3, Hochschulschriften, Tausch, Geschenk grant number [01125435-9987].

References

Arrazola, P. J., & Özel, T. (2009). Finite element modeling of machining processes. In T. Özel & J. P. Davim (Eds.), *Intelligent machining: Modelling and optimization of the machining processes and systems* (pp. 125–163). London: ISTE and Wiley.

Arrazola, P. J., & Özel, T. (2010). Investigations on the effects of friction modeling in finite element simulation of machining. *International Journal of Mechanical Sciences, 52*, 31–42.
Arrazola, P. J., Özel, T., Umbrello, D., Davies, M., & Jawahir, I. S. (2013). Recent advances in modelling of metal machining processes. *CIRP Annals – Manufacturing Technology, 62*, 695–718.
Astakhov, V. P. (2006). *Tribology of metal cutting*. Oxford: Elsevier.
Attanasio, A., Ceretti, E., Fiorentino, A., Cappellini, C., & Giardini, C. (2010). Investigation and FEM-based simulation of tool wear in turning operations with uncoated carbide tools. *Wear, 269*, 344–350.
Attanasio, A., & Umbrello, D. (2009). Abrasive and diffusive tool wear FEM simulation. *International Journal of Material Forming* [Internet], *2*, 543–546. doi:10.1007/s12289-009-0475-z
Aurich, J. C. C., & Bil, H. (2006). 3D finite element modelling of segmented chip formation. *CIRP Annals – Manufacturing Technology, 55*, 47–50.
Bonnet, C., Valiorgue, F., Rech, J., & Hamdi, H. (2008). Improvement of the numerical modeling in orthogonal dry cutting of an AISI 316L stainless steel by the introduction of a new friction model. *CIRP Journal of Manufacturing Science and Technology, 1*, 114–118.
Buchkremer, S., Wu, B., Lung, D., Münstermann, S., Klocke, F., & Bleck, W. (2014). FE-simulation of machining processes with a new material model. *Journal of Materials Processing Technology, 214*, 599–611.
Calamaz, M., Coupard, D., & Girot, F. (2008). A new material model for 2D numerical simulation of serrated chip formation when machining titanium alloy Ti–6Al–4V. *International Journal of Machine Tools and Manufacture, 48*, 275–288.
Chagas, G. M. P., Barbosa, P. A., Barbosa, C. A., & Machado, I. F. (2013). Thermal analysis of the chip formation in austenitic stainless steel. *Procedia CIRP, 8*, 293–298.
Davim, J. P., & Maranhão, C. (2009). A study of plastic strain and plastic strain rate in machining of steel AISI 1045 using FEM analysis. *Materials and Design, 30*, 160–165.
De Oliveira Junior, C., Diniz, A., & Bertazzoli, R. (2014, October). Correlating tool wear, surface roughness and corrosion resistance in the turning process of super duplex stainless steel. *Brazilian Society of Mechanical Sciences and Engineering, 36*, 775–785.
Ducobu, F., Rivière-Lorphèvre, E., & Filippi, E. (2014). Numerical contribution to the comprehension of saw-toothed Ti6Al4V chip formation in orthogonal cutting. *International Journal of Mechanical Sciences, 81*, 77–87. doi:10.1016/j.ijmecsci.2014.02.017
Gilat, A., & Wu, X. (1997). Plastic deformation of 1020 steel over a wide range of strain rates and temperatures. *International Journal of Plasticity, 13*, 611–632.
Guo, Y. B., & Yen, D. W. (2004). A FEM study on mechanisms of discontinuous chip formation in hard machining. *Journal of Materials Processing Technology, 155–156*, 1350–1356.
Guo, Z. L., Saunders, N., Miodownik, A. P., & Schille, J. P. (2007). Quantification of high temperature strength of nickel-based superalloys. *Materials Science Forum, 546–549*, 1319–1326.
Gupta, A., Singh, H., & Aggarwal, A. (2011). Taguchi-fuzzy multi output optimization (MOO) in high speed CNC turning of AISI P-20 tool steel. *Expert Systems with Applications, 38*, 6822–6828.
Haddag, B., & Nouari, M. (2013). Tool wear and heat transfer analyses in dry machining based on multi-steps numerical modelling and experimental validation. *Wear, 302*, 1158–1170.
He, A., Xie, G., Zhang, H., & Wang, X. (2014). A modified Zerilli-Armstrong constitutive model to predict hot deformation behavior of 20CrMo alloy steel. *Materials and Design, 56*, 122–127.
Heisel, U., Storchak, M., & Krivoruchko, D. (2013). Thermal effects in orthogonal cutting. *Production Engineering* [Internet], *7*, 203–211. doi:10.1007/s11740-013-0451-9
Hou, Q. Y., & Wang, J. T. (2010). A modified Johnson–Cook constitutive model for Mg-Gd-Y alloy extended to a wide range of temperatures. *Computational Materials Science, 50*, 147–152.
Iraola, J., Rech, J., Valiorgue, F., & Arrazola, P. J. (2012). Characterization of friction coefficient and heat partition coefficient between an austenitic steel AISI304L and a tin -coated carbide cutting tool. *Machining Science and Technology, 16*, 189–204.
Kalpakjian, S., & Schmit, S. R. (2006). *Manufacturing engineering and technology* (5th ed.). Upper Saddle River, NJ: Pearson Education.
Keleştemur, O., Arıcı, E., Yıldız, S., & Gökçer, B. (2014). Performance evaluation of cement mortars containing marble dust and glass fiber exposed to high temperature by using Taguchi method. *Construction and Building Materials, 60*, 17–24.

Klocke, F. (2011). *Manufacturing processes 1: Cutting*. Heidelberg: Springer.
Klocke, F., Lung, D., & Buchkremer, S. (2013). Inverse identification of the constitutive equation of inconel 718 and AISI 1045 from FE machining simulations. In *Procedia CIRP* (Vol. 8, pp. 212–217), Turin, Italy.
Koné, F., Czarnota, C., Haddag, B., & Nouari, M. (2011). Finite element modelling of the thermo-mechanical behavior of coatings under extreme contact loading in dry machining. *Surface and Coatings Technology, 205*, 3559–3566.
Koné, F., Czarnota, C., Haddag, B., & Nouari, M. (2013). Modeling of velocity-dependent chip flow angle and experimental analysis when machining 304L austenitic stainless steel with groove coated-carbide tools. *Journal of Materials Processing Technology, 213*, 1166–1178.
Koyee, R. D., Heisel, U., Eisseler, R., & Schmauder, S. (2014, October). Modeling and optimization of turning duplex stainless steels. *Journal of Manufacturing Processes, 16*, 451–467.
Koyee, R. D., Schmauder, S., & Eisseler, R. (2013). Machining of stainless steels: A comparative study. In A. Maffei & A. Archenti (Eds.), *3rd International Conference on Advanced Manufacturing Engineering and Technologies* (pp. 53–62). Stockholm, Sweden.
Krolczyk, G., Legutko, S., & Gajek, M. (2013). Predicting the surface roughness in the dry machining of duplex stainless steel (DSS). *Metalurgija, 52*, 259–262.
Li, H. Y., Li, Y. H., Wang, X. F., Liu, J. J., & Wu, Y. (2013). A comparative study on modified Johnson Cook, modified Zerilli-Armstrong and Arrhenius-type constitutive models to predict the hot deformation behavior in 28CrMnMoV steel. *Materials and Design, 49*, 493–501.
Liu, R., Melkote, S., Pucha, R., Morehouse, J., Man, X., & Marusich, T. (2013). An enhanced constitutive material model for machining of Ti–6Al–4V alloy. *Journal of Materials Processing Technology, 213*, 2238–2246. doi:10.1016/j.jmatprotec.2013.06.015
Lorentzon, J., & Järvstråt, N. (2008). Modelling tool wear in cemented-carbide machining alloy 718. *International Journal of Machine Tools and Manufacture, 48*, 1072–1080.
Lorentzon, J., Järvstråt, N., & Josefson, B. L. (2009). Modelling chip formation of alloy 718. *Journal of Materials Processing Technology, 209*, 4645–4653.
Maranhão, C., & Paulo, D. J. (2010). Finite element modelling of machining of AISI 316 steel: Numerical simulation and experimental validation. *Simulation Modelling Practice and Theory, 18*, 139–156.
Nasr, M., Ng, E.-G., & Elbestawi, M. (2007). Effects of workpiece thermal properties on machining-induced residual stresses – Thermal softening and conductivity. *Proceedings of the Institution of Mechanical Engineers, Part B: Journal of Engineering Manufacture, 221*, 1387–1400. doi:10.1243/09544054JEM856
Niesłony, P., Grzesik, W., Chudy, R., & Habrat, W. (2014). Meshing strategies in FEM simulation of the machining process. *Archives of Civil and Mechanical Engineering*. Retrieved from http://dx.doi.org/10.1016/j.acme.2014.03.009
Nomani, J., Pramanik, A., Hilditch, T., & Littlefair, G. (2013). Machinability study of first generation duplex (2205), second generation duplex (2507) and austenite stainless steel during drilling process. *Wear, 304*, 20–28.
Odeshi, A. G., Al-ameeri, S., & Bassim, M. N. (2005). Effect of high strain rate on plastic deformation of a low alloy steel subjected to ballistic impact. *Journal of Materials Processing Technology, 162–163*, 385–391.
Olovsjö, S., Hammersberg, P., Avdovic, P., Ståhl, J.-E., & Nyborg, L. (2012). Methodology for evaluating effects of material characteristics on machinability – Theory and statistics-based modelling applied on Alloy 718. *The International Journal of Advanced Manufacturing Technology, 59*, 55–66.
Outeiro, J. C., Umbrello, D., & M'Saoubi, R. (2006). Experimental and numerical modelling of the residual stresses induced in orthogonal cutting of AISI 316L steel. *International Journal of Machine Tools and Manufacture, 46*, 1786–1794.
Paro, J., Hänninen, H., & Kauppinen, V. (2001). Tool wear and machinability of HIPed P/M and conventional cast duplex stainless steels. *Wear, 249*, 279–284.
Pu, Z., Umbrello, D., Dillon, O. W., Lu, T., Puleo, D. A., & Jawahir, I. S. (2014). Finite element modeling of microstructural changes in dry and cryogenic machining of AZ31B magnesium alloy. *Journal of Manufacturing Processes, 16*, 335–343.
Pujana, J., Arrazola, P. J., M'Saoubi, R., & Chandrasekaran, H. (2007). Analysis of the inverse identification of constitutive equations applied in orthogonal cutting process. *International Journal of Machine Tools and Manufacture, 47*, 2153–2161.

Puls, H., Klocke, F., & Lung, D. (2012). A new experimental methodology to analyse the friction behaviour at the tool-chip interface in metal cutting. *Production Engineering, 6*, 349–354.

Quiza, R., & Davim, J. P. (2011). Computational methods and optimization. In J. P. Davim (Ed.), *Mach hard mater* (pp. 191–208). London: Springer-Verlag.

Rao, R. V. (2013). *Decision making in manufacturing environment using graph theory and fuzzy multiple attribute decision making methods*. London: Springer.

Rech, J., Arrazola, P. J., Claudin, C., Courbon, C., Pusavec, F., & Kopac, J. (2013). Characterisation of friction and heat partition coefficients at the tool-work material interface in cutting. *CIRP Annals – Manufacturing Technology, 62*, 79–82.

Rhim, S. H., & Oh, S. I. I. (2006). Prediction of serrated chip formation in metal cutting process with new flow stress model for AISI 1045 steel. *Journal of Materials Processing Technology, 171*, 417–422.

Riedmiller, M., & Braun, H. (1993). A direct adaptive method for faster back propagation learning: The RPROP algorithm. In *Proceedings of IEEE International Conference on Neural Networks* (pp. 586–591), San Francisco, CA.

Rosochowska, M., Balendra, R., & Chodnikiewicz, K. (2003). Measurements of thermal contact conductance. *Journal of Materials Processing Technology, 135*, 204–210.

Samantaray, D., Mandal, S., & Bhaduri, A. K. (2009). A comparative study on Johnson Cook, modified Zerilli–Armstrong and Arrhenius-type constitutive models to predict elevated temperature flow behaviour in modified 9Cr–1Mo steel. *Computational Materials Science, 47*, 568–576.

Saunders, N., Guo, Z., Li, X., Miodownik, A. P., & Schille, J. P. (2004). Modelling the material properties and behaviour of Ni-based superalloys. *Superalloys, 2004*, 849–858.

SFTC. (2010). DEFORM v10.2.1 User's manual. Columbus, OH, USA.

Shrot, A., & Bäker, M. (2012). Determination of Johnson–Cook parameters from machining simulations. *Computational Materials Science, 52*, 298–304.

Sima, M., & Özel, T. (2010). Modified material constitutive models for serrated chip formation simulations and experimental validation in machining of titanium alloy Ti–6Al–4V. *International Journal of Machine Tools and Manufacture, 50*, 943–960. doi:10.1016/j.ijmachtools.2010.08.004

Smolenicki, D., Boos, J., Kuster, F., Roelofs, H., & Wyen, C. F. (2014). In-process measurement of friction coefficient in orthogonal cutting. *CIRP Annals – Manufacturing Technology, 63*, 97–100.

Song, W., Ning, J., Mao, X., & Tang, H. (2013). A modified Johnson–Cook model for titanium matrix composites reinforced with titanium carbide particles at elevated temperatures. *Materials Science and Engineering: A, 576*, 280–289.

Ulutan, D., & Özel, T. (2013). Determination of tool friction in presence of flank wear and stress distribution based validation using finite element simulations in machining of titanium and nickel based alloys. *Journal of Materials Processing Technology, 213*, 2217–2237.

Umbrello, D., M'Saoubi, R., & Outeiro, J. C. (2007). The influence of Johnson–Cook material constants on finite element simulation of machining of AISI 316L steel. *International Journal of Machine Tools and Manufacture, 47*, 462–470.

Vaziri, M. R., Salimi, M., & Mashayekhi, M. (2011). Evaluation of chip formation simulation models for material separation in the presence of damage models. *Simulation Modelling Practice and Theory, 19*, 718–733.

Wang, X., Huang, C., Zou, B., Liu, H., Zhu, H., & Wang, J. (2013). Dynamic behavior and a modified Johnson–Cook constitutive model of Inconel 718 at high strain rate and elevated temperature. *Materials Science and Engineering: A, 580*, 385–390. doi:10.1016/j.msea.2013.05.062

Xue, Q., Liao, X. Z., Zhu, Y. T., & Gray, G. T. (2005). Formation mechanisms of nanostructures in stainless steel during high-strain-rate severe plastic deformation. *Materials Science and Engineering: A, 410–411*, 252–256.

Yang, X., & Liu, C. R. (2002). A new stress-based model of friction behavior in machining and its significant impact on residual stresses computed by finite element method. *International Journal of Mechanical Sciences, 44*, 703–723.

Yang, X. S. (2010). *Nature-inspired metaheuristic algorithms*. London: Luniver Press.

Zorev, N. (1963). Inter-relationship between shear processes occurring along tool face and on shear plane in metal cutting. *ASME International Research in Production Engineering*, 42–49.

3

Developing location indicators for Agricultural Service Center: a Delphi–TOPSIS–FAHP approach

Morteza Zangeneh[a]*, Asadolah Akram[a], Peter Nielsen[b] and Alireza Keyhani[a]

[a]Department of Agricultural Machinery Engineering, Faculty of Agricultural Engineering and Technology, School of Agriculture & Natural Resources, University of Tehran, Karaj, Iran;
[b]Department of Mechanical and Manufacturing Engineering, Aalborg University, Aalborg, Denmark

The main goal of this paper is to construct objectives and attributes of the service centers location problem. This problem is often found within production management as a location problem, when designing for example supply chains or manufacturing layouts. The main contribution of this paper is constructing related location indicators. Since there was no similar literature in this discipline, so a Delphi survey applied to quantify expert's attitudes about location problem of Agricultural Service Center (ASC) and construct location selection attributes and also ASC objectives. A TOPSIS survey is done to rank extracted attributes to import in fuzzy analytical hierarchy process (AHP) study. Then a fuzzy AHP technique is applied to compute the weight of these most important attributes using four objectives, which obtained by Delphi technique too. In the simplest form with this assumption that all objectives have the same priority, the results illustrated that the service, cost, speed, and ASC profit are first to last important objectives, respectively. At the end of this paper, the multi-choice goal programming method recommended to use when the priorities of objectives are not the same. Finally, the weight of ASC location attributes computed to consider in ASC location problem.

Keywords: Agricultural Service Centers (ASC); location; objectives; attributes; service

1. Introduction

Each member of supply chain needs to perform specific functions or activities in value addition process. Performance of supply chain could be improved if supply chain is integrated and the concerned activities are properly coordinated (Shukla, Garg, & Agarwal, 2014). Energy, labor, input costs, markets conditions, crop yielding and prices, machinery service availability, pest and disease infection, environmental conditions, and even economic and political strategies change the agricultural productivity in the supply chain (Lak & Almasi, 2011). Among these dynamic variables, providing agricultural services in the right way can have a strategic role to improve the agricultural productivity. Several services can be given to the whole of agricultural supply chain. These problems can be seen as analog to facility location problems in supply chains or design problems inside physical manufacturing systems. Some services are for farms, specifically. In following section, four types, including input supply, mechanization services,

*Corresponding author. Emails: mzangeneh@ut.ac.ir, morteza@m-tech.aau.dk

advisory services, and financial services, which are considered in this research, are introduced.

A wide range of location problems effect today's firm and organizational objectives and decision-makings. The interest and service level, economic profits, quality of service, security, environmental protection, availability, and optimum performance are some of the many criteria and objectives of the facility location problem; and hence, different methods and approaches have been employed in the literature to maximize the achievement of these objectives (Hosseini-Nasab, Tavana, & Yousefi, 2014).

1.1. Input supply service

Unfortunately after market liberalization, a majority of the farm input supply companies remains concentrated in urban areas or central rural zones. Due to these changes millions of poor farmers in rural areas do not have access to agricultural inputs on time and in right way, such as improved seeds, chemical fertilizers which are needed to help them improve their productivity. So poor development and weak performance of rural agricultural input markets explain the current low productivity of small holder farmers (Dorward & Chirwa, 2011).

1.2. Agricultural mechanization service

The manufacture, distribution, repair, maintenance, management and utilization of agricultural tools, implements, and machines are covered under mechanization services (Lak & Almasi, 2011). The important point in mechanization services is that how supply mechanization services to the farmer in an efficient and effective manner. If mechanization is implemented in the right way, it will have a considerable effect on agricultural productivity improvement (Lak & Almasi, 2011). In many countries, agricultural mechanization has made a significant contribution to the agricultural and rural development. After applying farm mechanization levels of production have increased, soil and water conservation measures were constructed, the profitability of farming improved, the quality of rural life enhanced, and development in the industrial and service sectors was stimulated so mechanization services are highly required to be supplied in an effective way (Bishop, 1997). Inns (1995) mentioned that agricultural mechanization development depends on the farmers' satisfaction and capability to identify opportunities for achieving sustainable benefits by improved and/or increased use of power and machinery, selecting the most worthwhile opportunity and carrying it through to successful implementation. Lack of consideration to the necessity of development in mechanization of agricultural sector, insufficient cooperation between industrial and agricultural sector, and unrealistic selection of goals and objectives and perhaps, more importantly misuse and poor management of resources could all be counted toward the considerable fall back in the agricultural sector (Bagheri & Moazzen, 2009). Sims and Kienzle (2009) have done three mechanization supply chain case studies in three countries. They presented five elements for mechanization supply chain (see Figure 1).

The role and activity of machinery hire service stakeholders should include following aspects: Coordination with other stakeholders; Business management; Quality control; Operator training; Maintenance and servicing; and closely fallow farmers' needs (Sims & Kienzle, 2009). The previous papers did not consider to efficient manner of supplying and distributing the mechanization services for customers. In current research the strategy of providing mechanization services with all other agricultural services is

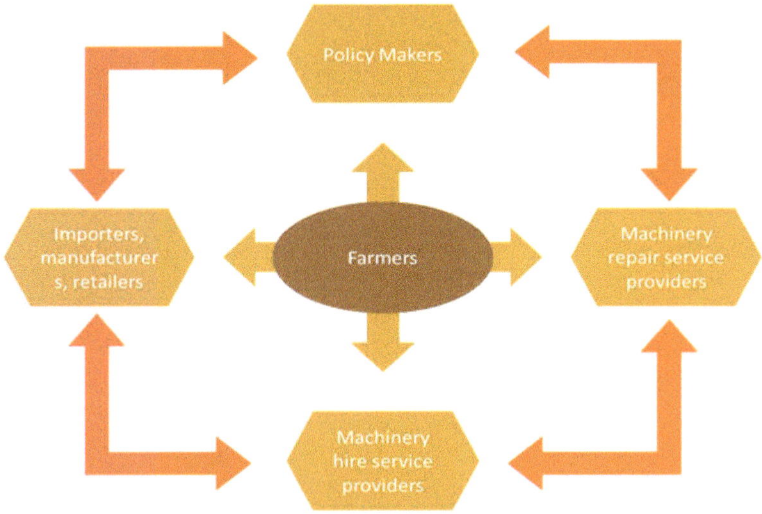

Figure 1. Farm power and machinery supply chain stakeholders (Sims & Kienzle, 2009).

considered. Therefore, supplying high quality mechanization services in coordination with the needs of farms by Agricultural Service Centers (ASCs) can lead to the improvement of productivity in whole supply chain of agricultural production.

1.3. Agricultural advisory service

Agricultural consulting services are known as activities that make new knowledge available to the agricultural producers and assist them to improve their farming and management skills. The services may include: Sharing the information, training and advice of farmers, testing new technologies on farms, and developing farm management tools. The basic indicators for success of a demand-driven advisory service system in agriculture are: (a) farmers have access to agricultural advisory services; (b) farmers use the advisory services; (c) advices lead to income's increase from the agricultural production; and also (d) competition among agricultural advisers (Chipeta, 2006). Altogether, the advisory is a critical need to the agricultural supply chain and the access to this service must be easy to encourage the farmers to use it. This can be obtained by locating service centers in accurate places. The demands from farmers in many cases are different. The advisory service demands which are formulated by various stakeholders of agricultural supply chain can be seen in Figure 2.

1.4. Financial service

Access to external financial resources in agriculture is constrained. This is due to low enterprise profitability in agriculture, accumulated debts, high inflation, risk and uncertainty, and collateral problems (Johan, Swinnen, & Gow, 1999). The limited access to financial resources may lead to several problems in agricultural production such as: disturb the time of farming operations (input supply, planting, trading, and etc.), decrease the input quality, etc. So increasing the access to financial services including loan and insurance can improve the productivity in agriculture.

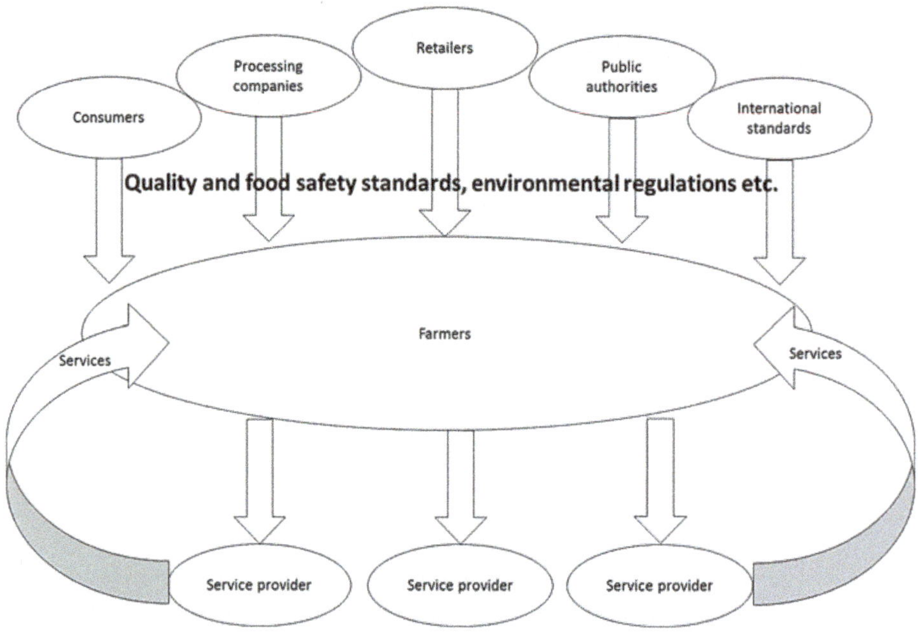

Figure 2. Demands are formulated by various stakeholders (Chipeta, 2006).

1.5. Location problem of ASCs

Agriculture is the only major sector that uses the land surface as an essential input into its production function. This wide geographical dispersion of agricultural production has the important economic consequence; transportation becomes essential. Output must be transported for consumption by others and inputs; such as modern seeds, fertilizer, pesticides, or machinery, and all required agricultural services must be transported to the farm to raise output (Timmer, Falcon, & Pearson, 1983). Therefore, to success in providing and distributing agricultural services for improvement of productivity in this sector, the right location of such service centers should be selected. In this regard, one of the initial steps to solve this location problem is finding related location objectives and attributes to use in multi-criteria decision-making (MCDM) location models. In following section the multi-attribute decision-making (MADM) and multi objective decision-making (MODM) techniques is reviewed and appropriate decision-making technique for ASC location problem is selected.

In a complementary study from the authors of this paper, a MADM approached developed, to use the attributes and objectives extracted in this study, and best candidate locations for establishing service centers for agricultural production identified (Zangeneh, Nielsen, Akram, & Keyhani, 2014).

1.6. MADM and MODM decision-making techniques

There are various techniques for ranking alternatives. In many real-world problems, the decision-maker likes to pursue more than one target or consider more than one factor or measure, which called MODM problem and MADM problem. There are many decision-making problems that their information is spatial and known as location decisions problems. Facility location is a branch of operation's research related to locating or positioning at least a new facility among several existing in order to optimize at least one objective function (Farahani, Steadieseifi, & Asgari, 2010).

There are many techniques which are used to solve the MADM problems. The most popular ones are as follows: lexicographic, permutation method, simple additive weighting (SAW), elimination and choice expressing reality (ELECTRE), technique for order preference by similarity to ideal solution (TOPSIS), linear programming techniques for multidimensional analysis of preference (LINMAP), and interactive SAW method and MDS with the ideal point (Hwang & Yoon, 1981a).

Also there are many techniques which are used for the MODM problems. The most popular ones are as follows: metric LP methods, bounded objective method, lexicographic method, goal programming (GP), goal attainment method, method of Zionts–Wallenius, the methods as step method (STEM) and related method, sequential multi-objective problem solving and sequential information generator for multi-objective problems method, GP STEM, and C-constraint method and adaptive search method (Hwang & Masud, 1979; Szidarovszky, Gershon, & Duchstein, 1986; Ulungu & Teghem, 1994; Zionts, 1979). Also, papers such as the work by Do et al. could be considered relevant as they combine simulation under uncertainty with heuristics to find suitable solutions.

The analytic hierarchy process (AHP) is widely used by authors to solve MADM problems. García et al. (2014) generated a multi-criteria and multi-attribute assessment model that allows selecting the ideal location for warehouses for perishable agricultural products. In another paper, Akıncı, Özalp, and Turgut (2013) determined suitable lands for agricultural use in Turkey by AHP. As García et al. (2014) newly reviewed the literatures, no papers has been seen, which focused on ASC location problem. A good paper on implementation of Fuzzy AHP and Delphi method in MADM problems is Cho and Lee (2013). They identified four decision areas and further prioritized the sixteen factors under a hierarchy model structured by Fuzzy AHP approach.

Location and the number of ASCs in each region is a key parameter to ensure the success of agricultural servicing. So the main scope of this study is constructing indicators for location selection of ASCs to improve the performance of agricultural supply chain viewpoint of productivity, quality, and competitiveness. Secondly, to give primary weight to the objectives and location attributes, Fuzzy AHP method is used. Also specifying the final weight of location attributes using GP method is provided at last step of this study. In other words an integrated approach which combines MADM and MODM is developed. First to third phase belongs to MADM and fourth phase is a MODM problem. Also GP is used to generalize proposed approach and make it utilizable for different/future real world applications. The managers can use developed approach to import their ideal values of agricultural supply chain goals to the model and get appropriate weight for the ASC location attributes. The results of current research are used in facility location problem of ASCs in another paper of the authors.

2. Material and methods

In this research, the FAHP has been used to assess the location indicators (both objectives and attributes) of ASCs. Using the FAHP method, the ASC location attributes is prioritized based on objectives, which assumed that they have same importance. But another phase added to the framework to consider the possible difference between objectives using multi-choice goal programming (MCGP), i.e. in the case that the priorities of objectives are not the same. So this research has four main phases for assessing the location attributes (see Figure 3):

(1) Constructing initial location attributes and refining them and also location objectives using Delphi method.
(2) Ranking initial ASC location attributes using TOPSIS technique.
(3) Developing the AHP hierarchal model containing the assessment of objectives and attributes, and determining the weights of objectives and attributes through FAHP.
(4) Prioritizing the attributes through GP.

2.1. Phase I

First of all, the location problem of ASCs is defined. Often the ASC location problem occur in developing countries, but may occurs in development ones where the strategic decision be taken to improve the productivity of their agricultural supply chain. However, the problem is also well known within production, where placing central maintenance or service facilities is often a critical issue when determining the cost of supporting an ongoing manufacturing environment. Services, however, play a particularly critical role in the agricultural sector. Tractor and farm machineries are needed, but in developing countries considerable number of them are old and their efficiency has been reduced, so need to be replaced. Also other services such as advisory, input supply

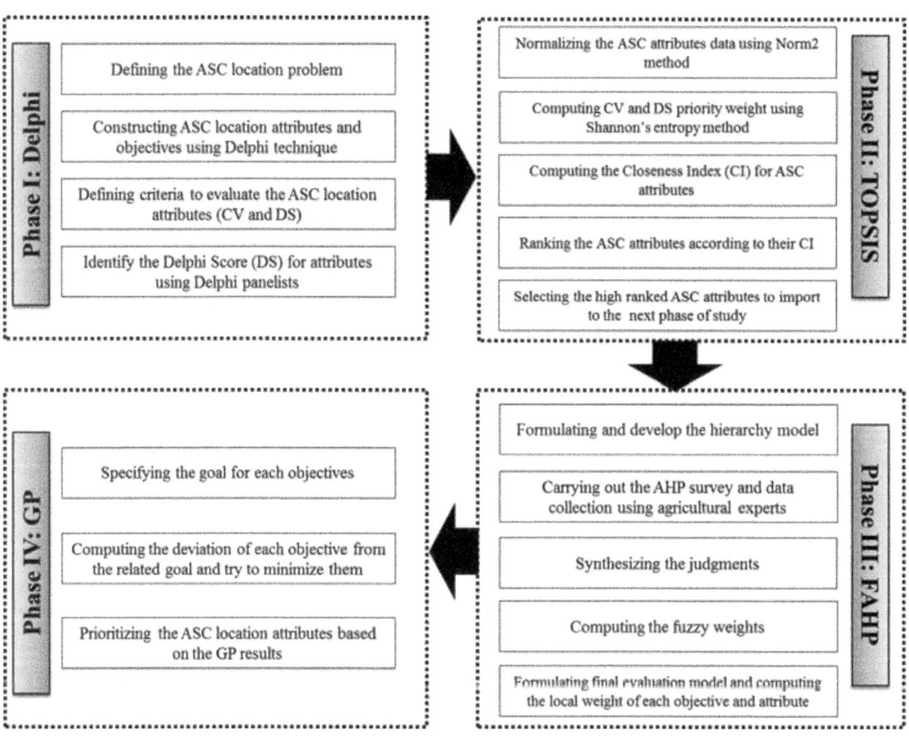

Figure 3. The schematic process of overall research framework.

and financial services can improve the efficiency of agricultural production. In some developing countries (for example Iran) governments have been concluded that establishing ASCs and keep them update in the aspects of tractors, machineries, and also other agricultural services will be improve the efficiency. So, the main issue for this policy is selecting the location of these centers. This problem has many aspects, but it is MADM problem and an AHP can solve it, primarily. It is common that in similar location problems, desired attributes be collected from literatures and similar studies. But in current problem no similar research study for the attributes found, so Delphi method was used to construct ASC location attributes.

2.1.1. Delphi method

Delphi is a technique of popular survey method which extract consensus of ideas among a set of experts or panelists by maintaining the unanimity among them. Delphi technique has been used for various purposes like setting goals, finding problems, developing system models, decision-making, etc. (Prusty, Mohapatra, & Mukherjee, 2010). In current research several rounds of Delphi survey is needed to construct the objectives and attributes of ASC location problem. In the following sections the process of Delphi survey has been described.

2.1.2. Delphi survey process

2.1.2.1. Design and administration of the initial questionnaire. There is no similar research about location of ASC in literatures, so the initial questionnaire was basically a brainstorming session among the selected Delphi panelists. Designer team was included four people who designed the open questions of first questionnaire and contributed in different steps of this research. The aim of this step of Delphi study was construction of the location selection indicators (both objectives and attributes) of ASC location problem. Each Delphi panelist was asked to answer two main questions as follows:

(1) What location can be appropriate to establish ASC? In other words what attributes should be considered to select a place for an ASC?
(2) What advantages can be achieved if an ASC located in best possible location (based on attributes you mentioned in previous question)? In other words what are the objectives of ASC location selection?

First question was designed to collect and construct ASC location attributes and the second to determine main objectives of ASC facility location problem. Eight experts replied to the first round questionnaire. The composition of the panelists in this round is given in Table 1.

The second questionnaire was designed with the objective of prioritizing the attributes raised in the first questionnaire and sent to the Delphi panelists. Attributes were rated with respect to priority criterion. The rating scale (see Table 2) was in the range of 1–9, with '1' representing 'not important' and '9' representing 'very highly important' meant that how much an attribute is important for each Delphi panelists participated in the survey to select best location of ASCs.

There are several methods to calculate the score of each attributes to determine the most important attributes to use in Fuzzy AHP method. The number of items for pairwise comparisons should be reduced in AHP to simplify judgment process and

Table 1. Composition of experts in the first round.

Participant type	Educations	Office	Number
Professor	Ph.D. with average background of 12 years in the field of Agricultural Engineering	University of Tehran, University of Zanjan	3
Ph.D. student	M.Sc. in the field of Agricultural Engineering	University of Tehran	3
Government expert	B.Sc. and M.Sc. with average background of 14 years in the field of Agricultural Mechanization Engineering	Agricultural Department of Zanjan and Alborz provinces	2

ensure the accuracy of results. So here coefficient of variation and Delphi score are used simultaneously in a TOPSIS ranking survey to reduce the number of items.

If the absolute dispersion is defined as the standard deviation, and the average is mean, the relative dispersion is called the coefficient of variation (CV) or coefficient of dispersion. The CV is attractive as a statistical tool because it apparently permits the comparison of varieties free from scale effects, i.e. it is dimensionless. The CV is defined as the ratio of the standard deviation σ to the mean μ (Brown, 1998) (Equation (1)):

$$\mathrm{CV} = \frac{\sigma}{\mu} \tag{1}$$

The lower value of CV has preference to select one attribute as ASCs location indicator, i.e. Delphi panelist opinions is similar to each other and average of them is high about an attribute, illustrate that all panelists want to consider that attribute in study.

Another method to analysis the Delphi panelist's scores to select best indicators is Delphi score, which proposed by Linstone and Turoff (2002) (Equation (2)):

$$\text{Delphi Score} = \frac{(\text{Lowest Score} + \text{Highest Score}) + (4 \times \text{Average Score})}{6} \tag{2}$$

Against the CV higher values of Delphi score is preferred. Each attribute which have the most Delphi score can be selected to include in ASCs location selection process.

2.2. *Phase II*

In this phase, the most important attributes will be selected according to their Delphi score and CV. Then the hierarchy model will be formulated, and the AHP survey can be run. Here, the TOPSIS technique is applied to rank attributes according to the value of CV and DS.

Table 2. Linguistic variables for the rating of attributes.

Linguistic variable	Value
Not important	1
Very low	2
Medium low	3
Low	4
Fair	5
Medium	6
High	7
Medium high	8
Very high	9

TOPSIS is a MCDM method which is initially developed by Hwang and Yoon (Hwang & Yoon, 1981b). The technique is based on the idea that the optimal solution should have the shortest distance from the positive ideal solution and the farthest from the negative ideal solution (Oztaysi, 2014). In current research, the maximum value of each attribute is considered as positive ideal solution and the minimum value as negative ideal solution. This value is mentioned for positive attributes where higher values of them are preferred for the location of ASC, e.g. population of candidate location, while for negative attributes is vice versa.

2.3. *Phase III*

The main aim of current research is estimating the weights of objectives and attributes from the FAHP model which is developed. So the AHP questionnaire distributed to the agricultural professors and experts to respond to judge about relative weight of each pairwise comparison between objectives and attributes. In the production environment this can easily be changed to managers and production technology experts.

2.3.1. FAHP

AHP is a structured technique for organizing and analyzing complex decisions based on paired comparisons of both projects and criteria (Saaty, 1986). The strength of the AHP method lies in its ability to construct complex, multi-person, multi-attribute, and multi-period problem hierarchically (Chin, Xu, Yang, & Lam, 2008). The judgments of conventional AHP method usually have ambiguity problem because the verbal attitudes of decision-makers evaluation process contain vague and multiplicity of meaning (Lee, 2010). Thus, in order to overcome the verbal ambiguity of linguistic variables, fuzzy set theory has applied to the judgment as an extension of AHP method. Fuzzy set theory was first introduced by Zadeh (1976) to deal with the uncertainty.

2.4. *Phase IV*

In this research, it is assumed that the importance of all objectives is the same, but for applying this approach to any agricultural region, the managers must consider their agricultural conditions and also their strategic decision to use ASC in their region. If the managers prefer to stress some objectives, in other words the managers follows specific level of each objective function to improve the performance of their ASC, so they strictly recommended prioritizing the objectives and then use GP technique (Chang, 2007) to compute the final weight of ASC location attributes. Then they can use these weights to define the process of MADM to find best location of ASCs. Traditional GP can be used in this case when there is only one goal level in each objective function, but MCGP is another version of GP which considers several maximum or minimum goals in (Chang, Chen, & Zhuang, 2014).

3. Results and discussion

3.1. *The results of Delphi survey*

The answers made by the Delphi panelists to the initial questionnaire covered a wide range of issues. According to the results, the panelists mentioned number of machinery available in the region, distances, future work condition of service centers in future, etc. Similarly, García et al. (2014) constructed attributes for perishable product warehouse which relatively is an ASC including: accessibility, security, needs of the agricultural product warehouse, social acceptance that the warehouse may have on the environment where it is supposed to be built, the costs of product transportation, wages, and salaries of workers and managers. After constructing the attributes various scores were obtained in the second questionnaire against each ASC location attributes. The title and scores of both evaluation methods (i.e. CV and DS) methods have been shown in Table 3.

Based on the main objectives of development of ASC establishment policy, four objectives was made for ASC location problem shown in Table 4, by consulting agricultural experts using Delphi method (the second question in first round questionnaire is related to the objectives). It can be said that the objectives constructed here is almost the main objectives of agricultural supply chain. The quality is the first and the most important objective in any supply chain management. All actions are done to improve the quality. So this study strictly focused on quality of services in accomplishing our research. The quality of services in agricultural operations such as mechanization and inputs can directly affect the quality and quantity of production. Competitiveness of agricultural production directly related to their quality and price. The farmers always try to reduce their costs and compete with other markets. Here the location attribute is

Table 3. Score of ASC location attributes.

No.	Attribute	CV	DS
C1	Surplus of tractor and machinery which concentrated in region	0.763	4.20
C2	Lack of tractor and machinery	0.393	6.50
C3	Easy reach of farmers to the ASC	0.447	5.73
C4	Easy reach to spare parts, repair, customer service center, and gas station	0.225	6.90
C5	Easy reach to main roads and byroads	0.553	5.27
C6	Possibility to develop the workspace in future	0.525	5.27
C7	Minimum distance to the covered regions	0.435	5.80
C8	Easy reach to its villages and covered regions	0.313	6.23
C9	Easy reach to water, electricity, gas, and telephone with minimum cost	0.484	5.67
C10	Existence of crop rotation with least fallow (irrigated be more than dry farming)	0.384	5.93
C11	Social acceptance by the farmers and bring their adoption	0.518	5.27
C12	Good climate for developing ASC operations and be safe from natural disasters	0.503	4.20
C13	Tractor and machinery being old and farmers could not be able replace them	0.479	5.20
C14	Maximum distance to other service centers or similar centers	0.334	5.60
C15	Possibility to extend new crop production in future	0.506	3.30
C16	The price of land for cheap extend	0.838	3.13
C17	Easy reach to required labor	0.569	4.23
C18	Enough security	0.594	4.63
C19	Visibility for farmers	0.859	4.27
C20	Minimum distance to residential areas	0.929	3.00
C21	Existence of progressive farmers	0.566	5.07
C22	Existence of good educations and tendency to modernism by farmers	0.536	4.70
C23	Good soil, good climate, less slope, big plot lands, and appropriate wind	0.516	5.00
C24	Existence of crowded population and vast cultivated area	0.316	6.60
C25	Minimum distance of villages to each other and large number of them (population being concentrated)	0.472	5.43
C26	Maximum number of crop types cultivated	0.503	5.47

Table 4. ASC location selection objectives made after first round Delphi survey.

Objectives	Objective Function (Max/Min)	Abbreviations
Quality of agricultural services	Max	Service quality (O1)
Cost of agricultural services paid by farmers instead of delivering services	Min	Service cost (O2)
Speed of service	Max	Service speed (O3)
Profitability of ASCs	Max	ASC profit (O4)

defined, which can minimize the agriculture service costs and consequently product cost. Locations which are near to their customers can reduce their transportation cost, so according to the objective of service cost, location attributes could be selected, which minimized the distance between service centers and the farms/farmers. Time is a critical factor in agricultural commodities, and extremely can influence the quality and price of agricultural production because of being perishable, dependence on whether, etc. The location and also capacity of service centers affects the service time. Therefore the attributes reflects these items can improve the location selection results. Service centers must be sustainable to provide appropriate services to the farms, so their profitability can be considered in this case.

3.2. Refinement of attributes by TOPSIS

Attributes were analyzed to refine and reduce their size, and also by selecting the most important ones. If too many factors be included in AHP model, then the number of pairwise comparisons will be increased, so the accuracy of AHP judgments may reduce. Also it is necessary to make them mutually exclusive and nonredundant. So to find five best attributes after the second round Delphi survey, a TOPSIS survey designed to rank ASC attributes, which have highest Delphi score and lowest CV simultaneously. Then they were selected as alternatives in the AHP model to make judgment easy and possible for experts and increase their accuracy. The results of TOPSIS technique are illustrated in Table 5 and Figure 4.

The highest rank ASC attributes has been introduced in Table 6.

3.3. Process of the AHP survey

The hierarchy model of ASC location problem formulated and developed as shown in Figure 5.

To find the weight of the ASC location selection attributes, an AHP questionnaire was designed with a 9-point scale and pairwise comparison format. The questionnaire was distributed to 10 agricultural experts. Approximately, 3 of 10 participants had more than 10 years of work experience in agricultural sector and also have related higher academic educations. Other experts almost are professor and PhD student of agricultural engineering. They were asked to make a pairwise comparison judgment and give the relative importance amongst the performance objectives and attributes. The pairwise judgment is conducted from the first level to the third level. Finally, each participant individually expressed their preference between each pair of elements.

Table 5. The results of TOPSIS survey.

Attribute	Normalized data		Distance to positive ideal solution	Distance to negative ideal solution	Closeness index to ideal solution	Rank
	CV	DS				
C1	0.208	0.666	0.345	0.116	0.252	23
C2	0.055	1.594	0.063	0.396	0.863	3
C3	0.072	1.240	0.160	0.304	0.655	7
C4	0.018	1.796	0.000	0.458	1.000	1
C5	0.110	1.047	0.220	0.244	0.526	15
C6	0.099	1.047	0.218	0.249	0.534	13
C7	0.068	1.269	0.152	0.313	0.673	6
C8	0.035	1.466	0.093	0.372	0.800	4
C9	0.084	1.212	0.170	0.293	0.632	9
C10	0.053	1.328	0.133	0.332	0.714	5
C11	0.096	1.047	0.217	0.250	0.535	12
C12	0.091	0.666	0.321	0.182	0.362	20
C13	0.082	1.020	0.222	0.251	0.530	14
C14	0.040	1.183	0.172	0.305	0.639	8
C15	0.092	0.411	0.391	0.158	0.287	22
C16	0.252	0.370	0.433	0.042	0.089	25
C17	0.116	0.676	0.321	0.168	0.343	21
C18	0.126	0.810	0.287	0.186	0.394	19
C19	0.264	0.687	0.357	0.102	0.223	24
C20	0.309	0.340	0.458	0.000	0.000	26
C21	0.115	0.969	0.242	0.225	0.482	17
C22	0.103	0.833	0.276	0.203	0.423	18
C23	0.096	0.943	0.245	0.228	0.482	16
C24	0.036	1.644	0.045	0.414	0.903	2
C25	0.080	1.114	0.196	0.272	0.582	11
C26	0.091	1.128	0.194	0.271	0.582	10
Shannon's weight	0.726	0.273				

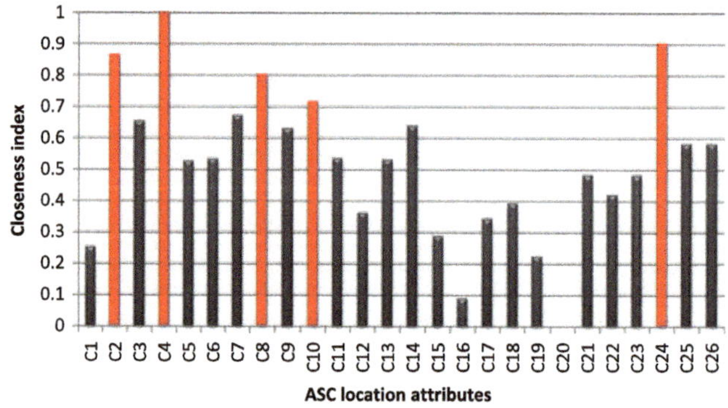

Figure 4. The value of closeness index in TOPSIS ranking method.

Table 6. The most important attributes based on Delphi survey.

No.	Attribute	New abbreviation
C2	Lack of tractor and machinery	Machinery lack (C1)
C4	Easy reach to spare parts, repair, customer service center, and gas station	Easy support (C2)
C8	Easy reach to its villages and covered regions	Easy reach (C3)
C10	Existence of crop rotation with least fallow (irrigated be more than dry farming)	Crop rotation (C4)
C24	Existence of crowded population and vast cultivated area	Population and area (C5)

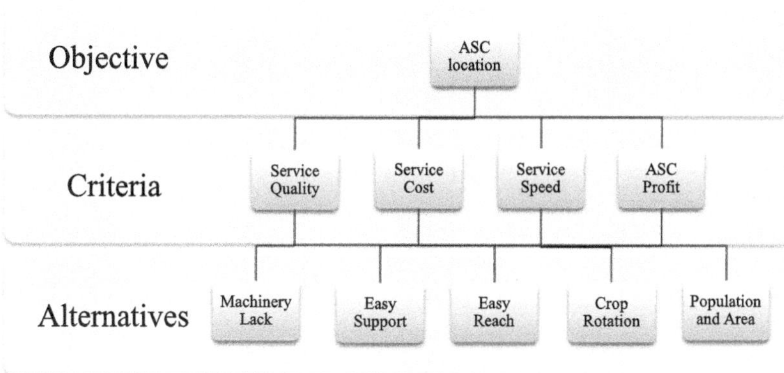

Figure 5. AHP model for location selection of the ASCs.

3.4. *FAHP model development*

3.4.1. *Computing the pairwise comparison matrices of objectives and attributes using fuzzy numbers*

By using triangular fuzzy numbers, the fuzzy pairwise comparisons matrix for the main objectives is constructed as follows in Table 7.

First, via pairwise comparison, the fuzzy synthetic extent values of the attributes were calculated in order to assess the priority weights of the main objectives. The values of the fuzzy synthetic extent of the four objectives were denoted $S_{\text{Service Quality}}$, $S_{\text{Service Cost}}$, $S_{\text{Service Speed}}$, and $S_{\text{ASC Profit}}$, respectively:

$$\begin{aligned} S_{\text{Service Quality}} &= (1.600, 5.661, 19.000) \otimes (6.169, 16.841, 72.000)^{-1} \\ &= (0.022, 0.336, 3.080) \end{aligned}$$

$$\begin{aligned} S_{\text{Service Cost}} &= (1.616, 4.521, 19.000) \otimes ((6.169, 16.841, 72.000))^{-1} \\ &= (0.022, 0.268, 3.080) \end{aligned}$$

$$\begin{aligned} S_{\text{Service Speed}} &= (1.476, 3.233, 18.000) \otimes ((6.169, 16.841, 72.000))^{-1} \\ &= (0.021, 0.192, 2.918) \end{aligned}$$

Table 7. Pairwise comparison matrix for objectives.

	O1	O2	O3	O4	Normalized weight
O1	(1.000, 1.000, 1.000)	(0.200, 1.415, 6.000)	(0.200, 1.711, 6.000)	(0.200, 1.533, 6.000)	0.257
O2	(0.166, 0.706, 5.000)	(1.000, 1.000, 1.000)	(0.200, 1.245, 7.000)	(0.250, 1.570, 6.000)	0.252
O3	(0.166, 0.567, 5.000)	(0.142, 0.785, 5.000)	(1.000, 1.000, 1.000)	(0.166, 0.880, 7.000)	0.245
O4	(0.166, 0.652, 5.000)	(0.166, 0.636, 4.000)	(0.142, 1.135, 6.000)	(1.000, 1.000, 1.000)	0.244
CR	0.001				

Table 8. Pairwise comparison matrix for service quality.

Service quality	C1	C2	C3	C4	C5	Normalized weight
C1	(1.000, 1.000, 1.000)	(0.125, 0.322, 4.000)	(0.142, 0.792, 4.000)	(0.142, 1.086, 5.000)	(0.142, 0.569, 5.000)	0.189
C2	(0.250, 3.101, 8.000)	(1.000, 1.000, 1.000)	(1.000, 3.282, 8.000)	(1.000, 3.702, 8.000)	(1.000, 2.907, 6.000)	0.225
C3	(0.250, 1.261, 7.000)	(0.125, 0.304, 1.000)	(1.000, 1.000, 1.000)	(0.125, 1.584, 5.000)	(0.125, 0.977, 6.000)	0.195
C4	(0.200, 0.920, 7.000)	(0.125, 0.270, 1.000)	(0.200, 0.630, 8.000)	(1.000, 1.000, 1.000)	(0.166, 0.812, 3.000)	0.190
C5	(0.200, 1.756, 7.000)	(0.166, 0.343, 1.000)	(0.166, 1.022, 8.000)	(0.333, 1.231, 6.000)	(1.000, 1.000, 1.000)	0.199
CR	0.007					

Table 9. Pairwise comparison matrix for service cost.

Service cost	C1	C2	C3	C4	C5	Normalized weight
C1	(1.000, 1.000, 1.000)	(0.142, 0.344, 2.000)	(0.142, 0.468, 3.000)	(0.166, 0.892, 4.000)	(0.200, 0.753, 4.000)	0.186
C2	(0.500, 2.905, 7.000)	(1.000, 1.000, 1.000)	(0.142, 1.492, 4.000)	(0.250, 2.334, 8.000)	(0.250, 1.614, 6.000)	0.214
C3	(0.333, 2.134, 7.000)	(0.250, 0.670, 7.000)	(1.000, 1.000, 1.000)	(0.333, 1.231, 6.000)	(0.333, 0.719, 6.000)	0.205
C4	(0.250, 1.120, 6.000)	(0.125, 0.428, 4.000)	(0.166, 0.812, 3.000)	(1.000, 1.000, 1.000)	(0.166, 0.748, 2.000)	0.192
C5	(0.250, 1.326, 5.000)	(0.166, 0.619, 4.000)	(0.166, 1.390, 3.000)	(0.500, 1.335, 6.000)	(1.000, 1.000, 1.000)	0.200
CR	0.01					

Table 10. Pairwise comparison matrix for service speed.

Service speed	C1	C2	C3	C4	C5	Normalized weight
C1	(1.000, 1.000, 1.000)	(0.111, 0.455, 3.000)	(0.111, 0.483, 6.000)	(0.200, 1.157, 3.000)	(0.200, 0.950, 4.000)	0.193
C2	(0.333, 2.197, 9.000)	(1.000, 1.000, 1.000)	(0.250, 0.821, 9.000)	(0.333, 2.099, 5.000)	(0.250, 1.876, 6.000)	0.209
C3	(0.166, 2.069, 9.000)	(0.111, 1.216, 4.000)	(1.000, 1.000, 1.000)	(0.200, 1.586, 9.000)	(0.142, 1.290, 6.000)	0.207
C4	(0.333, 0.863, 5.000)	(0.200, 0.476, 3.000)	(0.111, 0.630, 5.000)	(1.000, 1.000, 1.000)	(0.333, 0.835, 2.000)	0.207
C5	(0.250, 1.052, 5.000)	(0.166, 0.533, 4.000)	(0.166, 0.774, 7.000)	(0.500, 1.196, 3.000)	(1.000, 1.000, 1.000)	0.191
CR	0.1					

Table 11. Pairwise comparison matrix for ASC profit.

ASC profit	C1	C2	C3	C4	C5	Normalized weight
C1	(1.000, 1.000, 1.000)	(0.200, 1.453, 7.000)	(0.200, 2.065, 7.000)	(0.200, 2.432, 8.000)	(0.166, 1.000, 4.000)	0.207
C2	(0.142, 0.688, 5.000)	(1.000, 1.000, 1.000)	(0.200, 1.048, 8.000)	(0.333, 2.334, 6.000)	(0.166, 0.679, 4.000)	0.201
C3	(0.142, 0.484, 5.000)	(0.125, 0.954, 5.000)	(1.000, 1.000, 1.000)	(0.250, 1.231, 8.000)	(0.142, 0.357, 5.000)	0.196
C4	(0.125, 0.411, 5.000)	(0.166, 0.428, 3.000)	(0.125, 0.812, 4.000)	(1.000, 1.000, 1.000)	(0.166,0.328, 1.000)	0.183
C5	(0.250, 1.000, 6.000)	(0.250, 1.472, 6.000)	(0.200, 2.794, 7.000)	(1.000, 3.047, 6.000)	(1.000, 1.000, 1.000)	0.210
CR	0.1					

$$S_{\text{AMSC Profit}} = (1.476, 3.424, 16.000) \otimes ((6.169, 16.841, 72.000))^{-1}$$
$$= (0.021, 0.203, 2.594)$$

The degree of possibility of S_i $(i \neq j)$ was determined.

$$V\left(S_{\text{Service Quality}} \geq S_{\text{Service Cost}}\right) = 1$$

$$V\left(S_{\text{Service Quality}} \geq S_{\text{Service Speed}}\right) = 1$$

$$V\left(S_{\text{Service Quality}} \geq S_{\text{AMSC Profit}}\right) = 1$$

$$V\left(S_{\text{Service Cost}} \geq S_{\text{Service Quality}}\right) = \frac{3.080 - 0.022}{(3.080 - 0.022) + (0.336 - 0.268)} = 0.978$$

$$V\left(S_{\text{Service Cost}} \geq S_{\text{Service Speed}}\right) = 1$$

$$V\left(S_{\text{Service Cost}} \geq S_{\text{AMSC Profit}}\right) = 1$$

$$V\left(S_{\text{Service Speed}} \geq S_{\text{Service Quality}}\right) = \frac{2.918 - 0.022}{(2.918 - 0.022) + (0.336 - 0.192)} = 0.952$$

$$V\left(S_{\text{Service Speed}} \geq S_{\text{Service Cost}}\right) = \frac{2.918 - 0.022}{(2.918 - 0.022) + (0.268 - 0.192)} = 0.974$$

$$V\left(S_{\text{Service Speed}} \geq S_{\text{AMSC Profit}}\right) = \frac{2.918 - 0.021}{(2.918 - 0.021) + (0.203 - 0.192)} = 0.996$$

$$V\left(S_{\text{AMSC Profit}} \geq S_{\text{Service Quality}}\right) = \frac{2.594 - 0.022}{(2.594 - 0.022) + (0.336 - 0.203)} = 0.950$$

$$V\left(S_{\text{AMSC Profit}} \geq S_{\text{Service Cost}}\right) = \frac{2.594 - 0.022}{(2.594 - 0.022) + (0.268 - 0.203)} = 0.975$$

$$V\left(S_{\text{AMSC Profit}} \geq S_{\text{Service Speed}}\right) = 1$$

The minimum degree of possibility was stated as follows:

$$\acute{d}(\text{Service Quality}) = \min(1.000, 1.000, 1.000) = 1.000$$

$$\acute{d}(\text{Service Cost}) = \min(0.978, 1.000, 1.000) = 0.978$$

$$\acute{d}(\text{Service Speed}) = \min(0.952, 0.974, 0.996) = 0.952$$

$$\acute{d}(\text{AMSC Profit}) = \min(0.950, 0.975, 1.000) = 0.950$$

Accordingly, the weight vector was derived as $\acute{W} = (1.000, 0.978, 0.952, 0.950)$ and the normalized weight was given by $W = (0.257, 0.252, 0.245, 0.244)^T$. By applying the same calculation process to the remaining attributes, priority weights could be calculated.

From the calculated results listed in Table 7, it can be concluded that, by comparing the weights of objective in Level 2, service quality and service cost are more important than service speed and ASC profit objective.

The local weights of the relative importance are computed to the attributes against the objectives; and also the global weights, which are the relative importance attributes against the goal. To derive the global weight of each attribute, its local weight was multiplied by the local weight of each corresponding objectives. In this study, all consistency ratios range from 0.001 to 0.1 and overall CR is 0.04, which fell within the acceptance level of 0.10 as recommended by Saaty (1994). This shows that the survey respondents have assigned their weights consistently after examining the priorities of success attributes of location selection for ASCs.

3.4.2. Local weights based on the synthesized judgments

Tables 8–11 hierarchically display local weights according to the Level 3, with the CR using a 5 × 5 matrix. In the Level 3 attributes of service quality, 'easy support' (local *W*: 0.225) and 'population and area' (local *W*: 0.199) attributes rank first and second, respectively. Also 'easy reach' (local *W*: 0.195), 'crop rotation' (local *W*: 0.190), and 'machinery lack' (local *W*: 0.189) rank the third to fifth, respectively. To visualize this result, Figures 6 and 7 show the local weights of the attributes under service quality and service cost objectives in a radiated diagram.

Similarly from Figures 8 and 9 the priority of attributes under 'Service Speed' and 'ASC Profit' can be seen.

3.4.3. Global weights based on the synthesized judgments

To determine the overall prioritization of the five attributes with respect to the goal of best location of ASCs, the global weights of the attributes were calculated. Table 12 shows the prioritization of the four objectives and five attributes in terms of the global weights based on the synthesized judgments of evaluators.

From the Level 2 objective weights, we can derive the prioritization of four objectives. Table 12 shows that 'Service Quality' (global *W*: 0.257) is most important

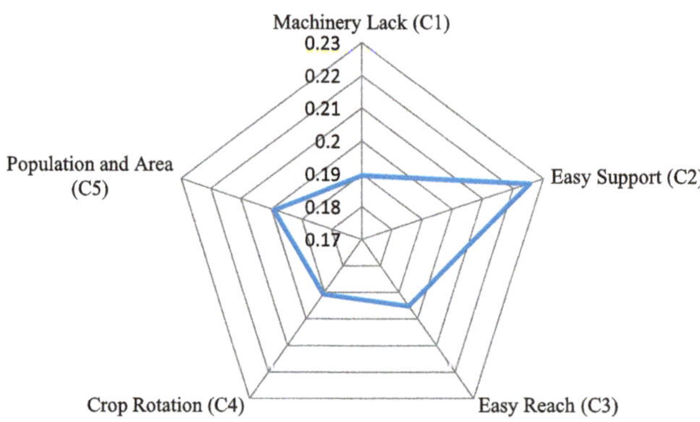

Figure 6. Radiated diagram of local weight of attributes under service quality objectives.

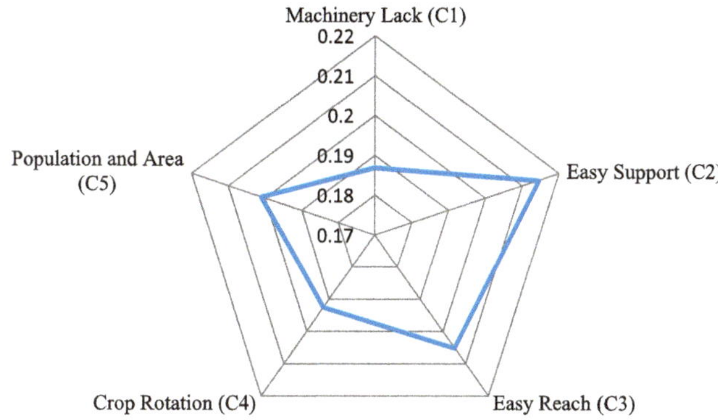

Figure 7. Radiated diagram of local weight of attributes under service cost objectives.

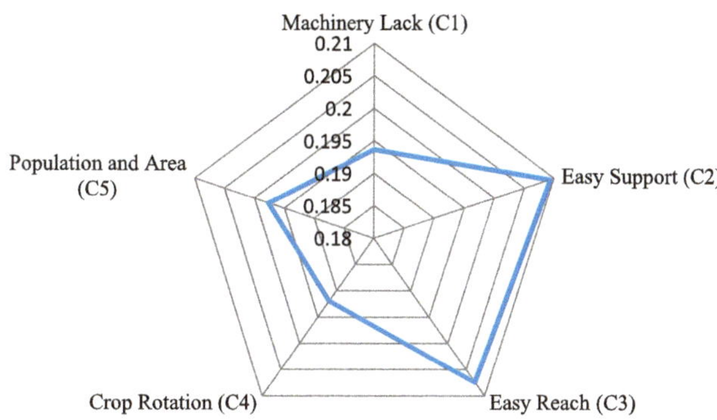

Figure 8. Radiated diagram of local weight of attributes under service speed objectives.

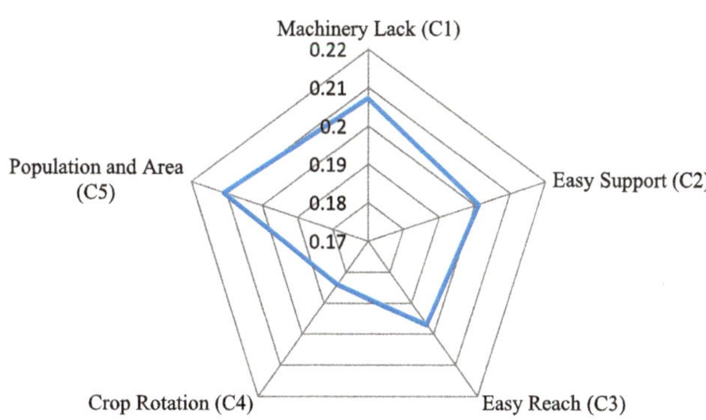

Figure 9. Radiated diagram of local weight of attributes under ASC profit objectives.

criterion for assessing the ASC location attributes. 'Service Cost' (global *W*: 0.252), 'Service Speed' (global *W*: 0.245) and 'ASC Profit' (global *W*: 0.244) are the second to fifth most important objectives, respectively, in this level.

For Level 3 perspective, the global weight of each attributes was calculated by multiplying the local weight of each attribute under each criterion to global weight of

Table 12. Global weight and prioritization of objectives and attributes.

Objectives	Global *W*	Prioritizations	Attributes	Global *W*	Prioritizations
Service quality (O1)	0.257	1	Easy support (C2)	0.212	1
Service cost (O2)	0.252	2	Population and area (C5)	0.202	2
Service speed (O3)	0.245	3	Easy reach (C3)	0.201	3
ASC profit (O4)	0.244	4	Machinery lack (C1)	0.194	4
			Crop rotation (C4)	0.189	5

related objectives and calculating sum of them for each attribute. As can be seen in Table 12 the attribute 'Easy Support' (global *W*: 0.212) has relatively higher value than other attributes. 'Population and area' (global *W*: 0.202) and 'Easy reach' (global *W*: 0.201) with less difference are the second and third important attributes. Also 'Machinery lack' (global *W*: 0.194) and 'Crop rotation' (global *W*: 0.189) are the less important attributes for location selection of ASC.

The attributes selected in this research are interested to use in facility location problem of ASCs. Service centers should be located on places which can reach to their required inputs to provide the services for farms. Easy support attribute means that candidate location is better to have access to fuel stations, agricultural input markets, human workers, etc. The second important attribute is population and area, which directly influence on the service cost and centers profit. Locations which are near to more customers can reduce their costs and get more profits. Accessibility to service centers is almost related to the distance between farmers and centers. But in agricultural areas the accessibility may related to centrality of centers location among villages, because farmers of around areas needed to travel between their farms/home and the center several times in each season. Easy reach to service center will increase the service lead time and center profit. Transportation and also distance are key factors in location problems. Since in this research, we have assumed same priority for objectives, the easy reach ranked third, but if in a situation the service cost and service speed be more important than quality and profit for service centers, then the easy reach will be the most important location attribute, because the transportation directly affect the cost and speed of service. The number of available tractors and machineries in any region can be considered as an attribute, because the service centers should provide mechanization services to the farms which have no machinery or needs some machinery, therefore the centers can be located there. The last attribute is crop rotation which means that service centers can locate on regions that several types of crops are cultivated. Agricultural operations are completely seasonal; it means that some operations in one month of a year are needed and its facilities will remain useless until next season. This phenomenon is not favorable for service centers economically. As our expert team selected the work for the facilities should be available in all times of season. It can be obtained where several crops cultivated in the region.

In this research, we assumed that all objectives have same importance in normal case, but if any manager/researcher want to use this approach with other assumption about objectives and make priority, they are recommended to use the GP method introduced through the paper. They can use some targets/goals for the objectives and then prioritize the location attributes of service centers. It can be very useful to manage the agricultural systems to be efficient.

4. Conclusions

The main contribution of this paper was to identify the list of objectives and attributes for assessing the best location for establishing ASCs. The paper elaborated upon a hierarchically assessment model and identified important objectives, and their weights to evaluate the relative importance of the ASC location selection problem. This paper contributes to experts by applying the TOPSIS and FAHP model with the Delphi method, which is considered one of the most structured techniques for industrial and managerial decision-making. Based on literature review and Delphi study, a TOPSIS, and an AHP model was generated by incorporating four decision objectives. In addition, the fuzzy approach has been used to make the conventional AHP model more sophisticated with much precise judgment. The results of the FAHP method indicated that agricultural service quality is the most important objective to assess the ASC location attributes. It can be concluded that the service quality is the main expectation from agricultural services. Further, several location attributes are adopted for ASC location and found that support for ASC is the main attribute. Tractors and agricultural machineries have too depreciations because of their working conditions. So these centers need continual repairs both in working season and off-season for next working season. This paper also proposed MCGP to prioritize the objective functions and define targets for them and then rank the attributes based on them. This method can be used to rank ASC location attributes, and this is another contribution of current research, because it is generalizable to other regions and other conditions. The studied problem closely relates to such problems found in manufacturing systems. The proposed methods and the conclusions reached can also be used for solving, e.g. maintenance facility placement problems in manufacturing or supply chain design problems. In future studies, extended research is needed to investigate the effect of current situation of services in agricultural supply chain on priority of location problem objectives in model developed in this paper into a variety of different areas. The main limitation of developed approach is related to the experts participated in this research. Since the methodology used in this paper relies on knowledge and experience of experts, thus it is so critical to use elite experts in such studies.

Acknowledgments

The financial support provided by University of Tehran is duly gratefully acknowledged. Thanks and appreciations of authors also go to colleagues in Aalborg University and people who have willingly helped us with their abilities.

References

Akıncı, H., Özalp, A. Y., & Turgut, B. (2013). Agricultural land use suitability analysis using GIS and AHP technique. *Computers and Electronics in Agriculture, 97*, 71–82.

Bagheri, N., & Moazzen, S. A. (2009). Optimum strategy for agricultural mechanization development in Iran. *Journal of Agricultural Technology, 6*, 225–237.

Bishop, C. (1997). *A guide to preparing an agricultural mechanization strategy.* Rome: Food and Agriculture Organization of the United Nations.

Brown, C. E. (1998). *Applied multivariate statistics in geohydrology and related sciences.* Berlin: Springer, Verlag Berlin Heidelberg.

Chang, C. T. (2007). Multi-choice goal programming. *Omega, 35*, 389–396.

Chang, C. T., Chen, H. M., & Zhuang, Z. Y. (2014). Integrated multi-choice goal programming and multi-segment goal programming for supplier selection considering imperfect-quality and price-quantity discounts in a multiple sourcing environment. *International Journal of Systems Sciences, 45*, 1101–1111.

Chin, K. S., Xu, D.-L., Yang, J. B., & Lam, J. P.-K. (2008). Group-based ER–AHP system for product project screening. *Expert Systems with Applications, 35*, 1909–1929.

Chipeta, S. (2006). *Demand driven agricultural advisory services.* Lindau: Neuchatel Group.

Cho, J., & Lee, J. (2013). Development of a new technology product evaluation model for assessing commercialization opportunities using Delphi method and fuzzy AHP approach. *Expert Systems with Applications, 40*, 5314–5330.

Dorward, A., & Chirwa, E. (2011). The Malawi agricultural input subsidy programme: 2005/06 to 2008/09. *International Journal of Agricultural Sustainability, 9*, 232–247.

Farahani, R. Z., Steadieseifi, M., & Asgari, N. (2010). Multiple criteria facility location poblems: A survey. *Applied Mathematical Modeling, 34*, 1689–1709.

García, J. L., Alvarado, A., Blanco, J., Jiménez, E., Maldonado, A. A., & Cortés, G. (2014). Multi-attribute evaluation and selection of sites for agricultural product warehouses based on an analytic hierarchy process. *Computers and Electronics in Agriculture, 100*, 60–69.

Hosseini-Nasab, H., Tavana, M., & Yousefi, M. (2014). A new heuristic algorithm for the planar minimum covering circle problem. *Production & Manufacturing Research, 2*, 142–155.

Hwang, C. L., & Masud, S. A. M. (1979). *Multiple objective decision making – Methods and applications: A state-of-the-art survey.* Berlin: Springer.

Hwang, C. L., & Yoon, K. (1981a). *Multiple attribute decision making – Methods and applications: A state-of-the-art survey.* New York, NY: Springer.

Hwang, C. L., & Yoon, K. P. (1981b). *Multiple attribute decision making methods and aplications.* New York, NY: Springer-Verlag.

Inns, F. M. (1995). *Selection, testing and evaluation of agricultural machines and equipment theory.* Rome: FAO Agricultural Services Bulletin.

Johan, F., Swinnen, M., & Gow, H. R. (1999). Agricultural credit problems and policies during the transition to a market economy in Central and Eastern Europe. *Food Policy, 24*, 21–47.

Lak, M. B., & Almasi, M. (2011). An analytical review of parameters and indices affecting decision making in agricultural mechanization Australia. *Journal of Agricultural Engineering, 2*, 140–146.

Lee, S. H. (2010). Using fuzzy AHP to develop intellectual capital evaluation model for assessing their performance contribution in a university. *Expert Systems with Applications, 37*, 4941–4947.

Linstone, H. A., & Turoff, M. (2002). *The Delphi method, techniques and applications.* Addison Wesley.

Oztaysi, B. (2014). A decision model for information technology selection using AHP integrated TOPSIS-Grey: The case of content management systems. *Knowledge Based Systems.* Retrieved from http://dx.doi.org/10.1016/j.knosys.2014.02.010

Prusty, S. K., Mohapatra, P. K. J., & Mukherjee, C. K. (2010). GOS tree (goal-objective-strategy tree) approach to strategic planning using a fuzzy-Delphi process: An application to the Indian Shrimp Industry. *Technological Forecasting & Social Change, 77*, 442–456.

Saaty, T. L. (1986). Axiomatic foundation of the analytic hierarchy process. *Management Science, 32*, 841–855.

Saaty, T. L. (1994). How to make a decision: The analytic hierarchy process. *Interfaces, 24*, 19–43.

Shukla, R. K., Garg, D., & Agarwal, A. (2014). An integrated approach of Fuzzy AHP and Fuzzy TOPSIS in modeling supply chain coordination. *Production & Manufacturing Research, 2*, 415–437.

Sims, G. B., & Kienzle, J. (2009). *Farm equipment supply chains; guidelines for policy-makers and service providers: Experiences from Kenya, Pakistan and Brazil.* Rome. FAO.

Szidarovszky, F., Gershon, M. E., & Duchstein, L. (1986). *Techniques for multi-objective decision making in systems management.* Amsterdam: Elsevier Science Publishers B.V.

Timmer, C. P., Falcon, W. P., & Pearson, S. R. (1983). *Food policy analysis*. Baltimore: The Johns Hopkins University Press.

Ulungu, E. L., & Teghem, J. (1994). Multi-objective combinatorial optimization problems: A survey. *Journal of Multicriteria Decisions Analysis, 3*, 83–104.

Zadeh, L. A. (1976). A fuzzy-algorithmic approach to the definition of complex or imprecise concepts. *International Journal of Man-Machine Studies, 8*, 249–291.

Zangeneh, M., Nielsen, P., Akram, A., & Keyhani, A. (2014). Agricultural service center location problem: Concept and a MCDM solution approach. In *APMS 2014 International Conference Advances in Production Management Systems "Innovative and knowledge-based production management in a global-local world" Ajaccio*. 20–24 September 2014.

Zionts, S. (1979). A survey of multiple criteria integer programming methods. *Annals of Discrete Mathematics, 5*, 389–398.

4

Development of a predictive model for determining mechanical properties of AA 6061 using regression analysis

Joseph Achebo*

Department of Production Engineering, University of Benin, Benin City, Nigeria

Aluminum alloys (AA) such as AA 6061 are difficult to weld, and their application to high-strength demands tends to be limited because of their inherently low-strength threshold levels when compared to other alloys. To effectively use this alloy to its full potential, its tensile strength, is investigated. In actual manufacturing settings, it has proven beneficial to attempt a prediction of the tensile properties of potential weld joints. To achieve this, a model was developed using the multiple linear regression analysis to predict these tensile properties such as the ultimate tensile strength, the yield strength (YS), and percentage elongation (% Elong). It was found from the scattered diagrams that the measured and predicted values were almost a perfect fit with a coefficient of determination of between 0.99 and 1.0. The analysis of variance further validated the adequacy of the model. The analysis showed that the claims observed in the study match with those of other investigators. The predictive model obtained is expected to help the welding community to pre-determine the tensile properties of AA 6061 weldment using selected values for each of the process parameters applied in this study.

Keywords: aluminum alloy 6061; mechanical properties; process parameters; regression analysis

1. Introduction

Materials can either be regarded as being ductile or brittle depending on their ability to undergo plastic deformation under the application of load without fracturing. Ductile materials are often used for engineering projects because they usually exhibit homogeneous load distribution. Brittleness is regarded as the opposite of ductility. In a brittle material, at critical conditions, fractures may likely occur suddenly when loads are applied because the YS equals the ultimate tensile strength (UTS). Therefore, because it is expected that most structural materials are bound to be subjected to repeat loading stretching over a long period, and at the same time, are expected to absorb impact energies without fracturing, the elastic property of these materials is of great importance in engineering applications, and ductility is therefore a highly prized quality.

Ductility could be measured by determining the % Elong and reduction in area of the structural material. Percent elongation as considered in this study shows the extent to which a material can endure the application of stress before a fracture can occur. The ability of an engineering material to withstand stresses cannot be overemphasized, in terms of equipment maintainability, coupled with the fact that replacement and

*Email: josephachebo@yahoo.co.uk

breakdown maintenance due to material failure could be very costly. Such unwarranted failures invariably lead to the resultant increase in idle time or downtime. Thus, in today's engineering landscape, structural materials are designed to carry even heavier loads, so that their service lives could exceed expectations with optimum efficiency. For instance, in bridge construction, structural materials are designed to serve for a very long period, as it would be quite uneconomical to sustain continued major routine overhauling processes.

Since ductility is a major factor to consider in selecting materials for structural engineering purposes, it has become imperative to develop a model to predict the tensile properties that define ductility. In this study, the ductile material considered is the AA 6061 aluminum alloy. Kumar, Kolhe, Morey, and Datta (2011) wrote that aluminum alloy 6061 (Al-Mg-Si) is widely used for the fabrication of food processing equipment, chemical containers, passenger cars, road tankers, and railway transport systems due to its high specific strength, excellent weldability, and resistance to corrosion. These aluminum alloy 6061 plates were welded with the gas metal arc welding (GMAW) equipment used to make weld deposits. Singla, Singh, and Deepak (2010) defined the GMAW as a process in which the heat source produces an arc between the consumable electrode and the workpiece.

Prediction of mechanical properties to optimize material production performance has been attempted by many investigators, such as Chiteka (2014). They aimed at using artificial neural networks in predicting UTS of a weldment, given rotational speed as well as the welding speed in a friction stir welding environment. Kim, Son, and Jeung (2001) applied novel algorithms to develop mathematical models for optimizing bead width for multi-pass welding using neural network as well as multiple regression methods. Xue et al. (2005) used the fuzzy regression method for predicting and controlling bead width in a robotic arc welding process. Mostafa and Khajavi (2006) developed a model for predicting the value of weld penetration using regression analysis. Karthikeyan and Mahadevan (2014) wrote on the modeling of the friction stir welded Al 6351 plates reinforced with SiC particles in the butt-joint region. Taguchi's experimental design was used to design the experiment matrix and a mathematical model was developed to predict the mechanical properties of the welded Al 6351 plates. The predicted results were compared with the actual experimental results Khorram, Ghoreishi, Yazdi, and Moradi (2011) employed the response surface methodology to model and predict weld bead geometry parameters. Palanivel, Mathews, and Murugan (2011) used mathematical model to predict the mechanical properties of AA 6351 aluminum alloy. Cavaliere, Campanile, Panella, and Squillace (2006) studied the effect of welding parameters on mechanical and microstructural properties of AA 6056 joints produced by friction stir welding. Sreeraj, Kannan, and Maji (2013) used a five-level five-factor full factorial design matrix based on central composite rotatable design technique for the mathematical development of the model used for predicting clad bead geometry of austenitic stainless steel deposited by GMAW. They also used a simulated annealing algorithm tool available in MATLAB 7 software which was efficiently employed for predicting this same clad bead geometry. Heidarzadeh, Mahmoudi, Khodaverdizadeh, and Nazari (2012) worked on tensile behavior of friction stir welded AA 6061-T4 aluminum alloy joints. Elangovan, Balasubramanian, and Babu (2009) predicted the tensile strength of friction stir welded AA 6061 aluminum alloy joints using mathematical model. Sen, Mukherjee, and Pal (2014) developed a mathematical model using a multiple linear regression analysis in MINITAB 13.1 to predict the weld bead geometry and the authors also checked for the adequacy of the model by conducting analysis of variance (ANOVA). It was found from their study that the predicted optimum bead geometry correlates very well with the confirmation test result.

In this study, a model was developed to predict tensile properties and to optimize the extent of ductile content in an aluminum alloy 6061 using the multiple linear regression analysis. Dhas and Satheesh (2012) acknowledged the claim made by other investigators that weld parameter optimization through experimental methods and mathematical models has grown substantially over time to achieve a common goal of improving higher weld efficiency. The model which was developed in this study, successfully predicted the tensile properties of AA 6061. The results obtained were almost perfect match with the corresponding experimental results obtained from the weldments produced using GMAW.

2. Materials and methods

2.1. Materials

Five tensile specimens each were prepared for 16 different clusters or combinations of process parameters, giving a total of 80 weld joints. These joints were prepared according to ASTM E8M-04 procedures (see Figure 1) using a CNC Lathe machine as shown in Figure 2. Tensile tests were carried out in a 100 kN computer-controlled Universal Testing Machine (see Figure 3) as used by Prasad, Rao, and Rao (2011). The specimens were loaded at a rate of 1.5 kN/min as per ASTM specifications, so that these tensile specimens could undergo the deformation process. From the stress strain curve, UTS, YS, and % Elong of the weld joints were evaluated and the average of the five test results is presented in Table 3.

2.2. Methods

The following steps were conducted:

(1) The 16 sets of welding process parameters (see Table 3) were applied, using a GMAW operation, to make weld deposits at five joints each on 60 mm × 140 mm × 10 mm AA 6061 aluminum plates (see Figure 4).
(2) Following the above GMAW welding operations, each of the 80 weld deposits was sawn off to form individual weldment.
(3) Each weldment was machined using the CNC Lathe machine resulting in individual tensile specimens according to ASTM E8M-04 procedures.

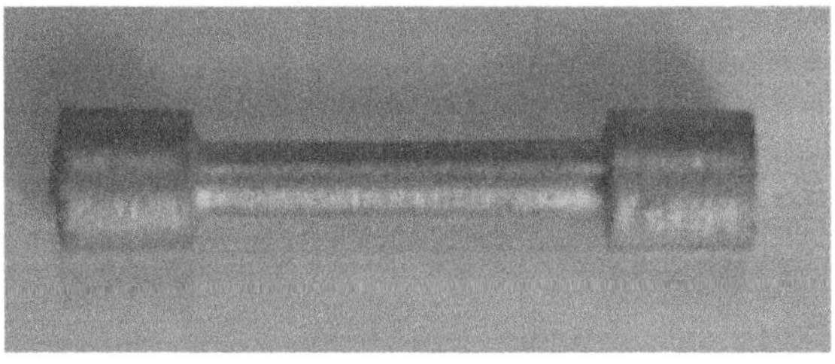

Figure 1. Tensile specimen.

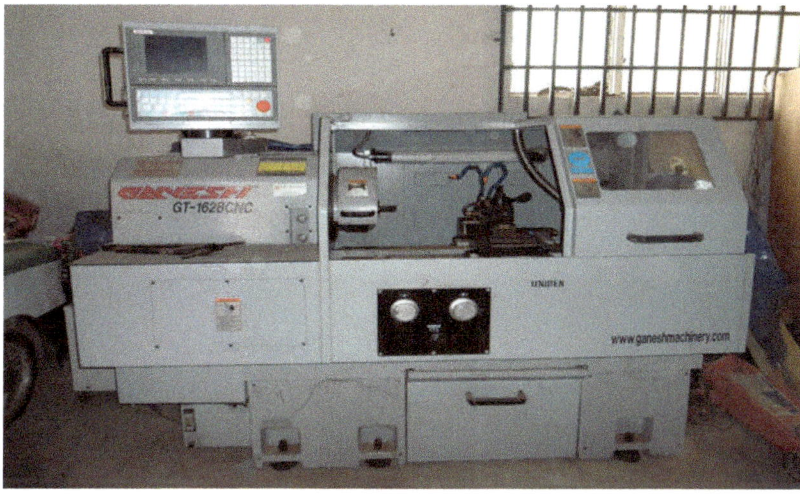

Figure 2. CNC Lathe machine.

Figure 3. 100 kN computer-controlled Universal Testing Machine.

(4) These machined tensile specimens were loaded into a 100 kN computer-controlled Universal Testing Machine.
(5) Each of the 80 loaded tensile specimens was measured for their UTS, YS, and % Elong.
(6) The results of these tensile tests were displayed in a digital screen.
(7) The average of the five tests conducted within each set for the 16 clusters of process parameters made was recorded.

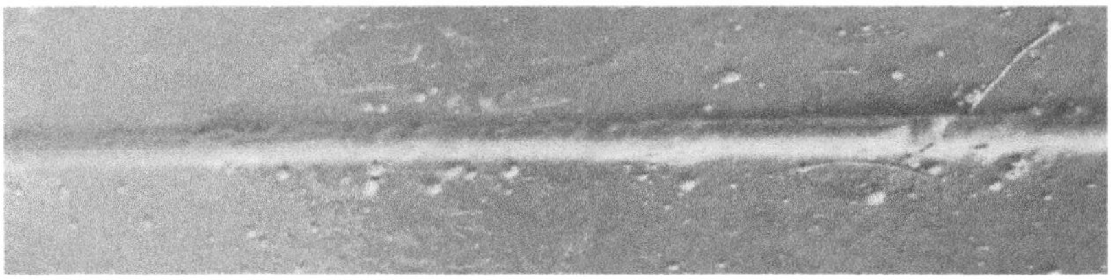

Figure 4. AA 6061 weldment sample.

(8) A statistical model was developed to predict the tensile properties of AA 6061 aluminum weldments.
(9) The effect of each of the 16 process parameters on the tensile properties of the aluminum weldments was examined.

3. Results and discussions

3.1. Results

Table 1 shows the design matrix containing both the process parameters (i.e. A, B, C, and D) and their interactions.

Table 2 shows the process parameters and their levels that were used for this study.

In Table 1, only the main parameters were considered, which comprises A, B, C, and D because linear multiple regression analysis would be used here to predict the tensile properties of aluminum plates treated in this study. The process parameters and their levels in Table 2 were substituted into the main parameters of the matrix design and used for welding the aluminum plates at their joints. Each weldment was tested for its UTS, YS, and % Elong and the measured values are recorded in Table 3.

Table 1. Design matrix.

	A	B	C	D	–AB	–BC	–CD	ABD	–AC	–BD	ABC	BCD	–ABCD	ACD	–AD
0	1	2	3	4	5	6	7	8	9	10	11	12	13	14	15
+	+	–	–	–	+	–	–	+	+	–	+	–	+	+	+
+	+	+	–	–	–	+	–	–	+	+	–	+	–	+	+
+	+	+	+	–	–	–	+	–	–	+	+	–	+	–	+
+	+	+	+	+	–	–	–	+	–	–	+	+	–	+	–
+	–	+	+	+	+	–	–	–	+	–	–	+	+	–	+
+	+	–	+	+	+	+	–	–	–	+	–	–	+	+	–
+	–	+	–	+	+	+	+	–	–	–	+	–	–	+	+
+	+	–	+	–	+	+	+	+	–	–	–	+	–	–	+
+	+	+	–	+	–	+	+	+	+	–	–	–	+	–	–
+	–	+	+	–	+	–	+	+	+	+	–	–	–	+	–
+	–	–	+	+	–	+	–	+	+	+	+	–	–	–	+
+	+	–	–	+	+	–	+	–	+	+	+	+	–	–	–
+	–	+	–	–	+	+	–	+	–	+	+	+	+	–	–
+	–	–	+	[illegible]	[illegible]	+	+	[illegible]	|	–	|	|	|	+	–
+	–	–	–	+	–	–	+	+	–	+	–	+	+	+	+
+	–	–	–	–	–	–	–	–	–	–	–	–	–	–	–

Table 2. The process parameters and their levels.

Process parameters	Unit	Notations	Levels	
			Low	High
Current	A	A (x_1)	320	390
Voltage	V	B (x_2)	18	20
Electrode diameter	mm	C (x_3)	1.6	3.2
Welding speed	mm/s	D (x_4)	80	120

Table 3. Process parameters with the corresponding measured and predicted (calculated) tensile properties.

x_1	x_2	x_3	x_4	Measured	Predicted	Measured	Predicted	Measured	Predicted
(*A*)	(*B*)	(*C*)	(*D*)	UTS (MPa)	UTS (MPa)	YS (MPa)	YS (MPa)	Elong (%)	Elong (%)
390	18	1.6	80	298	318.96	265	301.23	4.74	5.10
390	20	1.6	80	342	311.61	316	288.38	5.8	5.27
390	20	3.2	80	260	270.77	238	246.56	3.7	4.75
390	20	3.2	120	160	261.14	143	238.77	5.2	4.83
320	20	3.2	120	216	181.24	198	153.51	4.7	4.12
390	18	3.2	120	155	268.49	136	251.61	4.9	4.66
320	20	1.6	120	248	222.07	226	195.35	5.2	4.64
390	18	3.2	80	320	278.12	308	259.39	3.8	4.66
390	20	1.6	120	175	301.98	168	280.60	3.2	5.35
320	20	3.2	80	158	190.87	132	161.30	4.6	4.03
320	18	3.2	120	324	188.58	297	166.37	3.6	3.95
390	18	1.6	120	310	309.33	279	293.44	4.7	5.18
320	20	1.6	80	269	231.71	240	203.13	4.8	4.55
320	18	3.2	80	210	198.21	188	174.14	4.3	3.87
320	18	1.6	120	336	229.42	314	208.19	5.8	4.47
320	18	1.6	80	220	239.05	202	215.97	5.1	4.39

3.1.1. Developing the regression analysis model

The regression analysis model as proposed by Achebo (2008) is expressed as follows:

$$\text{SSE} = \sum_{i=1}^{n} (y_i - \bar{y})^2 = \sum_{i=1}^{n} (y_i - b_o - b_1x_1 - b_2x_2 - b_3x_3 - b_4x_4) = 0 = \text{minimum} \quad (1)$$

where SSE is the sum of square error

$$\text{but}\, y_i = b_o + b_1x_1 + b_2x_2 + b_3x_3 + b_4x_4 \quad (2)$$

$$\text{therefore, } b_0 = y - b_1x_1 - b_2x_2 - b_3x_3 - b_4x_4 \quad (3)$$

Differentiating (1) with respect to b_0

$$\frac{\partial \text{SSE}}{\partial b_0} = -2x_1 \sum_{i=1}^{n} (y - b_0 - b_1x_1 - b_2x_2 - b_3x_3 - b_4x_4) = 0 \quad (4)$$

From (4), it is deducted that

$$\sum_{i=1}^{n} y - b_0 - b_1 \sum_{i=1}^{n} x_1 - b_2 \sum_{i=1}^{n} x_2 - b_3 \sum_{i=1}^{n} x_3 - b_4 \sum_{i=1}^{n} x_4 = 0 \tag{5}$$

Making b_0 the subject of the equation

$$b_0 = \sum_{i=1}^{n} y - b_1 \sum_{i=1}^{n} x_1 - b_2 \sum_{i=1}^{n} x_2 - b_3 \sum_{i=1}^{n} x_3 - b_4 \sum_{i=1}^{n} x_4 \tag{6}$$

Differentiating (6) with respect to b_1

$$\frac{\partial \mathrm{SSE}}{\partial b_1} = -2x_1 \sum_{i=1}^{n} (y - b_0 - b_1x_1 - b_2x_2 - b_3x_3 - b_4x_4) = 0 \tag{7}$$

Rearranging (7), we have:

$$\sum_{i=1}^{n} \left(yx_1 - b_o x_1 - b_1 x_1^2 - b_2 x_1 x_2 - b_3 x_1 x_3 - b_4 x_1 x_4\right) = 0 \tag{8}$$

Rearranging (8), we have

$$\sum_{i=1}^{n} (yx_1) - b_o \sum_{i=1}^{n} x_1 - b_1 \sum_{i=1}^{n} x_1^2 - b_2 \sum_{i=1}^{n} x_1x_2 - b_3 \sum_{i=1}^{n} x_1x_3 - b_4 \sum_{i=1}^{n} x_1x_4 = 0 \tag{9}$$

Differentiating (1) with respect to b_2

$$\frac{\partial \mathrm{SSE}}{\partial b_2} = -2x_2 \sum_{i=1}^{n} (y - b_0 - b_1x_1 - b_2x_2 - b_3x_3 - b_4x_4) = 0 \tag{10}$$

Rearranging (10), we have:

$$\sum_{i=1}^{n} (yx_2) - b_o \sum_{i=1}^{n} x_2 - b_1 \sum_{i=1}^{n} x_1x_2 - b_2 \sum_{i=1}^{n} x_2^2 - b_3 \sum_{i=1}^{n} x_2x_3 - b_4 \sum_{i=1}^{n} x_2x_4 = 0 \tag{11}$$

Differentiating (1) with respect to b_3

$$\frac{\partial \mathrm{SSE}}{\partial b_3} = -2x_3 \sum_{i=1}^{n} (y - b_0 - b_1x_1 - b_2x_2 - b_3x_3 - b_4x_4) = 0 \tag{12}$$

Rearranging (12), we have:

$$\sum_{i=1}^{n} (yx_3) - b_o \sum_{i=1}^{n} x_3 - b_1 \sum_{i=1}^{n} x_1x_3 - b_2 \sum_{i=1}^{n} x_2x_3 - b_3 \sum_{i=1}^{n} x_3^2 - b_4 \sum_{i=1}^{n} x_3x_4 = 0 \tag{13}$$

Differentiating (1) with respect to b_4

$$\frac{\partial \mathrm{SSE}}{\partial b_4} = -2x_4 \sum_{i=1}^{n} (y - b_0 - b_1x_1 - b_2x_2 - b_3x_3 - b_4x_4) = 0 \tag{14}$$

Rearranging (14), we have:

$$\sum_{i=1}^{n} (yx_4) - b_o \sum_{i=1}^{n} x_4 - b_1 \sum_{i=1}^{n} x_1x_4 - b_2 \sum_{i=1}^{n} x_2x_4 - b_3 \sum_{i=1}^{n} x_3x_4 - b_4 \sum_{i=1}^{n} x_4^2 = 0 \tag{15}$$

From (9), we have that

$$b_1 = \frac{\sum_{i=1}^{n}(yx_1) - b_o\sum_{i=1}^{n}x_1 - b_2\sum_{i=1}^{n}x_1x_2 - b_3\sum_{i=1}^{n}x_1x_3 - b_4\sum_{i=1}^{n}x_1x_4}{\sum_{i=1}^{n}x_1^2} \quad (16)$$

From (11), we have that

$$b_2 = \frac{\sum_{i=1}^{n}(yx_2) - b_o\sum_{i=1}^{n}x_2 - b_1\sum_{i=1}^{n}x_1x_2 - b_3\sum_{i=1}^{n}x_2x_3 - b_4\sum_{i=1}^{n}x_2x_4}{\sum_{i=1}^{n}x_2^2} \quad (17)$$

From (13), we have that

$$b_3 = \frac{\sum_{i=1}^{n}(yx_3) - b_o\sum_{i=1}^{n}x_3 - b_1\sum_{i=1}^{n}x_1x_3 - b_2\sum_{i=1}^{n}x_2x_3 - b_4\sum_{i=1}^{n}x_3x_4}{\sum_{i=1}^{n}x_3^2} \quad (18)$$

From (15), we have that

$$b_4 = \frac{\sum_{i=1}^{n}(yx_4) - b_o\sum_{i=1}^{n}x_4 - b_1\sum_{i=1}^{n}x_1x_4 - b_2\sum_{i=1}^{n}x_2x_4 - b_3\sum_{i=1}^{n}x_3x_4}{\sum_{i=1}^{n}x_4^2} \quad (19)$$

Process parameters values in Table 3 were substituted into Equations (6), (16)–(19) and the model coefficients in Table 4 were obtained.

The developed predictive models as contained in Table 4 are expressed below:

(1) UTS, is
MPa $= -0.0099 + 1.1415x_1 - 3.6731x_2 - 25.5238x_3 - 0.2408x_4$

(2) YS, is
MPa $= -0.7846 + 1.2179x_1 - 6.4208x_2 - 26.1426x_3 - 0.1946x_4$

(3) % Elongation, is
% Elong. $= -0.0089 + 0.0101x_1 + 0.0844x_2 - 0.3251x_3 + 0.0021x_4$

The above predictive models were used to predict the various ductile properties in Table 3.

3.1.2. *Confirmation tests*

Confirmation tests were done to ascertain the adequacy of the developed model. Five weldments were tested for their tensile properties using the computerized tensile test machine. For each weldment, five tensile test results were obtained. The average of the five results was recorded for each of the tensile properties and this process was repeated for the five weldments. The developed model was used to predict the values of these tensile properties to confirm the adequacy of the model by comparing its values with those of the corresponding measured ones, as presented in Table 5.

Table 4. Regression model coefficients.

Regression coefficients	UTS (MPa)	YS (MPa)	Elong (%)
b_0	–0.0099	–0.7846	–0.0089
b_1	1.1415	1.2179	0.0101
b_2	-3.6731	–6.4208	0.0844
b_3	–25.5238	–26.1426	–0.3251
b_4	–0.2408	–0.1946	0.0021

3.1.3. Effect of process parameters on tensile properties

The effects of the process parameters on the tensile properties of aluminum alloy plates were studied as shown in Figures 5–8.

Figure 5 depicts the observed effects of welding current on tensile properties of each specimen. It can be seen from the graph that the curves representing UTS, YS, and % Elong increase following graduated increments of the current applied. This indicates that a small increase in current can alter the tensile property of the aluminum alloy. The atoms that make up this aluminum alloy were still within their elastic limits until the current reached 350 A. At this current, the atoms become more agitated and the atomic collision increases. When this molten aluminum alloy cools under controlled conditions, the material experiences some strain hardening effects. This means that as it cools down, the atoms experience less agitation and dislocation. This resistance to dislocation formation makes the aluminum alloy more plastic, thereby establishing the strength properties of the tensile specimen.

As the current is increased from 360 to 380 A, the parameters reduced. This is because as more current is applied the molten aluminum drops in a process known as necking down. This process negatively affects UTS, YS, and % Elong properties of the aluminum alloy leading to a reduction in these properties. Achebo (2012) was of the opinion that welds at certain high temperatures tend to absorb moisture rapidly. Moisture lowers the cooling rate. At low cooling rate, more time is available for the molten aluminum alloy to mix with moisture. This action leads to the coarsening of the weld grains by making the grains angular in shape. This angular shape of the grains creates voids. The presence of voids in welds lowers the quality of such welds. Prasad et al. (2011) observed that the formation of coarser grains in the fusion zone is responsible for lower UTS of the welded joints.

As the current reaches 380 A, the strain hardening effect of the aluminum alloy becomes plastic. This is because at this current, the temperature also increases, further altering the atomic nature of the molten aluminum alloy, making it less susceptible to moisture interference while cooling. As a result of the reduction in the effect of moisture, the grain formation of the aluminum alloy weldment becomes less coarse or finer, thereby increasing its plastic nature. This plastic nature increases the ability of the alloy to absorb energy and this eventually increases the alloy's UTS and YS. The increase in percentage elongation indicates that the material is still ductile and at the same time slightly malleable but has almost exceeded its proportionality limits. This observation compares well with the claim made by Prasad et al. (2011), who emphasized that rapid cooling due to the decrease in current leading to decrease in heat input could lead to the formation of finer grains which eventually increases the UTS.

Table 5. Measured and predicted results of mechanical properties.

x_1	x_2	x_3	x_4	Measured UTS (MPa)	Predicted UTS (MPa)	Measured YS (MPa)	Predicted YS (MPa)	Measured Elong (%)	Predicted Elong (%)
340	19	1.6	110	251	250.99	227	228.07	4.74	4.72
360	18	1.6	80	286	284.71	266	264.69	4.79	4.81
330	20	3.2	100	201	197.47	168	169.59	4.18	4.23
380	18	3.2	90	265	264.30	242	245.27	4.50	4.54
390	18	1.6	85	319	317.75	301	300.25	5.11	5.09

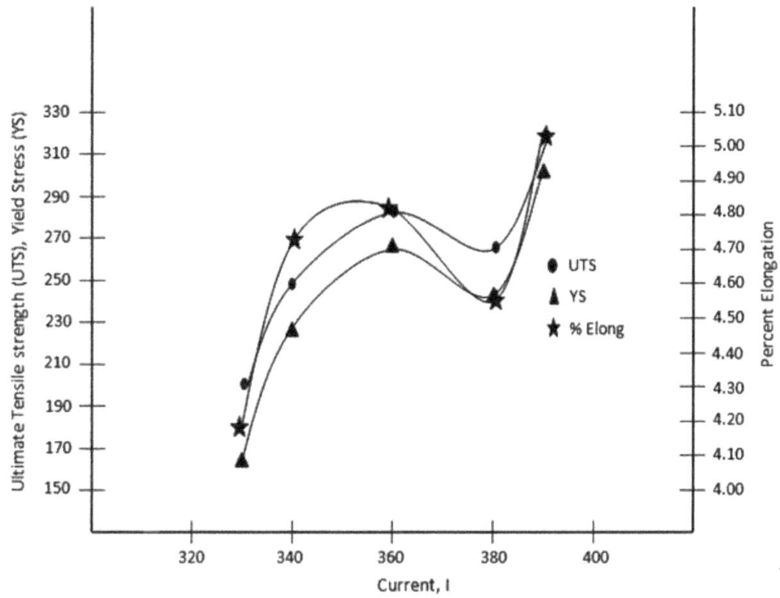

Figure 5. Effect of current on UTS, YS, and % elongation.

Figure 6 shows the relationship between tensile properties and the welding voltage. It is observed that at a constant welding voltage of 18 V, UTS, YS, and % Elong increased steeply. However, as the voltage increases from 18 V, the UTS, YS, and % Elong reduced drastically. This indicates that at 18 V, the aluminum weld was still at its solid–plastic transition state where the grains are assumed to have disintegrated repeatedly and eventually become finer and denser, but as the voltage increases, the weld transits from plastic to liquid. The hot liquid weld metal absorbs moisture and becomes coarser. This action reduces the quality of the weld which negatively affects the weld's UTS, YS, and eventually its % Elong. Its reduction in percentage elongation indicates that the weldment is not as ductile. Bahman and Alialhosseini (2010) showed that as the arc voltage increases, hardness, UTS, and YS decrease. Yang, Chandel, and Bibby (1992) said that at high voltage, bead widths are large and large bead widths are associated with a poor-quality welding process, which invariably negatively affects the UTS, YS, and % Elong.

Figure 7 shows the relationship between tensile properties and the electrode diameter. It is observed that when 1.6-mm electrode diameter was used for the welding process, UTS, YS, and % Elong increased steeply and as the electrode diameter increases, the UTS, YS, and % Elong decrease. This indicates that the 1.6-mm electrode diameter gave a more localized welding process where spatter was eliminated. This theoretically shows that at this electrode diameter, the weld dilution and deposition rate would have been high with adequate and controlled weld penetration, which is assumed to have occurred due to the Marangoni flow process. However, as the diameter of the electrode exceeds 1.6 mm, the increase in the size of the electrode produces beads with large widths forming spatter, which oxidizes rapidly and forms coarse grains, eventually reducing the UTS, YS and % Elong of the aluminum weld. Too much of an oxidization process of the spatter causes the weld to contract.

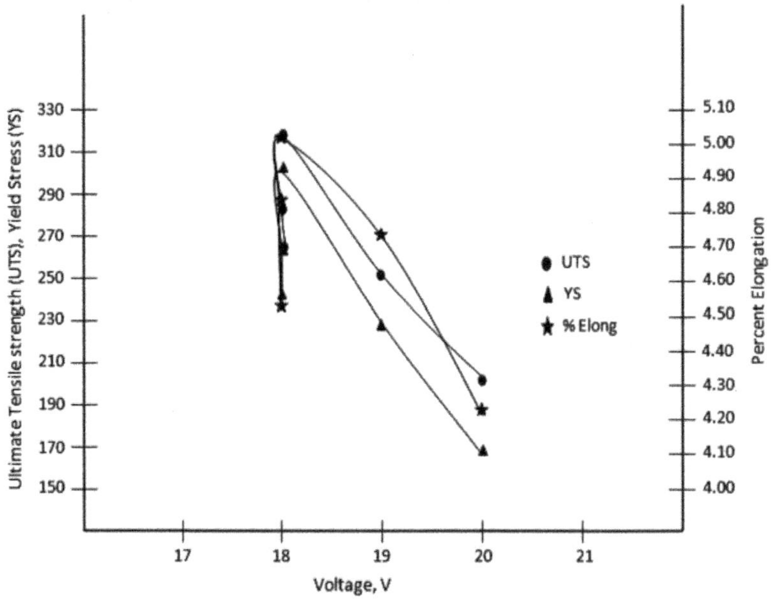

Figure 6. Effect of voltage on UTS, YS, and % elongation.

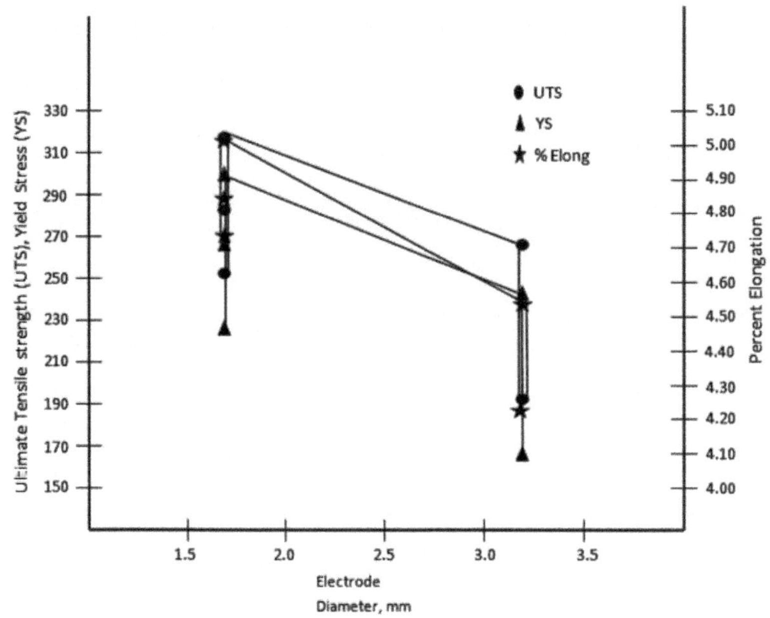

Figure 7. Effect of electrode diameter on UTS, YS, and % elongation.

Figure 8 shows the relationship between tensile properties and the welding speed. It is observed that as the welding speed increases from 80 to 85 mm/s, the UTS, YS, and % Elong also increase. However, when the welding speed increased above 85 mm/s, the UTS, YS, and % Elong decreased. These tensile properties also increased when the

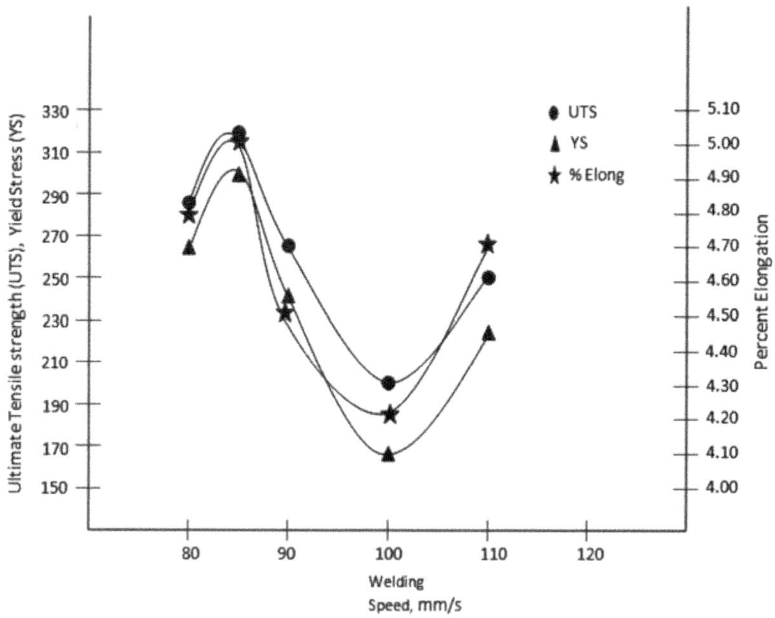

Figure 8. Effect of welding speed on UTS, YS, and % elongation.

welding speed increased from 100 to 110 mm/s. The tensile behavior of the weld indicates that the increase in welding speed, reduces the time spent in welding, this leads to a smaller heat affected zone (HAZ) and a small bead width, and as a result, less heat is utilized for the welding process. This process is used to achieve increased UTS and YS of weldment. These observations confirm the claim made by Bahman and Alialhosseini (2010) that increasing the welding speed, increase hardness, YS, and UTS of weld metal and the situation occurs when a low heat input is achieved. Conversely, prolonged heating at incremental welding speeds produces welds with a large HAZ, this eventually reduces the tensile properties of the weld.

3.1.4. Validation of the developed model

The adequacy of the developed model tested by making a scattered diagram of the results in Table 5, to show the correlation between the measured and the predicted (calculated) UTS of the aluminum alloy studied, as shown in Figure 9.

The adequacy of the developed model tested by making a scattered diagram of the results in Table 5, to show the correlation between the measured and the predicted (calculated) YS of the aluminum alloy studied, as shown in Figure 10.

The adequacy of the developed model tested by making a scattered diagram of the results in Table 5, to show the correlation between the measured and the predicted (calculated) % Elong of the aluminum alloy studied, as shown in Figure 11.

3.1.5. Analysis of variance

Table 6 presents the ANOVA at 95% confidence level. In addition, the coefficient of determination was also determined for each of the measured and predicted (calculated) tensile property.

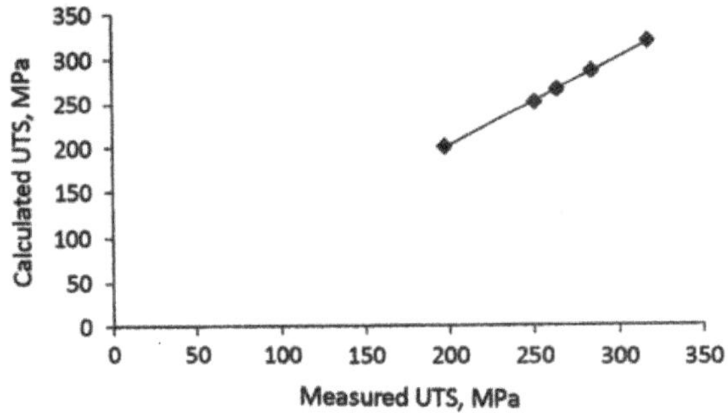

Figure 9. Scattered diagram of the UTS.

3.2. Discussion

A predictive model is developed for determining the tensile properties of AA 6061. These tensile properties include UTS, YS, and % Elong. Process parameters as shown in Table 2, and explicitly distributed in Table 3, were used to conduct 16 welding experiments. Each welding experiment was used to produce five weld samples, the five weld samples were subjected to tensile tests and the average of the responses from the tensile test results obtained such as UTS, YS, and % Elong were recorded for the 16 welding experiments conducted.

These test results were used to develop a model for predicting the tensile properties that define ductility. These mechanical properties are UTS, YS, and % Elong. In this study, the measured UTS is in the range of 155–342 MPa, the predicted value UTS is within the range of 180.24–317.96 MPa. The measured YS is in the range of 132–316 MPa, the predicted YS is found to be in the range of 153.51–301.23 MPa whereas, the measured % Elong is in the range of 3.2–5.8%, the predicted % Elong is within the range of 3.87–5.35%. Lakshminarayanan and Balasubramanian (2009) in their study measured AA 7039 weld joint whose tensile strength was obtained to be within the range of 170–317 MPa using friction stir welding process. Venkateswarlu, Mandal, Mahapatra, and Harsh (2013) determined UTS and % Elong of AA 7039 alloy, the measured UTS is in the range of 125.67–267.05 MPa, the predicted UTS is 116.49–261.97 MPa. Also these researchers determined the measured % Elong of AA 7039 to

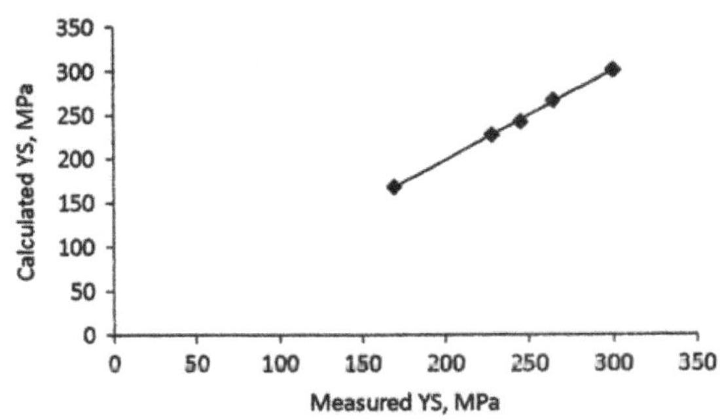

Figure 10. Scattered diagram of the yield strength.

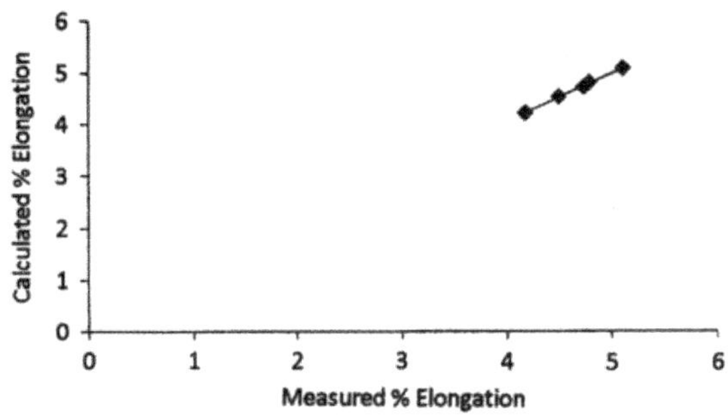

Figure 11. Scattered diagram of the % elongation.

Table 6. ANOVA and coefficient of determination.

Mechanical properties		SSE	F_{test}	F_{table}	r^2
UTS (MPa)	Predicted	0.000049	1.1767×10^{10}		1
	Measured	16.198869	35594.88209		0.99
YS (MPa)	Predicted	0.000016	3.039×10^{10}	230	1
	Measured	16.664116	29183.65826		0.99
Elong (%)	Predicted	0.000046	3940066.116		0.99
	Measured	0.005146	35220.17903		0.99

be between 1.84 and 5.9%, whereas the predicted value of the % Elong is within the range of 1.72–6.32%. Kumar et al. (2011) applied the modified Taguchi method to optimize process parameters. In their research work, they determined UTS and % Elong of AA 6061, the experimental results obtained for UTS are within the range of 256.40–386.07 MPa, with a corresponding predicted UTS of 389.31 MPa. The measured % Elong results are within the range of 7.84–12.94%, whereas the predicted % Elong is 12.98%. Hsieh and Chen (2008) determined the tensile strength of AA 6061 T-6 alloy to be within the range of 225.4–247.9 MPa and the % Elong of 5.5–6.8%. Palanivel et al. (2011) used regression model to predict UTS, YS, and % Elong of AA 6351 aluminum alloy. They obtained for measured UTS, a range of 193–219 MPa, YS within the range of 166–190 MPa and % Elong of between 3.61 and 5.24%, whereas the predicted values for UTS are within the range of 199.70–216.54 MPa, predicted YS is within the range of 172.92–187.29 MPa, and predicted % Elong is within the range of 3.52–5.08%. The mechanical properties obtained in this study falls within the range of values obtained in other reported literature. From the correlation analysis, it is shown that the measured and predicted (calculated) values are nearly a perfect fit with coefficient of determination of between 0.99 and 1.0. The effects of the process parameters used for the welding process on the tensile properties of the AA 6061 were examined. It was proposed that a prolonged increase in current could have a negative effect on the tensile properties of the weld. The increased current also increases in the HAZ profile of the weld. It was found that the increase in voltage could lead to a tremendous reduction in the UTS, YS, and % Elong. The 1.6-mm electrode diameter appears to produce welds with better mechanical properties than the 3.2 mm ones. In Figure 4, it is shown that the increase in welding speed also increases tensile properties.

An ANOVA was done to determine the significance of the level of tensile properties to the quality of the weld at 95% confidence. In each case, the Fisher test was conducted for both measured and predicted (calculated) weld properties. The predicted (calculated) and measured Fishers test values were in each case greater than the table value. This shows that the predictive model has great potential when compared with the measured values.

4. Conclusion

A predictive model using linear regression analysis was painstakingly developed along with a detailed methodology, showing a step-by-step approach. The developed model has proved to be very potent and quite beneficial. It has been used to predict the mechanical properties of AA 6061 with a focus on its UTS, YS, and Fisher Test . These tensile properties are known to be major determinant factors regarding the ductility level of the said alloy considered in this study. The proposed predictive models for determining these tensile properties are expected to ensure that engineers are equipped with a reliable means of pre-determining these tensile properties to ensure that weld operations are properly planned and concomitantly efficient. The correlation analysis and ANOVA conducted confirmed the adequacy of the model. The confirmation tests showed that the variances between the measured and calculated tensile properties were not significant. Also the measured and predicted values of these tensile properties were compared with those obtained in related literature. It was found that values obtained in this study compare well with those from other investigators. This also further confirms the potency of the derived predictive models.

Disclosure statement

No potential conflict of interest was reported by the author.

References

Achebo J. I. (2008). *Development of aluminium welding flux from local raw materials* (PhD Thesis). Department of Production Engineering, University of Benin, Nigeria.

Achebo J. I. (2012, July 4–6). *Optimization of fluence energy in relation to weld properties based on vogel approximation method.* World Congress on Engineering: International Conference on Mechanical Engineering, 3. London.

Bahman, A. R., & Alialhosseini, E. (2010). Change in hardness, yield strength and ultimate tensile strength of welded joints produced in st 37 grade steel. *Indian Journal of Science and Technology, 3*, 1162–1164.

Cavaliere, P., Campanile, G., Panella, G., & Squillace, E. (2006). Effect of welding parameters on mechanical and microstructural properties of AA6056 joints produced by friction stir welding. *Journal of Materials Processing Technology, 180*, 263–270. doi:10.1016/j.jmatprotec.2006.06.015

Chiteka, K. (2014). Artificial neural networks in tensile strength and input parameter prediction in friction stir welding. *International Journal of Mechanical Engineering and Robotics Research, 3*, 145–150.

Dhas, R. E. J., & Satheesh, M. (2012). Multiple objective optimization of submerged arc welding process parameters using grey based fuzzy logic. *Advances in Production Engineering & Management, 7*, 5–16.

Elangovan, K., Balasubramanian, V., & Babu, S. (2009). Predicting tensile strength of friction stir welded AA6061 aluminium alloy joints by a mathematical model. *Materials & Design, 30*, 188–193. doi:10.1016/j.matdes.2008.04.037

Heidarzadeh, A., Mahmoudi, A., Khodaverdizadeh, H., & Nazari, E. (2012). Tensile behavior of friction stir welded AA 6061-T4 aluminum alloy joints. *Materials & Design, 37*, 166–173. doi:10.1016/j.matdes.2011.12.022

Hsieh, H., & Chen, J. L. (2008). Influence of welding parameters on mechanical properties of friction stir welded 6061-T6 launch box. *Materials Transactions, 49*, 2179–2184.

Karthikeyan, P., & Mahadevan P. K. (2014, August 17). *A study of modeling on the mechanical properties of friction stir welded al 6351 plates added with* sic *particles in the butt-joint region*. Proceedings of Eleventh IRF International Conference, 62–65, Chennai. ISBN: 978-93-84209-47-6

Khorram, A., Ghoreishi, M., Yazdi, M. R. S., & Moradi, M. (2011). Optimization of bead geometry in CO_2 laser welding of Ti 6Al 4 V using response surface methodology. *Engineering, 3*, 708–712. doi:10.4236/eng.2011.37084

Kim, I., Son, S., & Jeung, Y. (2001). Control and optimization of bead width for multipass welding in robotic arc welding process. *Australian Welding Journal, 46*, 43–47.

Kumar, P., Kolhe, K. P., Morey, S. J., & Datta, C. K. (2011). Process parameters optimization of an aluminium alloy with pulsed gas tungsten arc welding (GTAW) using gas mixtures. *Materials Sciences and Applications, 2*, 251–257. doi:10.4236/msa.2011.24032

Lakshminarayanan, A. K., & Balasubramanian, V. (2009). Comparison of RSM with ANN in predicting tensile strength of friction stir welded AA7039 aluminium alloy joints. *Transactions of the Nonferrous Metals Society of China, 19*, 9–18. doi:10.1016/S1003-6326(08)60221-6

Mostafa, N. B., & Khajavi, M. N. (2006). Optimization of welding parameters for weld penetration in FCAW. *Journal of Achievements in Materials and Manufacturing Engineering, 16*, 132–138.

Palanivel, R., Mathews, K. P., & Murugan, N. (2011). Development of mathematical model to predict the mechanical properties of friction stir welded AA6351 aluminium alloy. *Journal of Engineering Science and Technology Review, 4*, 25–31.

Prasad K. S., Rao C. S., & Rao D. N. (2011). Optimizing pulsed current micro plasma arc welding parameters to maximize ultimate tensile strength of inconel 625 nickel alloy using response surface method. *International Journal of Engineering Science and Technology, 3*, 226–436. doi: http://dx.doi.org/10.4314/ijest.v3i6.18

Sen M., Mukherjee, M., & Pal, T. K. (2014, December 12–14). *Prediction of weld bead geometry for double pulse gas metal arc welding process by regression analysis*. 5th International & 26th All India Manufacturing Technology, Design and Research Conference, 814-6, IIT Guwahati, Guwahat.

Singla, M., Singh, D., & Deepak, D. (2010). Parametric optimization of gas metal arc welding processes by using factorial design approach. *Journal of Minerals & Materials Characterization & Engineering, 9*, 353–363.

Sreeraj, P., Kannan, T., & Maji, S. (2013). Prediction and control of weld bead geometry in gas metal arc welding process using simulated annealing algorithm. *International Journal of Computational Engineering Research, 3*, 213–222.

Venkateswarlu, D., Mandal, N. R., Mahapatra, M. M., & Harsh, S. P. (2013). Tool design effects for FSW of AA7039. *Welding Journal, 92*, 41s–47s.

Xue, Y., Kim, I. S., Son, J. S., Park, C. E., Kim, H. H., Sung, B. S., … Kang, B. Y. (2005). Fuzzy regression method for prediction and control the bead width in the robotic arc-welding process. *Journal of Materials Processing Technology, 164–165*, 1134–1139. doi:10.1016/j.jmatprotec.2005.02.174

Yang, L. J., Chandel, R. S., & Bibby, M. J. (1992). The effects of process variables on the bead width of submerged-arc weld deposits. *Journal of Materials Processing Technology, 29*, 133–144. doi:10.1016/0924-0136(92)90430-Z

5

Inventory classification based on decoupling points

Joakim Wikner*,[1] and Eva Johansson

Department of Industrial Engineering and Management, School of Engineering, Jönköping University, Jönköping, Sweden

The ideal state of continuous one-piece flow may never be achieved. Still the logistics manager can improve the flow by carefully positioning inventory to buffer against variations. Strategies such as lean, postponement, mass customization, and outsourcing all rely on strategic positioning of decoupling points to separate forecast-driven from customer-order-driven flows. Planning and scheduling of the flow are also based on classification of decoupling points as master scheduled or not. A comprehensive classification scheme for these types of decoupling points is introduced. The approach rests on identification of flows as being either demand based or supply based. The demand or supply is then combined with exogenous factors, classified as independent, or endogenous factors, classified as dependent. As a result, eight types of strategic as well as tactical decoupling points are identified resulting in a process-based framework for inventory classification that can be used for flow design.

Keywords: decoupling points; inventory classification; planning and control; materials management

1. Introduction

Increasing customer requirements for customization, cost issues, and ever shortening product life cycles result in a dynamic behaviour and structural complexity of the logistics system that are very demanding. The decisions are not only a complex matter due to the intricacy of each individual decision but also the sheer volume of decisions adds weight to the burden. In addition, the reduction of inventories in the logistics systems leads to stronger dependencies across the material flows, which increases the risk for knock-on effects of disturbances as pointed out by, among others, Goldratt and Cox (1984). Considering these characteristics in combination, it is obvious that logistics managers need tools to simplify and automate decision-making. Important concepts in this context are abstraction, in order to retain only information that is relevant for a particular decision problem, and aggregation, to focus on groups of entities where the same decision can be implemented.

From a decision-maker's perspective, the implications of applying abstraction (selecting relevant information) and aggregation (grouping of information) are significant. Abstraction is usually applied through hierarchical decision-making. At the lowest level, detailed decisions are usually made with a local and short-term focus. At higher levels, issues of more strategic character are handled. Typical examples at the strategic

*Corresponding author. Email: joakim.wikner@ju.se

[1]The paper is a revised and extended version of Wikner, Johansson, and Persson (2009).

level are to focus on constraining resources and materials critical to customer service. Aggregation can be applied by aggregating decision problems and making a decision based on a key property of the aggregate, such as a product group, and then more or less automatically applying the decision to all products of that product group. The terminology used in this paper refers to this type of aggregation as classification. By classifying an item as a specific type due to some of its properties, a group of items sharing this property is created.

These two concepts, abstraction and aggregation, are the theme of this paper. Item and inventory classifications of different types are very common in logistics management (see, e.g. Broeckelmann, 1999), and our ambition is to further emphasize the importance of a process perspective in this context. Previously, much focus has been on classifications using item properties as a point of departure, e.g. ABC classification, but with an additional process perspective, further support can be provided to logistics management. The purpose of the research presented in this paper is to develop a process-based framework for inventory classification. The application of the framework in planning and control is discussed in order to show how the results can be used.

The paper is structured as follows. Next, a general overview of the inventory management context is provided in terms of decoupling points. The concept of supply system is introduced and some key properties of decoupling points related to replenishment and withdrawal are defined. Thereafter, a generic structural model for logistics systems is introduced to identify four key delivery strategies and their respective positions of the two critical decoupling points of strategic importance. Finally, a process-based approach to inventory classification is derived from the discussion on strategic and tactical decoupling points, followed by managerial implications for planning and scheduling, and some ideas for further research.

2. Decoupling points and material flows

The context of material flows is here referred to as a supply system that can be described as a set of resources transforming raw materials into finished goods. Throughout the material flow, resources interact in the transformation process to create value in terms of form, time, and place. In an ideal state, the material flow is continuous and one piece (Ohno, 1988, p. 96) meaning that there is a steady and even flow where all the resources are synchronized and performing with a minimum of balancing loss in the system. This ideal state is, however, rarely possible to realize in the complete supply system. Usually, this type of material flow is limited to a set of resources performing a number of steps continuously but in a context of only a manufacturing cell or a line. In the context of place transformation, the only case where continuous flow can be achieved, except for in pipeline transportation, is through aggregation of several material flows to, for instance, a truck load, thus providing one-piece flow for each separate material flow. Using common lean terminology, this kind of well-defined section of one-piece flow is referred to as a process (Rother & Shook, 2003, p. 18). Each process then represents transformation and replenishes a point of inventory downstream and withdraws materials from one or more other points of inventory upstream. At a ‘micro’ level, the material flows can therefore be modelled as a set of stock points, i.e. inventories, each connected with processes that withdraw materials and processes that replenish materials. Each inventory can hence be seen as a point that decouples two processes (replenishment and withdrawal) and enables them to act somewhat independently. This kind of decoupling point is referred to as a DP (this definition is in line with the general perception of the concept, see, e.g. Blackstone (2008). The supply system can hence be modelled as a set of DPs that interact through replenishment and withdrawal processes. Note that a process that is a replenishment process for one DP can be a withdrawal process for another DP. The only exceptions are the replenishment processes related to supply external to the supply system (e.g. external suppliers) and withdrawal processes that are external to the supply process (e.g. external customers).

2.1. Decoupling points categorization

The concept of DP is a core component in the management of material flows. Considering different combinations of unexpected and expected events, related to replenishment and withdrawal, for each and every DP results in a very demanding challenge for management since the number of DPs in an enterprise or supply chain is staggering. It is very difficult in all realistic settings, at least with today's technology, to control each and every DP to any greater extent. To handle this complexity, the DPs can be differentiated into categories. In particular three categories play a key role in materials management; they are here referred to as operational, tactical, and strategic:

- *Operational DPs* connect transformation points in the physical material flows and are, for example, known as queues in front of resources.
- *Tactical DPs* are related to materials given a unique identifier, usually referred to as item number or part number. Introducing this identity code enables item-level materials management.
- *Strategic DPs* are a subset of all tactical DPs that play a role of critical importance to the interface of the supply system and its context.

2.2. Decoupling points and planning and scheduling

Strategic DPs relate to items that are exposed to external influence on the withdrawal side, such as customer orders, or on the replenishment side, such as purchase orders. External demand is not possible to derive from demand for other items and is hence usually unpredictable in terms of both quantity and timing. Classifying items as exposed to external influence or only as internally affected corresponds to the classification in terms of dependent and independent demand, which is the fundamental principle behind materials planning (Vollmann, Berry, Whybark, & Jacobs, 2005, p. 134). Demand for an item that can be derived from demand for other items is in this context referred to as dependent demand (DD). In its most straightforward application, demand for a component is calculated as the demand for the parent item multiplied by the explosion quantity. Demand that cannot be calculated in this way but is directly affected by external demand is thus classified as independent demand (ID). This classification of demand has a long history and is particularly associated with material requirements planning, MRP (Orlicky, 1975), of which backward scheduling is a key concept. As this scheduling approach is performed, the demand for the next level of items is calculated in the bill-of-material (BoM) explosion process. Many items with DD may also have a certain proportion of ID, if they, for example, are sold separately as spare parts besides being used in the production of another parent item. In summary, DD items are exposed to internal demand and hence of a less strategic nature. ID items, on the other hand, are exposed to external demand, e.g. customer orders, and due to their strategic nature correspond to strategic DPs. The similarities between independent and strategic DPs have so far been covered from a demand-based perspective but, as will be shown below, the corresponding similarities can also be identified from a supply-based perspective. Integrating the demand-based and the supply-based perspectives results in a more comprehensive approach to the classification of DPs.

3. The logistics system

The basic supply system outlined above corresponds to an inventory subsystem related to one DP connected with a replenishment process and a withdrawal process. In isolation, this is a relatively simple entity where item-based inventory management can be applied. An inventory subsystem is, however, usually a part of a larger logistics system consisting of a set of connected inventory subsystems. As the complete logistics system is considered, a range of dependencies is introduced, which increases the complexity of the decision problem. Instead of dealing with single-item inventory management, the decision-maker is facing a scenario with a network of inventory subsystems that interact with each other, and to some extent also interact with the context of the complete logistics system. This type of system also corresponds to the logical entity introduced by Wikner (2014) as a supplement to the traditional actor-based approach to flow design. Such a complex decision problem is very difficult to handle with a monolithic model and therefore requires some kind of hierarchical approach to system analysis providing the system analyst with an appropriate abstraction level for each decision. Above, the lower abstraction level in terms of one individual DP was introduced. Next, the higher abstraction level is introduced involving the key concepts of strategic DPs and strategic supply systems. In particular, we are interested in DPs exposed to exogenous influence. Exogenous influence refers to external influence, of which the customer order is the most important. Internal influence, on the other hand, is dependent on other internal events and is thus endogenous.

3.1. Demand and supply categories

To discriminate between different DPs with the objective of identifying the vital few, it is necessary to identify the DPs exposed to exogenous influence. Dependent and independent demands (DD and ID) were described earlier and are classified in Table 1. To the authors' knowledge, there is no corresponding general conceptual framework for bottom-up approaches where the supply is the driver in the process. More specific approaches do, however, exist, such as the forward flow scheduling approach of process flow scheduling (Taylor & Bolander, 1994). Taking supply as the point of reference would result in employing a forward scheduling approach to generate supply for the next higher level in the BoM. Note that a common denotation of levels in a BoM is that

Table 1. Summary of demand and supply categories.

	Supply	Demand
Independent	Supply that must be estimated (purchase orders)	Demand that must be estimated (forecast and customer orders)
Dependent	Supply that can be calculated (forward scheduling)	Demand that can be calculated (backward scheduling)

the top level is 0 and for each subsequent lower level the level number increases. In analogy with the demand-based concept, this is referred to as dependent supply (DS). Using the concept of forward scheduling, the supply of material on level i in the BoM would be dependent on supply on the next lower level, level $i + 1$, during the scheduling phase. In a backward scheduling scenario, the demand for items on level i would be dependent on the demand for items on the next higher level, level $i - 1$. The key property of DD and DS is that they both can be calculated based on the available endogenous information of the logistics system, such as lead times and available inventory.

The description above shows the clear analogy between DS and DD. The next step is to identify independent supply (IS) as an analogy to ID. This analogy is, however, not as clear-cut, but we define supply from a source outside the logistics system as IS since it is outside the strict control of the logistics system. Control can cover a complete manufacturing network or just a single site. The important aspect is that within the controlled supply system there is no need for traditional purchasing orders; instead internal orders (sometimes referred to as distribution orders) can be used covering both the demand process from one DP and the supply process of another DP. In summary, four types of drivers can be identified in the material flows as in Table 1, i.e. the intersections between supply/demand and independent/dependent.

Based on these two types of independent drivers (IS and ID), two types of corresponding DPs can be identified; they are discussed in the following two sections.

3.2. Demand-based decoupling point

Delivery strategies such as make-to-stock (MTS) and make-to-order (MTO) play a key role in providing an appropriate balance between efficiency and responsiveness in meeting market demand (see, e.g. Chopra & Meindl, 2007). The baseline for designing the appropriate strategy in this perspective is the relation between the delivery lead time (D), requested by the market, and the total supply lead time (S) within which the system needs to produce the output. The ratio between these two lead times is critical; it was first identified by Shingo (1981) as the D:P relation. Mather (1984) changed the relation to the inverse for practical reasons and since then it is known as the P:D ratio or P:D relation. Shingo used P as he was referring to product lead time. In a similar vein, the more general supply lead time is used and thus referred to as S (see, e.g. Wikner, 2014). S more explicitly involves, for example, also purchasing and distribution lead times. If S is shorter than D, all supply activities can be performed while the customer is waiting. But when S is longer than D some activities must be performed in advance to create semi-processed parts to be used when the customer order arrives. The supply system is thus split into two parts corresponding to two subsystems. The subsystem supplying semi-processed parts is performed before the customer order arrives and is therefore forecast-driven. The second subsystem uses the semi-processed parts in stock to deliver on the customer order. This is sometimes referred to as customer-order-driven (see, e.g. Hoekstra & Romme, 1992) but also as demand-driven (see, e.g. Christopher, 1998). The semi-processed parts are kept in stock at a decoupling point that is known as the customer order decoupling point, CODP, (see, e.g. Giesberts & van der Tang, 1992) but also as order penetration point, OPP (Sharman, 1984) or simply as decoupling point (Hoekstra & Romme, 1992). The CODP has been defined in a number of different ways but, as summarized by Wikner (2014, p. 195), '[t]he customer order decoupling point (CODP) separates decisions about initiating flow based on speculation on future customer orders from commitment against actual customer orders'. Using these

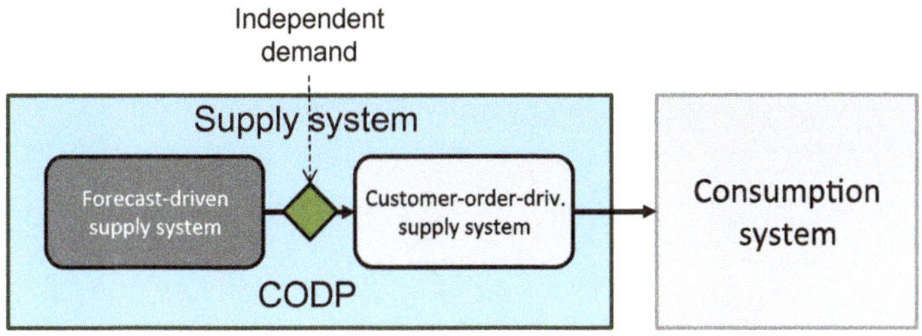

Figure 1. The CODP perspective of the supply system.

three building blocks (forecast-driven supply system, CODP, and customer-order-driven supply system), the supply system can be depicted as in Figure 1, where a consumption system has been introduced as an origin of demand and to represent the customers of the supply system. The diamond is used to represent the CODP in line with the guidelines suggested by Wikner (2014).

Considering a customer order arriving in the supply system of Figure 1, the customer must accept to wait during the time it takes to perform the customer-order-driven supply system (during the lead time D). This system lead time could involve activities outside manufacturing (see, e.g. Rudberg & Wikner, 2004). To replenish the CODP, some kind of forecast is necessary since customer orders are assumed to arrive in time to perform only the customer-order-driven supply system in Figure 1. The forecasting can be performed at the end-item level but if any parts are shared between different end items it is easier, due to aggregation effects reducing the coefficient of variation in demand, to forecast at the CODP on semi-processed parts. This corresponds to defining the CODP parts as subject to ID as shown in Figure 1. Activities downstream of the CODP may be performed under certainty of demand since a customer order is driving the activities. The parts at the CODP must, however, be produced to forecast since they are expected to be available at the moment the customer order arrives.

3.3. Supply-based decoupling point

The CODP emphasizes the relation between the delivery lead time D and the supply lead time S, and hence investigates the demand-related properties of S but does not provide any further information on what actually constitutes supply. Taking a supply perspective, an important concept is to what extent the complete supply system is integrated. In this case, the focus is on inventory and materials management and therefore also the level of integration is defined based on the level of control that can be exercised on the supply system. In an integrated supply system, the complete set of supply activities would be controlled as one entity. If, however, the supply system is not integrated, the supply system could be separated into two subsystems, here referred to as one controlled subsystem and one uncontrolled subsystem, see Figure 2. A hexagon is used to illustrate the purchase order decoupling point (PODP) in line with Wikner (2014). The uncontrolled subsystem is out of control reach for the controlled subsystem.

An example is a component supplier of a company that supplies parts per order and where there is no further integration. A purchase order is sent to replenish the inventory when required. In this case the supplier would be modelled as an uncontrolled system including not only the supplier but also all the supplier's suppliers, etc. Using this

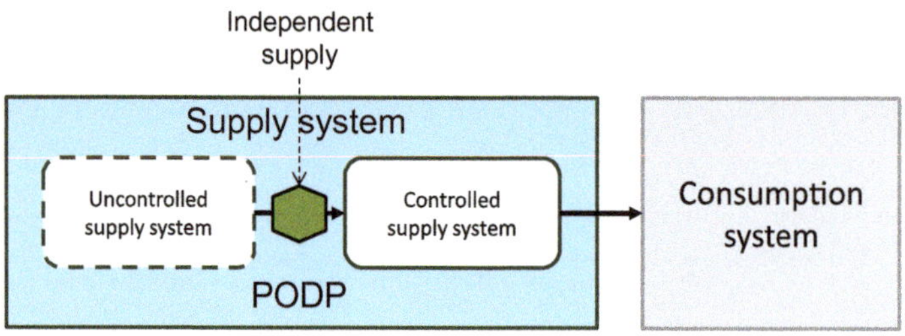

Figure 2. The PODP perspective of the supply system.

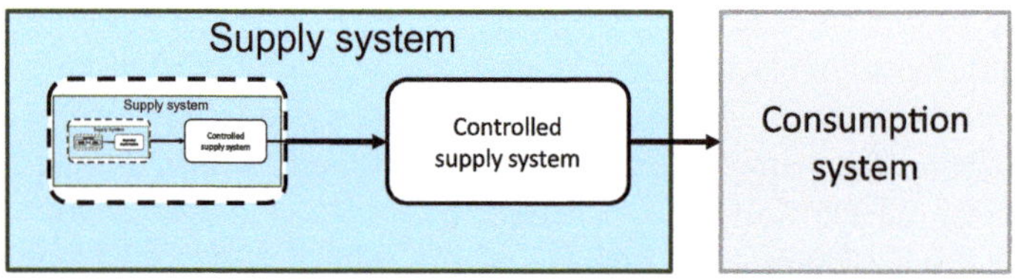

Figure 3. Recursive model of a supply system.

approach, a recursive systems definition is obtained where the controlled supply system and how it interacts with the consumption system is the focus. Remember that from the uncontrolled supply system perspective, the controlled supply system would represent the consumption system as depicted in Figure 3. This is because all upstream activities not covered by the controlled supply system are included in the uncontrolled supply system.

Within the controlled supply system, there is full transparency in terms of planning and execution which enables a cost-efficient allocation of decoupling points, i.e. buffers, from a system's perspective. The uncontrolled supply system does, however, not provide this level of transparency by definition. Since control cannot be exercised on this system by the controlled supply system, the only way to manage unexpected events occurring in this system is to introduce a buffer explicitly targeting these properties and decoupling the controlled system to protect it. As shown by Lee (2002), this relation must be carefully designed and alternative strategies for minimizing unexpected events should be considered, but if uncertainty prevails it must be handled with some type of buffer, otherwise customer service and/or cost efficiency will suffer. Since the relation with the uncontrolled supply system usually is through some kind of purchase order, we introduce the term PODP for this kind of buffer, as in Figure 2. Activities within the controlled supply system are usually coordinated by production orders and distribution orders but when it comes to coordinating with an external system, the relation is usually handled through a purchase order. This model is also related to the lead time based analysis performed by Wallin, Rungtusanatham, and Rabinovich (2006). They related the delivery lead time D to production time P in the spirit of Shingo (1981), but then made a distinction between the part of P covered by the supplier and the part covered by the focal company. Using this approach, different strategies for inventory management of purchased material were identified, but here the emphasis is on the complete supply system structure and not only purchased materials at the PODP. In Wallin et al.

(2006), the focal company is the unit of analysis. Here, however, the unit of analysis is a supply system, and to emphasize what specific supply system is considered it may be referred to as the focal supply system. In Figure 2 the focal supply system would correspond to the controlled supply system. A similar approach to dividing the supply system into subsystems based on exogenous influence has also been investigated by Wikner (2005), but then from the perspective of efficiency and responsiveness.

The PODP can hence be defined as follows: '[t]he PODP separates decisions about delimiting flow based on what is external to the logical entity from what is internal and hence controllable' (Wikner, 2014, p. 199). This indicates that the uncontrolled supply system provides input to the controlled supply system. This input contains more uncertainty compared to what is perceived from the controlled supply system and thus requires some kind of estimation (forecasting), which also is the key property of IS, see Table 1.

3.4. Modelling the logistics system based on delivery strategies

The fundamental building blocks of the logistics system have been defined above and are summarized in Figures 1 and 2. Basically, the properties of the supply systems highlighted are type of driver, related to CODP, and level of control, related to PODP. Each particular logistics system has its own unique characteristics based on its history, market conditions, available resources, etc. There are, however, some generic configurations that can provide a baseline for illustrating how the constructs introduced above can be employed to represent an enterprise from a logistics perspective. The two fundamental configurations used here are the completely customer-order-driven system (MTO) and the completely forecast-driven system (MTS). The third system (assemble-to-order, ATO) represents a compromise system with elements of both MTO and MTS (Wemmerlöv, 1984). The fourth system, finally, involves also the possibility of outsourcing not just forecast-driven activities to uncontrolled supply systems but also customer-order-driven activities; this is referred to as purchase-and-make-to-order (PMTO) (Hoekstra & Romme, 1992).

3.4.1. Make-to-stock (MTS)

When the MTS strategy is used, all activities are forecast-driven (with the possible exception of some picking and delivery activities). In this case, the CODP is positioned as far downstream in the material flow as possible. No activities are customer-order-driven and

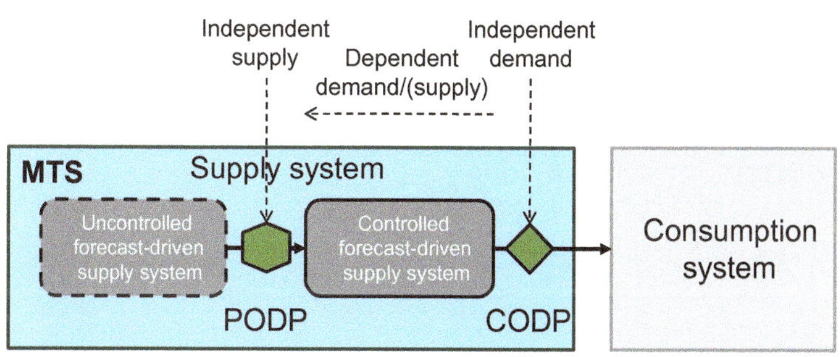

Figure 4. Modelling make-to-stock (MTS) supply systems.

usually the master schedule (see, e.g. MRPII) or pacemaker (see, e.g. rate-based master scheduling as in lean) is positioned at the end-item level, where ID is present. This allows for a materials planning approach based on DD (backward scheduling), which is the most common approach in this setting, for all items down to the PODP. The PODP hence represents the extension of the integrated planning and control approach as shown in Figure 4. In Figures 4–7, the dashed vertical arrows represent exogenous input to the supply system. The dashed horizontal lines represent endogenous propagation of demand and supply respectively, in the direction indicated by the arrow.

3.4.2. Make-to-order (MTO)

The second fundamental configuration corresponds to the MTO strategy, where all activities of the controlled system are customer-order-driven. When the customer order is received, the customer waits until the product is finished. The planning approach can be based on DD (as in MRP), where the end item is master scheduled and DD is calculated for all other items down to the CODP. A preferred scenario would, however, in some cases be to start with the CODP items availability and then to forward schedule all items (DS) up to the end item to provide, for example, information about delivery capability. This process would then preferably also involve finite loading considering any finite capacity to provide reliable delivery dates. Since the CODP here is positioned at the beginning of the controlled supply system, it coincides with the PODP as shown in Figure 5.

3.4.3. Assemble-to-order (ATO)

A preferred strategy in many cases is to establish a level in the BoM that allows for modularization. Different modules can then be used in a multitude of configurations enabling a mass customization approach. Modules are then produced to forecast using a standard MTS 'sub-strategy', whereas the final assembly is customer-order-driven and similar to the MTO approach. Setting the two models of MTS (Figure 4) and MTO (Figure 5) in tandem results in an ATO scenario as in Figure 6. The CODP is positioned between the forecast-driven and the customer-order-driven supply systems, and the PODP is also in this case positioned at the interface between the uncontrolled and the controlled supply systems.

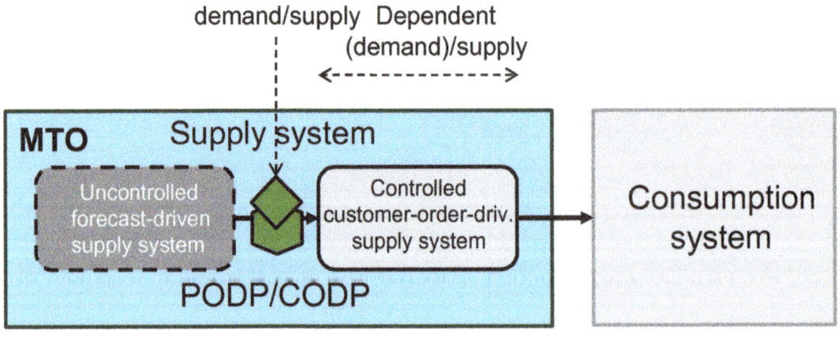

Figure 5. Modelling make-to-order (MTO) supply systems.

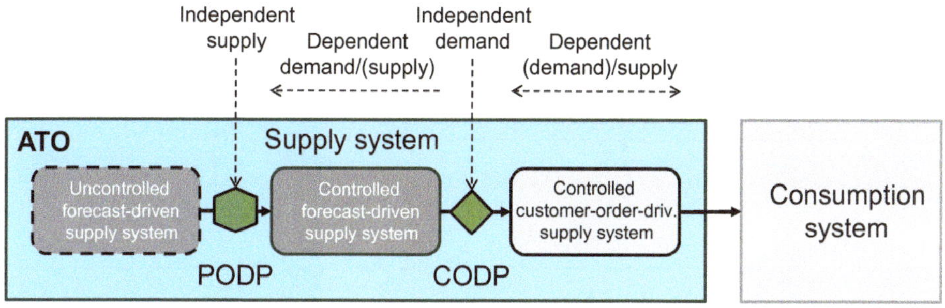

Figure 6. Modelling assemble-to-order (ATO) supply systems.

3.4.4. *Purchase-and-make-to-order (PMTO)*

If only the CODP is considered, the three strategies identified above, MTS, MTO, and ATO, are key scenarios. Each strategy consists of a unique combination of forecast-driven and customer-order-driven supply systems. Taking a broader perspective and also considering that the delivery lead time D may extend beyond the lead time of the controlled supply system provide leeway for a fourth scenario. This scenario involves an external supplier (here referred to as an uncontrolled supply system) that is customer-order-driven, as illustrated in Figure 7. There are now two customer-order-driven supply systems separated by the PODP, and the CODP is positioned upstream of these two supply systems. This is a possible, if not always desirable, strategy as some activities performed while the customer, i.e. the consumption system, is waiting are performed beyond the control of the supply system responsible for delivery to the customer. A possible approach in this case would be to seek to integrate, i.e. 'in-source', the uncontrolled customer-order-driven supply system and, if this is not possible, to establish a safety lead time at the PODP, unless all uncertainties between the CODP and the PODP can be eliminated.

3.4.5. *Summary of the four logistics system configurations*

To conclude the discussion on logistics systems configured based on the concepts of customer-order/forecast-driven and uncontrolled/controlled, the configurations are summarized in Table 2. The four concepts provide four possible configurations of supply systems as represented by the columns. In addition, it is also required, by definition, that

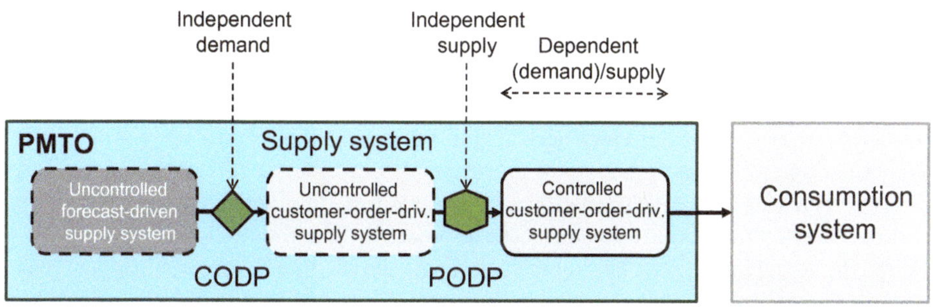

Figure 7. Modelling purchase-and-make-to-order (PMTO) supply systems.

Table 2. Summary of the four possible configurations of the logistics supply system.

	Uncontrolled forecast-driven	Controlled forecast-driven	Uncontrolled customer-order-driven	Controlled customer-order-driven
Config. MTS	X	X	–	–
Config. ATO	X	X	–	X
Config. MTO	X	–	–	X
Config. PMTO	X	–	X	X

certain precedence constraints are fulfilled. A forecast-driven supply system should not be positioned downstream of a customer-order-driven supply system. This is because the supply systems are positioned along a time axis and if it is possible to be customer-order-driven, this is assumed to always be the preferred option. Why forecast-driven if actual demand is available? The second precedence constraint is that uncontrolled supply systems are positioned upstream from controlled supply systems. One exception to this logic is when certain operations within a routing are outsourced. This precedence constraint is nevertheless assumed here.

To verify that there are no other fundamental configurations regarding these constraints, a simple analysis can be performed. To begin with, it is assumed that there is always the possibility of having uncontrolled forecast-driven activities which corresponds to purchasing materials to forecast. This means that the configurations of the leftmost column of Table 2 are valid (uncontrolled forecast-driven). Configurations MTS and ATO in Table 2 both involve a forecast-driven supply system that also is controlled. This means that there can be no uncontrolled customer-order-driven activities without invalidating any of the precedence rules. The only option available is thus to include or exclude controlled customer-order-driven activities, which results in MTS and ATO. If there are no controlled forecast-driven activities, as in configurations MTO and PMTO, that is only possible with two different configurations. Either all customer-order-driven activities are controlled (MTO) or parts of them are uncontrolled (PMTO). If all activities were uncontrolled, the investigated system would be empty and here considered as void.

These four configurations of logistics supply systems highlight the structural importance of the two types of decoupling points: CODP and PODP. Based on this insight, we return to the complete set of DPs to introduce a classification that takes this insight as a point of departure.

4. Process-based inventory classification using decoupling points

Inventory classification is usually based on some properties of the items in inventory such as value, weight, or perishability. In the case of classification based on, e.g. withdrawal frequency or supplier characteristics, more emphasis is placed on a process perspective but one of the most important properties of processes is still not considered. As processes represent all the value-adding and non-value-adding activities performed in an enterprise, a key factor is how and why these processes are initiated, i.e. what triggers a process. In the analysis of the logistics system above, two important triggers were identified. Basically, the triggers can be characterized as exogenous or endogenous. An exogenous trigger refers to external influence, of which the customer order is the most important, and an endogenous trigger refers to internal influence, corresponding to DD and DS. Finally, a process is also vulnerable when it crosses organizational boundaries, and the relation to external suppliers is of particular interest since it is a consequence of a make/buy decision and represents the reach of what is within process control of the business. In all, there are four process-based concepts that can be used in classifying the DPs and hence inventory. They are related to dependent or independent, which can be combined with a supply or demand perspective in four different ways. These four concepts (dependent, independent, demand, and supply) are combined below and then classified as strategic (related to the choice of delivery strategy and level of outsourcing) or tactical (related to other item-based DPs of importance when synchronizing the material flows).

4.1. Strategic process-based decoupling points

The customer order triggers the processes in the customer-order-driven supply system. From an inventory perspective, the DP of outstanding importance is the CODP which is subject to ID and crucial to the service level that can be achieved. The CODP can be seen as the incarnation of the delivery strategy since the position of the CODP is based on the delivery lead time. The positioning of the CODP has been identified as a key strategic decision, for example, in terms of the CODP itself (Olhager, 2003), as a consequence of balancing efficiency and responsiveness (Chopra & Meindl, 2007), as the point where lean meets agile (leagility) (Mason-Jones, Naylor, & Towill, 2000), or where the physical and innovative parts of the supply chain meet (Olhager, Selldin, & Wikner, 2006).

The positioning of the PODP is also strategic in the sense that it represents the selected level of vertical integration, i.e. the level of outsourcing. It is also crucial for buffering external disturbances and hence for providing the internal flows with the best possible conditions. From an inventory perspective, a second important property of the PODP is the amount of control that can be exercised on a process, and it is assumed that uncontrolled subsystems constitute a risk and may have a negative impact on the controlled supply system, which is downstream from the PODP.

The inventory at the CODP enables the forecast-driven supply system to focus on efficiency since the inventory at the CODP buffers against seasonal variations as well as short-term fluctuations related to demand. The dimensioning of the CODP inventory is usually quantity based, as it is the impact of the demand variations on the service level that is in focus. At the PODP, on the other hand, the situation is more concerned with the reliability of the supply lead times and hence one is more inclined to use safety lead time at this DP. In addition, the PODP also handles the lot-size inventory due to the delivery frequency from the external suppliers.

In summary, the strategic DPs are either driven by downstream exogenous factors (related to external customers through customer orders) or driven by upstream exogenous factors (related to external suppliers through purchase orders). The strategic DPs

Table 3. Strategic process-based decoupling points.

	No independent demand (NID)	Independent demand (ID)
No independent supply (NIS)	–	NIS–ID
Independent supply (IS)	IS–NID	IS–ID

are categorized in Table 3 based on how they are related to these two types of exogenous triggers. If the trigger is only related to ID with no independent supply (NIS), the decoupling point is of the type NIS–ID, better known as CODP (see Figure 1). The reverse, IS but no independent demand (NID), is indicated as IS–NID and defined above as PODP (see Figure 2). These two types of strategic DPs may be separate but still present in the same flow as shown in Figures 4, 6, and 7. However, in some particular cases, a strategic DP can be exposed to both independent supply and independent demand (IS–ID), which represents the case where a DP is both a CODP and a PODP (see Figure 5). The fourth case finally, with NIS and NID, is not included in terms of strategic DPs but will be further analysed next at the tactical level.

4.2. Tactical process-based decoupling points

Tactical decisions should follow as a consequence of the decided strategy. In the same way the positioning of DD and DS is a direct consequence of ID and/or IS. The strategic DPs create a structural framework based on the delivery strategy and the outsourcing strategy. The objective of the positioning of DD and DS DPs is to support this strategy. These DPs are endogenously triggered to support the ID of all related items using the bill-of-material/distribution as the road map. In all, three DP classes can be identified at the tactical level as shown in Table 4. DPs with no dependent supply (NDS) but with DD, i.e. type NDS–DD, are backward scheduled based on the demand downstream, and DPs with DS but no dependent demand (NDD), i.e. type DS–NDD, are forward scheduled based on the supply upstream. Since backward and forward scheduling in practice are difficult to combine at the same DP, the combination of dependent supply and dependent demand (DS–DD) is rarely established. An example could however be when one item is part of two different products one of which is made to customer order and the other is sold from finished goods, i.e. a hybrid of MTS (see Figure 4) and MTO (see Figure 5) for that particular item. The combination involving no dependent supply and no dependent demand (NDS–NDD) is not included at the tactical level since a tactical driver is not present. DPs classified as tactical are generally used during the planning phase to establish the appropriate level of inventory over time at these decoupling points.

4.3. Summary of process-based inventory classifications

Three types of DPs were identified above as strategic and related to ID (CODP-based) and IS (PODP-based). If NID and NIS were present (upper left cell of Table 3), the DPs would be classified as tactical and could still be differentiated as exposed to DD or DS as in Table 4. In total, these two tables highlight six different types of DPs. By combining these two tables, two of the strategic DPs, IS–NID and NIS–ID, may be further differentiated. Compared to Table 3, NID is differentiated into no demand (ND) and DD. If there is NID there could still be DD. NIS is differentiated into no supply

Table 4. Tactical process-based decoupling points.

	No dependent demand (NDD)	Dependent demand (DD)
No dependent supply (NDS)	–	NDS–DD
Dependent supply (DS)	DS–NDD	(DS–DD)

Table 5. Process-based inventory classifications.

	No demand (ND)	Dependent demand (DD) (backward scheduled)	Independent demand (ID) (CODP based)
No supply (NS)	–	NS–DD	NS–ID
Dependent supply (DS) (forward scheduled)	DS–ND	(DS–DD)	DS–ID
Independent supply (IS) (PODP based)	IS–ND	IS–DD	IS–ID

(NS) and DS in a corresponding way as in Table 5. This results in five different types of strategic DPs, as highlighted in Table 5. In combination with the three types of DPs in Table 4, there are in total eight types of DPs (inventory) that have been identified using the process-based approach suggested here, with a foundation in planning and control. Five of them are of a strategic character involving either IS or ID. Three classes of DPs are of a tactical character involving NIS and NID as summarized in Table 5. Table 5 is a combination of Table 3 (mainly represented by the white area) and Table 4 (represented by the shaded area). 'No Demand (ND)' in Table 5 represents that the use of inventory in the DP is not directly driven by demand. Correspondingly, 'No Supply (NS)' refers to the fact that the replenishment of the DP is not driven by supply pushing items into the DP. A DP with a combination of ND and NS would be completely disconnected from the system and is therefore not classified in Table 5.

In essence, Table 5 represents eight (nine) classes of decoupling points, i.e. inventory points. Hence, the table constitutes a classification of inventory based on the intersection between the demand perspective (the columns) and the supply perspective (the rows). Note that these DPs are tactical or strategic meaning that all DPs are related to items. From this perspective, Table 5 can be seen as providing eight different boxes where all items in an enterprise can be allocated to at least one box. In some cases, one individual item can be related to two boxes if it, for example, is used in two different end products where one is MTS and the other is MTO. It is important to note that Table 5 does not provide any information on if an item is forecast-driven or customer-order-driven, except for the CODP items which are forecast-driven since they are expected to be in stock when the customer order arrives. The table does not provide a set of possible systems but rather the building blocks that can be used for designing any type of system. The NS–ND category should be empty since all items are either driven by demand or supply.

5. Managerial implications

The eight different classes of DPs in Table 5 represent an abstraction in the sense that these are the DPs that should be handled at a strategic and at a tactical level. They also represent an aggregation in the sense that each cell in Table 5 represents a type, or group, of DPs that have certain prerequisites concerning, for example, planning and control. Using the terminology of the ABC classification (see, e.g. Vollmann et al., 2005), the process-based classification in Table 5 can be summarized as generating three classes of inventory (DPs) to focus on. Class A is considered as the most important class and is fundamental for the order fulfilment process. Class B is the second in rank and relates to exogenous uncertainty that is difficult to control, and Class C, finally, is of least importance since it is within the controlled system:

- Class A: CODP inventory (column 3 in Table 5), since it is key to the delivery strategy chosen and a foundation for the delivery reliability to the customer.
- Class B: PODP inventory (row 3 in Table 5), since it is key to the sourcing strategy chosen and based on the delivery reliability of the suppliers.
- Class C: Tactical DPs (grey cells in Table 5), as they represent the link between PODPs, CODPs, and the customers.

Interpreting the different types of DPs in terms of planning and control methods provides some further insights into the significance of the eight types of DPs. The details of this mapping are not the subject of this paper but some guidelines have already been outlined above and some further insights are summarized below.

The framework suggested above and summarized in Table 5 provides a baseline for materials management but does not in itself provide any methods for planning and control. It does, however, highlight some key properties of when a particular planning and control method is appropriate. A key aspect of the framework is process triggering; in materials management this aspect has received considerable interest and in general a process is assumed to be triggered in basically two ways (Hopp & Spearman, 2000, p. 340):

- A customer to the process demands that the process should be performed (frequently referred to as pull). The customer can either be an external customer or an internal customer represented by another process or a point of inventory.
- A schedule demands that the process should be performed (frequently referred to as push). In some instances it is assumed that the schedule in push is based on forecasted demand (see, e.g. Marchwinski, Shook, & Schroeder, 2008), but this is a confusing and misleading assumption in this context. How should, for example, customer-order-driven manufacturing using scheduling software be classified under this assumption? The key issue here is rather that the mode is proactive and the schedule can be based on a wide range of information, whereas the pull approach operates in a reactive mode.

The terminology used in these two cases is somewhat different but still from a functional view they are perfectly in line with the framework introduced here, to which process triggering is central. To provide further support for this conclusion, three examples are presented of what kind of scenarios three different approaches are designed to handle. First, the standard MRPII approach is discussed, followed by takt-based planning and control, and finally an example from the process industry is presented.

The DPs NS–DD and NS–ID correspond to Orlicky's (1975) framework for MRP, where NS–ID is referred to as ID and should be master scheduled, whereas NS–DD is referred to as DD and should be handled by the MRP logic. In case there is NDD (only single-level bill-of-materials), the NS–ID could be managed using traditional inventory control methods such as an order point system. Another example is the classical ATO scenario where the order promising process considers stock availability of modules (NS–ID) at the CODP and the order promising logic checks availability downstream using forward scheduling (DS–ND) to determine possible delivery dates.

In the lean environment, the interpretation of Table 5 requires a distinction between how the system is dimensioned during planning (takt time analysis, line balancing, estimation of number of kanbans, and size of FIFO buffers, etc. (see, e.g. Duggan, 2013) and how it actually works within those boundaries during execution. The pacemaker is frequently positioned at the CODP (see, e.g. Rother & Shook, 2003). Downstream from the pacemaker, the DPs would then be considered as DS–ND and upstream as NS–DD during the planning phase. During the execution phase, all DPs would act as NS–ID responding to pull signals from downstream processes considered basically as ID.

In some process industries, the incoming raw material is not controllable but rather a consequence of long-term contracts, such as in farming. The livestock may be supplied to the abattoir at a certain age and must be processed without delay and therefore represents IS–ND and requires a planning and control approach based on divergent product profiles. Referring to process flow scheduling (Taylor & Bolander, 1994), the DPs covered by reverse flow scheduling (referred to as backward scheduled here) would be NS–DD except for the final stage, which would be NS–ID. Forward flow scheduling (forward scheduled) would be DS–ND except for the first stage, which would be IS–ND or NS–ID depending on whether demand or supply drives the flow. Mixed flow scheduling finally would then be a combination with some central point in the flow being NS–ID and then upstream NS–DD and downstream DS–ND.

These examples highlight how inventory can be classified and related to different scheduling and control techniques. As can be seen from these examples, the approach for the operational DPs here follows from the approach selected for the corresponding item DP (which is not generally necessary if control can be differentiated on an operations level). This means that if the item DPs are of the type NS–DD, then also all the operations to produce that item are assumed to be DD (backward scheduled).

As shown above, the classification framework introduced can be used in a contextual way for a wide range of environments to classify different types of DPs. The approach is hence generic and not only applicable in certain environments. The approach rests on a set of fundamental system properties and can therefore be applied in most environments.

6. Conclusions and further research

Materials management is a comprehensive topic usually involving several thousands of items and a multitude of aspects to analyse such as risk, investment, durability, etc. Using abstraction and aggregation as outlined in this paper, the management of inventory can be more explicitly targeted at focusing on the inventory of strategic importance, i.e. the strategic DPs. Process-based classification enables evaluation of inventory management across different delivery strategies (different positions of the CODP), which are of particular interest to companies with hybrid strategies where, for example, both MTO and MTS products are present. From a customer service perspective the management of the CODP inventory is critical to all kinds of delivery strategies. In combination with evaluation of how the inventory at the PODP is managed, a comprehensive view of management of strategic inventory can be obtained. The CODP is also the key starting point for most planning and control methods for manufacturing and logistics. The framework does not in itself provide a new approach to materials management but it provides a structured way to better understand the role of different DPs, i.e. inventory points, and their role in materials management.

With this classification as a point of reference, there are many possible lines of research to pursue. One of the most important is probably that this classification may guide researchers on inventory and materials management in terms of which DPs to focus on. Also, in case of evaluation of manufacturing strategies across different delivery strategies, the strategic DPs may serve as a point of reference for delineation between lean and agility (see, e.g. Mason-Jones et al., 2000). Giesberts and van der Tang (1992) pointed out that in many cases companies have 'hybrid production situations' and they showed how information systems may take this into account. The impact of this classification scheme on hybrid strategies remains an area of interesting opportunities for further research that could also take the framework suggested by Wikner (2014) into consideration. In the same spirit, the practitioners can use this classification to focus their investment on inventory and materials management where it gives the best leverage. The classification may also work as a tool for categorizing different concepts such as planning and control methods to provide guidelines on how to implement these methods based on strategic as well as tactical decoupling points.

Acknowledgements

This research has been performed in collaboration with six companies in the projects KOPeration and KOPtimera.

Disclosure statement

No potential conflict of interest was reported by the authors.

Funding

The research has been funded by the Swedish Knowledge Foundation (KKS), Jönköping University and the participating companies.

References

Blackstone, J. H. (Ed.). (2008). *APICS dictionary.* Chicago, IL: APICS The Association for Operations Management.

Broeckelmann, R. G. (1999). *Inventory classification innovation: Paving the way for electronic commerce and vendor managed inventory.* Boca Raton, FL: St. Lucie Press.

Chopra, S., & Meindl, P. (2007). *Supply chain management: Strategy, planning, and operation.* Upper Saddle River, NJ: Pearson Prentice Hall.

Christopher, M. (1998). *Logistics and supply chain management: Strategies for reducing cost and improving service* (2nd ed.). London: Financial Times/Prentice Hall.

Duggan, K. J. (2013). *Creating mixed model value streams: Practical lean techniques for building to demand* (2nd ed.). Boca Raton, FL: CRC Press.

Giesberts, P. M. J., & van der Tang, L. (1992). Dynamics of the customer order decoupling point: Impact on information systems for production control. *Production Planning & Control, 3*, 300–313.

Goldratt, E. M., & Cox, J. (1984). *The goal: A process of ongoing improvement.* Croton-on-Hudson, NY: North River Press.

Hoekstra, S., & Romme, J. (1992). *Integrated logistics structures: Developing customer oriented goods flow*. London: McGraw-Hill.
Hopp, W. J., & Spearman, M. L. (2000). *Factory physics* (2nd ed.). New York, NY: McGraw-Hill.
Lee, H. L. (2002). Aligning supply chain strategies with product uncertainties. *California Management Review, 44*, 105–119.
Marchwinski, C., Shook, J., & Schroeder, A. (Eds.). (2008). *Lean lexicon – A graphical glossary for lean thinkers* (4th ed.). Cambridge, MA: Lean Enterprise Institute.
Mason-Jones, R., Naylor, B., & Towill, D. R. (2000). Engineering the leagile supply chain. *International Journal of Agile Management Systems, 2*, 54–61.
Mather, H. (1984). *Attack your P:D ratio*. APICS Conference Proceedings, Falls Church, VA, 84–88.
Ohno, T. (1988). *Toyota production system: Beyond large-scale production*. Cambridge, MA: Productivity Press.
Olhager, J. (2003). Strategic positioning of the order penetration point. *International Journal of Production Economics, 85*, 319–329.
Olhager, J., Selldin, E., & Wikner, J. (2006). Decoupling the value chain. *International Journal of Value Chain Management, 1*, 19–32.
Orlicky, J. (1975). *Material requirements planning: The new way of life in production and inventory management*. New York, NY: McGraw Hill.
Rother, M., & Shook, J. (2003). *Learning to see: Value stream mapping to create value and eliminate muda*. Cambridge, MA: Lean Enterprise Institute.
Rudberg, M., & Wikner, J. (2004). Mass customization in terms of the customer order decoupling point. *Production Planning & Control, 15*, 445–458.
Sharman, G. (1984). The rediscovery of logistics. *Harvard Business Review, 62*, 71–79.
Shingo, S. (1981). *Study of 'Toyota' production system from industrial engineering viewpoint*. Tokyo: Japan Management Association.
Taylor, S. G., & Bolander, S. F. (1994). *Process flow scheduling: A scheduling framework for flow manufacturing*. Falls Church, VA: APICS.
Vollmann, T. E., Berry, W. L., Whybark, D. C., & Jacobs, F. R. (2005). *Manufacturing planning and control systems for supply chain management*. New York, NY: McGraw-Hill.
Wallin, C., Rungtusanatham, M. J., & Rabinovich, E. (2006). What is the 'right' inventory management approach for a purchased item? *International Journal of Operations & Production Management, 26*, 50–68.
Wemmerlöv, U. (1984). Assemble-to-order manufacturing: Implications for materials management. *Journal of Operations Management, 4*, 347–368.
Wikner, J. (2005). *Hantering av osäkerhet i försörjningskedjor med utgångspunkt i frikopplingspunkter* [Management of uncertainty in supply chains from a decoupling point perspective]. Proceedings of Produktionslogistik 2005 – Plans forsknings-och tillämpningskonferens, Borås, 221–236.
Wikner, J. (2014). On decoupling points and decoupling zones. *Production & Manufacturing Research, 2*, 167–215.
Wikner, J., Johansson, E., & Persson, T. (2009). *Process based inventory classification*. The 21st Annual Nofoma Conference, Jönköping Sweden.

6

An extension for dynamic lot-sizing heuristics

Fabian G. Beck*, Eric H. Grosse and Ruben Teßmann

Carlo and Karin Giersch Endowed Chair "Business Management: Industrial Management", Technische Universität Darmstadt, Hochschulstr. 1, 64289 Darmstadt, Germany

This paper presents an efficient procedure to extend dynamic lot-sizing heuristics that has been overlooked by inventory management literature and practice. Its intention is to show that the extension improves the results of basic heuristics significantly. We first present a comprehensive description of the extension procedure and then test its performance in an extensive numerical study. Our analysis shows that the extension is an efficient tool to improve basic dynamic lot-sizing heuristics. The results of the paper may be used in inventory management to assist researchers in selecting dynamic lot-sizing heuristics and may be of help for practitioners as decision support.

Keywords: EOQ; dynamic lot-sizing; heuristics

1. Introduction

Changes in the competitive environment, such as high time pressure and variations in the demand pattern of the customers, induce companies to simultaneously reduce the costs and increase the service level and quality of production and manufacturing processes. For this reason, the management of inventories is essential as inventories directly influence product and raw material availability as well as production lead time. It has a high cost impact and is therefore one of the main operational activities in the manufacturing industry. The management of inventories requires determining economic batch sizes by balancing inventory holding and setup/order costs with the objective of providing a high service level at minimal total costs. For its relevance in industry, it is not surprising that the lot-sizing problem has lost none of its attention in research and practice since the publishing of the first decision model to determine the economic order quantity (EOQ) (Harris, 1913).

The EOQ model is a simple, robust and efficient tool for companies to resolve the conflict between ordering and inventory costs (Dobson, 1988). While a plethora of extensions of the EOQ model exists and continues to be developed with a wide pallet of theoretical orientations from multi-stage models to incentives and productivity issues (see, for reviews, Brahimi, Dauzere-Prese, Najid, & Nordli, 2006; Glock, 2012; Glock, Grosse, & Ries, 2014; Jans & Degraeve, 2008), the basic case with deterministic time-varying demand, also known as dynamic lot-sizing problem, was solved more than 50 years ago (Wagner & Whitin, 1958). A typical dynamic lot-sizing decision is to determine the amount and timing of replenishment of items, with known but varying

*Corresponding author. Email: beck@bwl.tu-darmstadt.de

demand and fixed ordering and linear holding costs. Although the Wagner–Whitin (WW) algorithm derives optimal solutions for this problem, it is only very infrequently used in practice where heuristics are applied more commonly. One may ask why anyone would use a heuristic when a practical optimal algorithm is available. This is partly due to the fact that the algorithm is not commonly known in practice and that heuristics are simpler and faster to compute (Boe & Yilmaz, 1983). However, proper computer implementation, meanwhile, makes it possible that the WW algorithm runs in linear time (Karimi, Fatemi Ghomi, & Wilson, 2003; Saydam & Evans, 1990). Nevertheless, dynamic lot-sizing heuristics still provide value for practical applications, research and educational exercises (Simpson, 2001). For example, material requirements planning modules in enterprise resource planning (ERP) software, to the best of the authors' knowledge, do not provide the WW method (Bahl & Bahl, 2009). Instead, some of the heuristic methods that may lead to poor results are covered. This is consistent with the findings that poorer rules are well known in educational texts, while other more efficient rules have not received attention in research and practice (Simpson, 2001).

The paper at hand builds on this line of thought and presents an efficient extension algorithm for dynamic lot-sizing heuristics published in 1997 but inexplicably overlooked by inventory management literature. We test the extension for two basic and popular heuristics and illustrate its performance and robustness in an extensive numerical study. Thus, the intention of this paper is to introduce the algorithm and to show that it improves existing heuristics significantly. In addition, we present a practical approach as we implement the simulation study using the environment of MS Excel. This paper may support researchers in selecting heuristics for dynamic lot-sizing problems, and may be of help for practitioners to support inventory management decisions.

The remainder of the paper is structured as follows: The next section reviews related literature. Subsequently, formulations for the dynamic lot-sizing models and algorithms under study are presented in Section 3. This is followed by an extensive numerical analysis in Section 4 to evaluate the performance of the heuristics, and managerial implications are deduced from its results. Section 5 concludes the paper and provides suggestions for future research.

2. Literature review

Reviewing the overall literature on dynamic lot sizing is not within the scope of this paper due to the sheer number of works available. The reader may refer to Robinson, Narayanan, and Sahin (2009) for a review and Andriolo, Battini, Grubbström, Persona, and Sgarbossa (2014) as well as Glock et al. (2014) for recent overviews and surveys on dynamic lot sizing. In this paper, we present an overview of works that concentrated on the comparison and evaluation of dynamic lot-sizing heuristics and algorithms.

To find an optimal solution in the dynamic lot-sizing problem, dynamic programing was used (Wagner & Whitin, 1958). However, several authors developed heuristic solution procedures to solve the dynamic lot-sizing problem. A description and classification scheme of popular heuristics, such as least unit cost (LUC) (Gorham, 1968), Silver–Meal (SM) (Silver & Meal, 1973) and part-period balancing (PP) (DeMatteis, 1968), was given in the review paper of De Bodt, Gelders, and Van Wassenhove (1984). The authors noted that choosing a suitable heuristic for a specific application is not straightforward as cost structure and variability of demand determine how heuristics perform. Bitran, Magnanti, and Yanasse (1984) analytically derived worst case error bounds for LUC and PP. Early comprehensive overviews and performance evaluations of dynamic lot-sizing procedures, including PP, SM, and WW, were developed by Blackburn and Millen (1980), Axsäter (1982, 1985), Baker (1989) and Ritchie and Tsado (1986). Blackburn and Millen (1985) added other algorithms, such as Groff's rule (GR) (Groff, 1979), to previous performance evaluations. Another early classification scheme of basic heuristics was developed by Maes and Van Wassenhove (1988). The authors confirmed that the selection of heuristics depends on environmental characteristics and that universal conclusions about the performance of an algorithm are difficult to draw. Another comparison of dynamic lot-sizing techniques (including LUC, PP, SM and GR) with extensive numerical evaluation can be found in Zoller and Robrade (1988). A worst case and performance analysis of dynamic lot-sizing heuristics, including LUC, PP, and SM, was studied in Vachani (1992). Gupta, Keung, and Gupta (1992) compared the performance of dynamic lot-sizing heuristics, such as LUC and GR, in a multi-stage system. A comprehensive sensitivity analysis of popular dynamic lot-sizing heuristics, among them LUC and SM, was conducted by Pan (1994). Ganas and Papachristos (1997) presented an analytical evaluation and comparison of SM and PP. Kazan, Nagi, and Rump (2000) extended and compared the performance of WW and SM to account for unexpected changes in previously set production and setup schedules. Jeunet and Jonard (2000) evaluated the robustness of lot-sizing techniques, including WW, PP and LUC, taking into account uncertain environments. Simpson (2001) extensively evaluated WW, GR, SM and LUC, among others, in a computational study and concluded that there exist some inconsistencies in dynamic lot-sizing heuristics and that these algorithms are equivalent in performance. The author noted that, curiously enough, some algorithms that are known to generate poorer results are more popular in research and practice than other algorithms that outperform the well-known heuristics. A comparison of dynamic lot-sizing techniques (among them WW and SM), taking into account the so-called end-effect, was presented by Van Den Heuvel and Wagelmans (2005). Gafaar (2006) used SM as benchmark to test the applicability of a genetic algorithm to the dynamic lot-sizing problem with the special case of batch ordering. More recently, Jans and Degraeve (2007) presented a summary and comparison of meta-heuristic solution algorithms for the dynamic lot-sizing problem. Ho, Solis, and Chang (2007) compared the performance of SM, PP, and least total cost (LTC) for the special case of deteriorating inventory. Teunter, Bayindir, and Van Den Heuvel (2006) and Schulz (2011) extended some heuristics, such as SM and LUC, to take into account product returns and remanufacturing considerations. Another modification of dynamic lot-sizing heuristics, among them SM, LUC, and PP, was presented by Toy and Berk (2013), who studied special 'warm/cold' production processes. In a current study, Baciarello, D'Avino, Onori, and Schiraldi (2013) presented a comparison of basic lot-sizing heuristics (among them PP, SM, and GR) in a computational study using the results of WW as benchmark. A recent comparison of dynamic lot-sizing algorithms, including SM and WW, for the special case of convex production costs and setups, can be found in Kian, Gürler, and Berk (2014).

From the literature review, we can deduce that most studies compared and tested the same popular heuristics, such as LUC, SM, or GR, while other heuristics and possible extensions have mainly been overlooked (Simpson, 2001). One unrecognized algorithm that extends basic dynamic lot-sizing heuristics is addressed in this paper. Leinz, Bossert, and Habenicht (1997) introduced a procedure which improves the results of stop rules using a combination of stop rules and cost comparisons. Interestingly enough, this extension algorithm has not received, to the best of the authors' knowledge, any attention in research and practice since the publishing of the working paper (in German) in 1997. We present a formal description of the algorithm, in the following denoted as LBH and test its performance in an extensive numerical study.

3. Dynamic lot-sizing algorithms under study

In this chapter, we give a short description of the dynamic lot-sizing algorithms under study. For our comparative analysis, we choose WW, LUC and GR as the literature review showed that these algorithms are popular in dynamic lot sizing. Special attention is paid to the LBH extension algorithm and how it can improve basic heuristics (here: LUC and GR).

3.1. Definitions

The following definitions are made for the single-level dynamic lot-sizing problem:

- discrete time periods of equal duration over a finite planning horizon;
- demand is deterministic with time-varying feature;
- full demand occurs at the beginning of each period;
- all period demands and costs are non-negative;
- lead time is constant (fixed and known);
- order costs are fixed for every order;
- replenishment rate is infinite;
- stock is empty at the beginning and the end of the planning horizon;
- lot-splitting is not allowed;
- no shortages are allowed;
- all products are considered separately;
- holding costs only have to be determined for a stock which is carried over from one period to another;
- holding costs for carrying inventory during the consumption period are not considered.

The following terminology will be used throughout the paper:

l	last period of a preliminarily considered order cycle
l^*	final resulting last period which is calculated by a heuristic; the related demand d_{l^*} in period l^* is added to the order in period b
b	first period of an order cycle
$\tau_{b,l}$	order cycle range including period b to l
$\Delta\tau$	difference in the average order cycle range between a heuristic and WW
y_t	binary tag in period t
$c_{b,l}$	total cost per unit
c_O	fixed cost per order
c_H	holding cost per unit and period
C_O	total ordering costs
ΔC_0	approximated marginal savings in ordering costs
C_H	total holding costs
ΔC_H	approximated marginal-added holding costs
d_t	demand in period t

t	index for time periods $t = 1, \ldots, T$
x_t	amount ordered in period t
n_i	order frequency for algorithm i until the planning horizon T is reached
d	demand rate
q	order quantity in the EOQ-model
I_t	inventory at the end of period t
C	total costs
C_t	total costs in period t
C^P	preliminary total costs for the first order
C_a	total costs of two aggregated orders, $a \in \{\psi, \phi, \theta\}$
pp	part-period quotient
κ	relative difference in the cost variance between a heuristic and the WW algorithm
$\bar{\kappa}_{\text{erratic}}$	arithmetical mean of κ that is required for the calculations with erratic demand structure
T	end of planning horizon
X, Y	random variables
μ	expected value for uniform and normal distribution
σ^2	variance for uniform and normal distribution

3.2. *Wagner–Whitin (WW)*

The WW algorithm (Wagner & Whitin, 1958) leads to optimal solutions for the problem under study and is used in this paper to calculate reference values to evaluate the heuristic algorithms. It can be denoted as follows:

$$C = \sum_{t=1}^{T} C_t(x_t, I_t) = \sum_{t=1}^{T} (y_t(x_t) \cdot c_O + I_t \cdot c_H) \rightarrow \text{Min!}$$

subject to

$$I_t = I_{t-1} + x_t - d_t$$

$$I_t, x_t \geq 0, \quad 1 \leq t \leq T,$$

where

$$y_t(x_t) = \begin{cases} 0, & \text{if} \quad x_t = 0 \\ 1, & \text{if} \quad x_t > 0. \end{cases}$$

Without loss of generality, we take $I_0 = I_T = 0$ (Baker, 1989).

Because the WW algorithm is established in the literature, the reader may refer to Wagner and Whitin (1958), De Bodt et al. (1984) and Gupta and Keung (1990), among others, for a detailed description of the algorithm.

3.3. *Least unit cost (LUC)*

LUC (Gorham, 1968) is a heuristic algorithm that is based on one of the characteristics of the EOQ model, which says that the order quantity is equal, regardless of whether the unit costs or the total costs are minimized. A proof is given in Appendix 1. Because of that characteristic, the LUC heuristic calculates the unit costs $c_{b,l}$ from the beginning period $b = 1$ and adds order quantities as long as the unit costs are decreasing. When the costs are increasing, for the first time, the cumulated order size is ordered and the calculation of the unit costs starts again in the period $b = l^* + 1$ which is the first one not considered in the order before. The heuristic ends when the planning horizon is reached. The algorithm is expressed as follows:

$$c_{b,l} = \frac{c_O + c_H \cdot \sum_{t=b}^{l} (t-b) \cdot d_t}{\sum_{t=b}^{l} d_t}$$

The stop rule $c_{b,l^*-1} \geq c_{b,l^*} < c_{b,l^*+1}$ selects the final resulting last period l^* which leads to the following order quantity in period b:

$$x_b = \sum_{t=b}^{l^*} d_t$$

3.4. *Groff's rule (GR)*

One of the heuristic algorithms that has proven to lead to good results for the dynamic lot-sizing problem is GR (Groff, 1979; Simpson, 2001). It uses another feature of the optimal order quantity in the EOQ-model. For the optimal order quantity, the marginal holding costs and the marginal ordering costs are equal according to amount. A proof is given in Appendix 2. The heuristic starts in period $b = 1$ and ends when the planning horizon is reached. The order quantity is increased as long as the approximated marginal savings in ordering costs ΔC_O from adding the l-th period's demand exceed the approximated marginal holding costs ΔC_H. These two marginal costs are calculated as follows:

$$\Delta C_O = \frac{c_O}{(l-b-1)} - \frac{c_O}{(l-b)} = \frac{c_O}{(l-b-1) \cdot (l-b)}$$

$$\Delta C_H = \frac{d_l \cdot c_H}{2}$$

As soon as the stop rule $\frac{c_O}{(l^*-b)\cdot(l^*-b+1)} - \frac{d_{l^*+1} \cdot c_H}{2} \leq 0$ is reached, the final resulting last period l^* is selected and the order quantity for period b can be computed. The related order quantity is $x_b = \sum_{t=b}^{l^*} d_t$. The calculations are continued in period $b = l^* + 1$ until $l = T$ and the stop rule is not reached.

3.5. *Leinz–Bossert–Habenicht (LBH)*

The LBH extension algorithm can be described in five steps (Leinz et al., 1997).

Let $b = 1$ and $l = 1$:

- Step 1: Determine the first two order cycle ranges, τ_{b,l^*} and $\tau_{l^*+1,l^{**}}$ by starting in period b and using a stop rule such as the ones of LUC or GR. These results are cumulated to a preliminary order cycle range $\tau_{b,l^{**}} = \tau_{b,l^*} + \tau_{l^*+1,l^{**}} = l^{**} - b + 1$.
- Step 2: Set $\psi = \tau_{b,l^{**}}$. The associated costs can be calculated using
  $C_\psi = c_O + c_H \cdot \sum_{t=b}^{b+\psi-1} [(t-b) \cdot d_t]$.

  An attempt is made to decrease these costs by dividing the preliminary order cycle range into two parts which leads to $\psi - 1$ possible combinations of two cycles. The resulting costs can be determined as follows:
  $C_\phi = c_O + c_H \cdot \sum_{t=b}^{b+\phi-1} [(t-b) \cdot d_t] + c_O + c_H \cdot \sum_{t=b+\phi}^{b+\psi-1} [(t-(b+\phi)) \cdot d_t]$
  with $\phi = 1, 2, \ldots, \psi - 1$.

  After these calculations, choose $C^* = \min_\phi \{C_\phi\}$ with $\phi = 1, 2, \ldots, \psi$ so that $\tau^*_{b,b+\phi^*-1} = \phi^*$. The associated preliminary total costs for the first order can then be calculated with
  $C^P = c_O + c_H \cdot \sum_{t=b}^{b+\phi^*-1} [(t-b) \cdot d_t]$.

  If $\phi^* \neq \psi$ respectively $\phi^* > 1$ go to step 3 otherwise go to step 4.

- Step 3: Another attempt is made to decrease the total costs by dividing the first calculated order cycle range $\tau^*_{b,b+\phi^*-1}$ into two orders. To this end, the following costs have to be calculated:
  $C_\theta = c_O + c_H \cdot \sum_{t=b}^{b+\theta-1}[(t-b)\cdot d_t] + c_O + c_H \cdot \sum_{t=b+\theta}^{b+\phi^*-1}[(t-(b+\theta))\cdot d_t]$
  with $\theta = 1, 2, \ldots, \phi^* - 1$.
  Determine $C^{**} = \min_\theta\{C_\theta\}$ with $\theta = 1, 2, \ldots, \phi^* - 1$.
- Step 4: Another case differentiation has to be made in this step.
  If $C^{**} \geq C^P$ or $\phi^* = \psi$ or $\phi^* = 1$, one gets the following order quantity $x_b = \sum_{t=b}^{b+\phi^*-1} d_t$.
  If $C^{**} < C^P$, one gets two different order quantities $x_b = \sum_{t=b}^{b+\theta-1} d_t$ and $x_{b+\theta} = \sum_{t=b+\theta}^{b+\phi^*-1} d_t$.
- Step 5: Set $b = b + \phi^*$ and $l = b + \phi^*$. Go back to step 1 and continue in this manner until the planning horizon is reached.

In the next section, an extensive simulation study is conducted to evaluate the performance of the algorithms.

4. Simulation study

4.1. Description

In the numerical simulation study, LUC, GR and the LBH extension based on LUC (LBH-LUC) as well as on GR (LBH-GR) are examined assuming different demand structures, i.e. constant, systematic and erratic demand (Leinz et al., 1997; Zoller & Robrade, 1988). The algorithms were programed using MS Excel and run on a laptop with Intel core i5 (2G) processer. The following assumptions for all tests are made:

- average demand is around 1000 [units] except all demands with trend structures (see I–VI below);
- *pp*-quotient $\left(pp = \frac{c_O}{c_H}\right)$ can attain the following values:
  $pp \in \{1000; 2500; 5000; 7500; 10,000; 15,000\}$;
- 100 periods of demand are considered ($T = 100$).

Using the assumptions above, six data-sets are created for the case of constant demand with $d_t = 1000$.

For the systematic demand, the trend and season are analyzed for eight systematic demand structures. The demand for each period will be calculated as follows:

(I)	linear increasing trend	$d_t = 1000 + 7.5 \cdot (t-1)$
(II)	progressive trend	$d_t = d_{t-1} \cdot \left[1 + \frac{5+5\cdot\left(\frac{t+99}{100}\right)}{1500}\right]$ with $d_1 = 1000$
(III)	saturation	$d_t = 2000 - \left[1000 \cdot e^{-\frac{(t-1)}{10}}\right]$
(IV)	linear decreasing trend	$d_t = 2000 + 7.5 \cdot (t-1)$
(V)	degressive increasing trend	$d_t = 1000 + \sqrt{5 \cdot d_{t-1} \cdot (t-1)}$ with $d_1 = 1000$
(VI)	degressive decreasing trend	$d_t = 2000 - \sqrt{5 \cdot d_{t-1} \cdot (t-1)}$ with $d_1 = 2000$
(VII)	additive constant season	$d_t = 1000 + s_t$
(VIII)	additive dynamic season	$d_t = 1000 + \frac{t}{25} \cdot s_t$

with $s_t = s_{t-5}; s_1 = -100; s_2 = 50; s_3 = 210; s_4 = 0; s_5 = -120$ for VII and VIII. In combination with the six different *pp*-values, there are 48 data sets to examine. The erratic demand is simulated by a uniform and a modified normal distribution (MND). The uniformly distributed random variables have an expected value of $\mu = 1000$ and fluctuation ranges of $\{\pm 20\%; \pm 40\%; \pm 60\%; \pm 80\%; \pm 100\%\}$, which lead to the variances $\sigma^2 \in \{13,333;\ 53,333;\ 120,000;\ 213,333;\ 333,333\}$. The modified normal distributed random variables X are generated out of normal distributed random variables Y with an expected value $\mu = 1000$ and a variance $\sigma^2 \in \{10{,}000;\ 100{,}000;\ 500{,}000\}$. The latter assumptions make negative demand values in Y possible, so in case a negative demand occurs, this value is set to zero in X. By this alteration, the initially normally distributed values are actually no longer normally distributed, which also means, that the distribution of X is no longer symmetrical. In fact, there is the following change to the distribution $P_{\text{MND}}(X = 0) = F_{X_{\text{ND}}}(0)$, where $F_{X_{\text{ND}}}(x)$ is the cumulative distribution function of the normal distribution (ND). The probability density function then shows a small rise at $x = 0$ and is no longer defined from $-\infty$ to $+\infty$ but is now defined from 0 to $+\infty$. With respect to the law of large numbers, we calculate 100 data sets for each erratic demand structure.

4.2. Results

The costs of the results obtained with a heuristic are compared to the results of the reference procedure WW (Leinz et al., 1997). The total cost variance is measured by the relative difference: $\kappa = \frac{C_{\text{Heuristic}} - C_{\text{WW}}}{C_{\text{WW}}} \times 100[\%]$. Because of the law of large numbers, 100 relative differences κ_i with $i = 1, \ldots, 100$ for each erratic demand structure are created and the arithmetical mean is needed for further calculations. The arithmetical mean can be computed as follows: $\bar{\kappa}_{\text{erratic}} = \frac{1}{100} \cdot \sum_{i=1}^{100} \kappa_i$.

In addition, the average order cycle range difference compared to WW can be calculated as follows: $\Delta\tau = \frac{T}{n_{\text{Heuristic}}} - \frac{T}{n_{\text{WW}}}$ [periods]. Note that results shown in Figures 1–8 are calculated using the arithmetical mean for all test sets with the specific indicated demand structures (in figure captions) and *pp*-values, i.e. $\bar{\kappa}$ and $\overline{\Delta\tau}$, respectively.

The results for the case of constant demand are summarized in Table 1. In this case, other values than $pp = 5000$ lead to no variances compared to WW and are thus not illustrated. However, our results show that *pp*-values of 5000 in combination with a fixed planning horizon lead to variance in total costs and order cycle range compared to WW. A rolling planning horizon, as common in practice, may be suitable to prevent this variance. As can be seen, LBH eliminates this variance as well.

Regarding systematic demand with trend (cf. Section 4.1, No. I to VI), our results show that LBH leads to substantial improvements in the basic heuristics as illustrated in Figure 1. As can be seen, GR generates the worst results in this case. However, even employing large *pp*-values, the cost variance is only slightly above 1%.

As for systematic seasonal demand (cf. Section 4.1, No. VII and VIII), we also observe that LBH performs well and, in this case, GR outperforms LUC. This relative advantage remains applying LBH as LBH-GR leads to better results than LBH-LUC.

Table 1. Results for constant demand.

Heuristic		GR	LUC	LBH-GR	LBH-LUC
$pp = 5000$	κ	.749	.749	0	0
	$\Delta\tau$	−.089	−.089	0	0

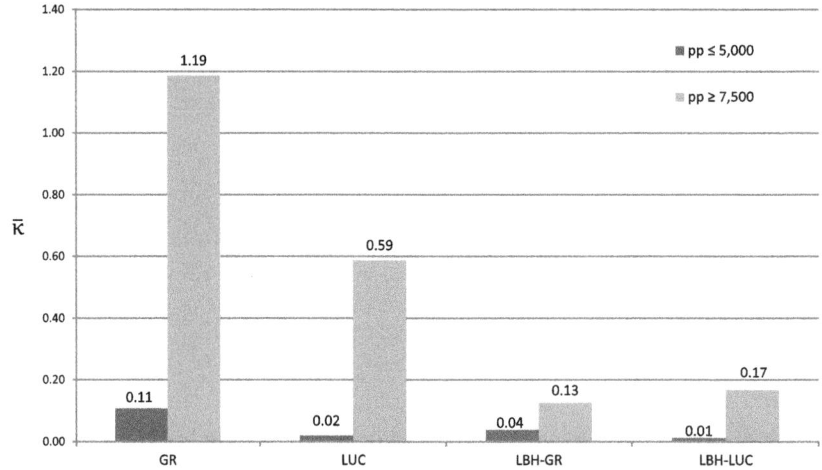

Figure 1. Results for systematic demand with trend structure (demand structures I–VI).

As can be seen, higher cost variances occur compared to the results of systematic demand with trend. This implies that inventory management for seasonal demand is a more challenging task in practice than for systematic demand with trend. However, as our results show, this situation could be enhanced significantly applying the LBH extension algorithm, in particular, LBH-GR as illustrated in Figure 2.

In the case of erratic demand, the results of our study show that higher *pp*-values lead to higher cost variances, except employing LUC, compare therefore Figure 3. Regarding LBH, the results show that average *pp*-values lead to the highest cost variances. As LUC calculates poor results with cost variance more than 6%, LBH-GR outmatches LBH-LUC accordingly.

All tested algorithms generate shorter average order cycle ranges compared to WW, as can be seen in Figure 4. One can observe a tendency that high variance in order

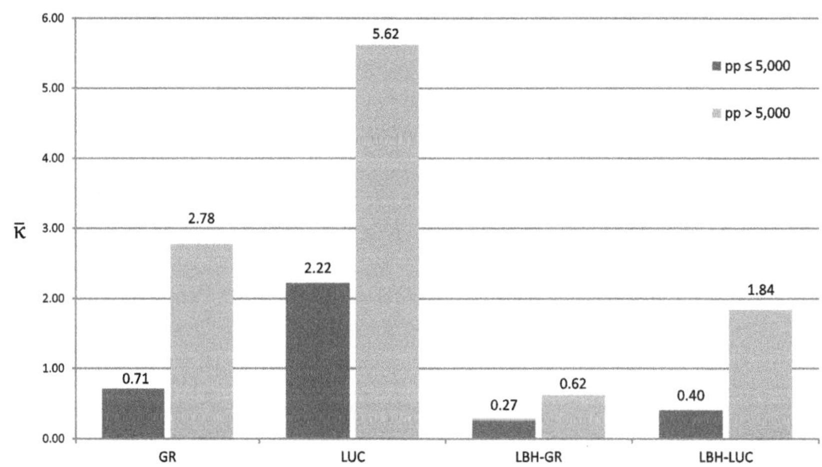

Figure 2. Results for systematic seasonal demand (demand structures VII–VIII).

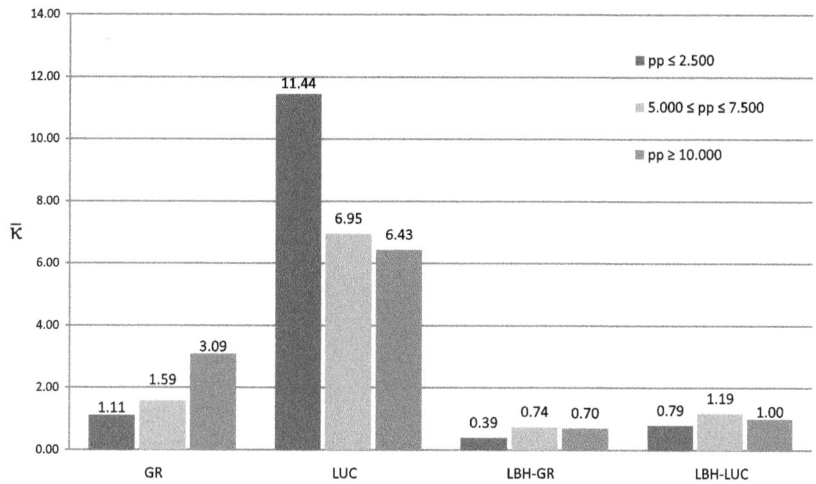

Figure 3. Results for erratic demand and varying *pp*-values.

cycle ranges may lead to high total costs variance. However, as can be seen for LUC ($pp \leq 2500$) in Figures 3 and 4, for example, a low order cycle range variance may result in high total cost variance. The results in Figure 4 also show that the LBH extension algorithm generates longer order cycle ranges that are then closer to the results of WW compared to the results of the basic heuristics.

In case of $\sigma^2 < 60{,}000$, all heuristics lead to small cost differences apart from LUC, which generates high cost variances. As can be seen in Figure 5, a significant improvement of LUC is possible applying LBH with $pp \leq 5000$. However, as the basic heuristic calculates poor results, the results of LBH-LUC are still quite far off the ones of WW. It is also notable that LBH-LUC leads to the same cost variances with different *pp*-values. Moreover, we can observe here, that LBH-GR generates the lowest cost

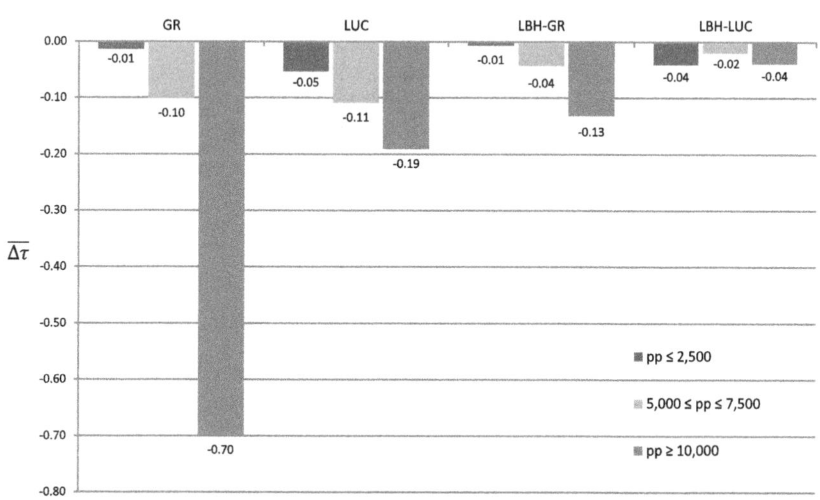

Figure 4. Order cycle range differences for erratic demand and varying *pp*-values.

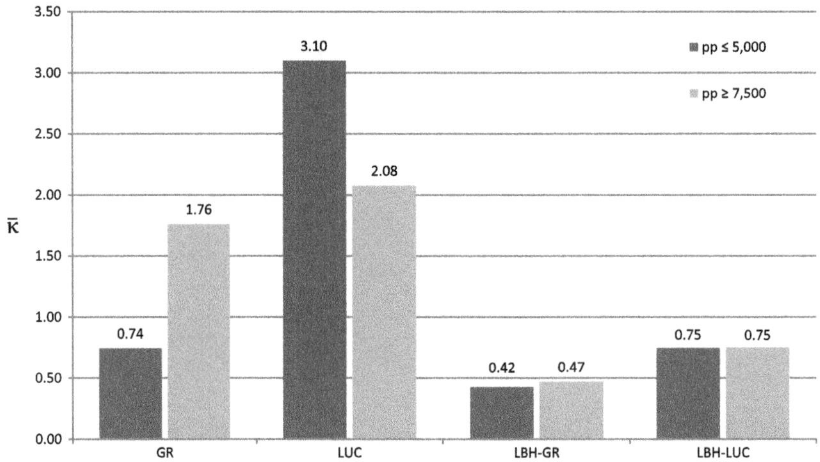

Figure 5. Cost variances for $\sigma^2 < 60{,}000$.

variances compared to WW with all *pp*-values. The additional costs of GR can be considerably reduced applying LBH even with higher *pp*-values.

Most procedures lead to shorter order cycle ranges compared to WW with low *pp*-values and $\sigma^2 < 60{,}000$. In turn, as illustrated in Figure 6, LBH-LUC generates longer order cycle ranges with high *pp*-values. We also observe the tendency that the closer the order cycle range generated by a basic heuristic and a heuristic extended with LBH gets to the order cycle range of WW, the better cost results are obtained (compare Figures 5 and 6). However, there are exceptions, such as LBH-GR with high *pp*-values. An explanation is probably that equal average order cycle ranges to WW are derived, but the order timing is different leading to cost variance of about .5%.

For $\sigma^2 > 200{,}000$, comparable results are obtained to the case of $\sigma^2 < 60{,}000$. However, cost variance is noticeably higher for $\sigma^2 > 200{,}000$, see therefore Figure 7. The only

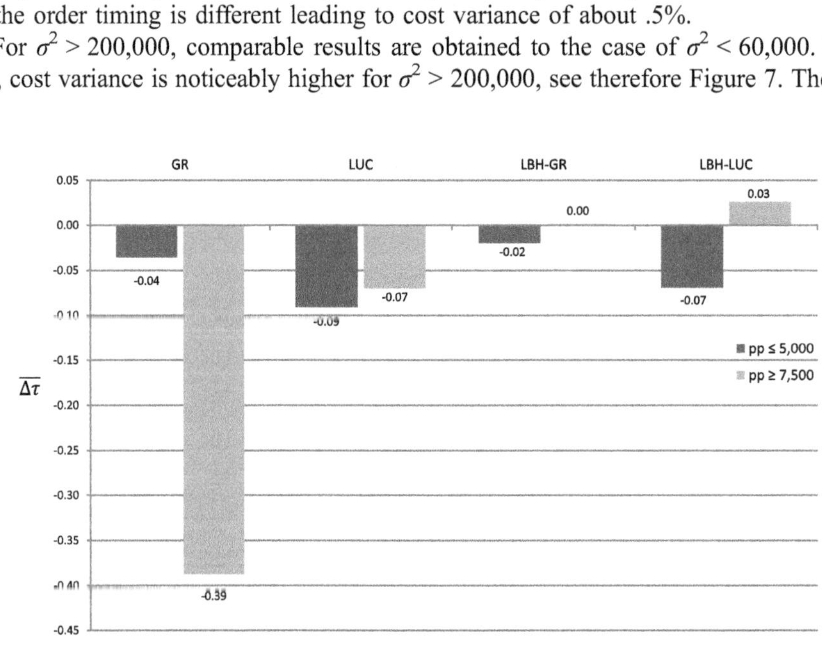

Figure 6. Order cycle ranges for $\sigma^2 < 60{,}000$.

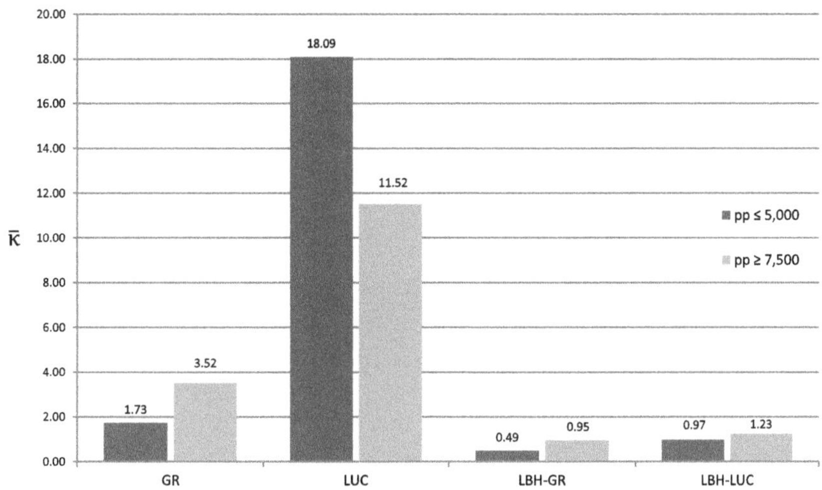

Figure 7. Cost variances for $\sigma^2 > 200{,}000$.

algorithm that leads to cost variances significantly lower than 1% compared to WW is LBH-GR with small *pp*-values. LBH-LUC is about 1% deviation. While GR generates acceptable cost variances with low *pp*-values, high *pp*-values lead in turn to significant poorer results. Obviously, LUC leads to the worst cost variances. Even though cost variance decreases with high *pp*-values, it is still very high. Particularly, in this scenario, one can see the strength of the LBH extension, which leads to considerable improvements in the results.

For the case of $\sigma^2 > 200{,}000$, we can again observe that average order cycle ranges calculated with all tested procedures are too short compared to WW, see Figure 8. Applying LBH leads to extended order cycle ranges. A peculiarity is that the cost results of LUC are better for high *pp*-values than for low *pp*-values (see Figure 7),

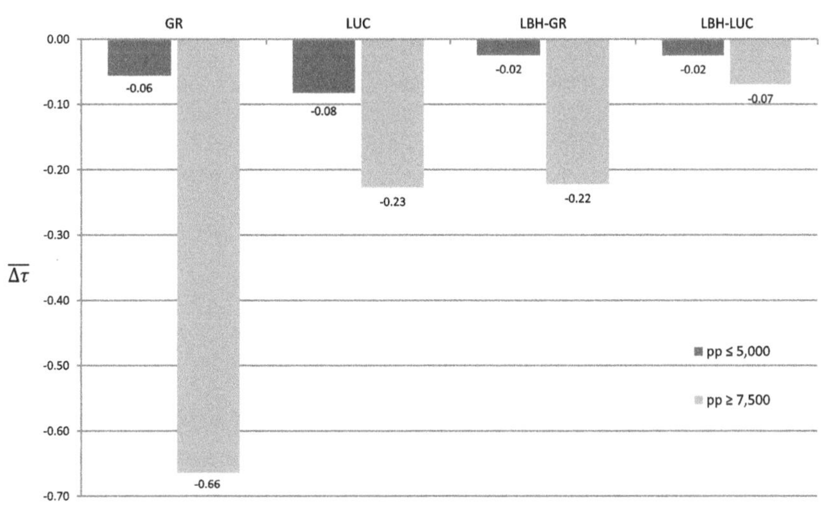

Figure 8. Order cycle ranges for $\sigma^2 > 200{,}000$.

although $\Delta\tau$ is nearer to zero, when *pp*-values are low (see Figure 8). However, we note that a certain variance in average order cycle ranges does not imply a certain goodness of an algorithm regarding cost performance (compare LUC with low *pp*-values).

The results of our study can be summarized as follows:

- LBH eliminates end-of-horizon effects for constant demand.
- All tested algorithms generate average order cycle ranges that are too short compared to WW.
- Average order cycle ranges that are close to WW can only be seen as an indication for good cost performance.
- LUC generates the highest cost variances compared to WW in most tested cases, particularly in the case of $\sigma^2 > 200{,}000$.
- High *pp*-values lead to higher cost variances for almost all tested scenarios in comparison to low *pp*-values. Only LUC leads to partly better results for high *pp*-values.
- The results of all tested heuristics can be significantly improved applying the LBH extension algorithm, and order quantities approximated the results of WW. However, poor results of a basic heuristic can only relatively be improved with LBH.

5. Conclusion

The intention of this paper was to present a comprehensive description of an extension for dynamic lot-sizing heuristics that has been overlooked by literature and practice. For this purpose, we conducted an extensive numerical study to evaluate its performance. The results of our study show that LBH leads to significant cost savings compared to the tested heuristics and may support researchers in improving existing dynamic lot-sizing heuristics. In addition, this paper may be of help for practitioners to support inventory management decisions. As ERP systems, to the best of the authors' knowledge, do not provide the WW algorithm, LBH represents a practical extension to improve the results of implemented procedures. As shown above for LUC and GR, LBH could be easy to integrate into ERP systems, as it may extend and improve any heuristic of the dynamic lot sizing problem significantly. The results of our numerical study showed, for example, that LBH-LUC reduced the cost variance of LUC compared to the optimal solution of WW from 11.4 to .79% (cf. Figure 3). We observed that LBH is adding about 20% computation time, whereas WW adds about 400% to the computation time of GR and LUC, which shows that LBH is suitable in practice. In addition, the results of this paper may be of help in education to advance existing heuristics in the literature.

This paper also has limitations. Generally, the improvements obtained with LBH depend on the results of the basic heuristic. The better the results of the heuristic, (i.e. near to the solution of WW) the better the result with LBH extension is. For example, the lot-for-lot heuristic (Orlicky, 1975) may be improved significantly applying LBH, but would still be far off the solution obtained with WW as LBH would solely consider two order periods, which means $\tau_{b,l^*} - 1$ and $\tau_{l^*+1,l^{**}} - 1$, for all order cycle ranges. In addition, we tested LBH solely within a range of possible heuristics. This and other limiting factors could be addressed in an extension of this paper.

References

Andriolo, A., Battini, D., Grubbström, R. W., Persona, A., & Sgarbossa, F. (2014). A century of evolution from Harris's basic lot size model: Survey and research agenda. *International Journal of Production Economics, 155*, 16–38.
Axsäter, S. (1982). Worst case performance for lot sizing heuristics. *European Journal of Operational Research, 9*, 339–343.
Axsäter, S. (1985). Performance bounds for lot sizing heuristics. *Management Science, 31*, 634–640.
Baciarello, L., D'Avino, M., Onori, R., & Schiraldi, M. M. (2013). Lot sizing heuristics performance. *International Journal of Engineering Business Management, 5*, 1–10.
Bahl, H. C., & Bahl, N. (2009). An empirical comparison of lot-sizing methods available in an ERP System. *California Journal of Operations Management, 7*, 77–83.
Baker, K. R. (1989). Lot-sizing procedures and a standard data set: A reconciliation of the literature. *Journal of Manufacturing and Operations Management, 2*, 199–221.
Bitran, G. R., Magnanti, T. L., & Yanasse, H. H. (1984). Approximation methods for the uncapacitated dynamic lot size problem. *Management Science, 30*, 1121–1140.
Blackburn, J. D., & Millen, R. A. (1980). Heuristic lot-sizing performance in a rolling-schedule environment. *Decision Sciences, 11*, 691–701.
Blackburn, J. D., & Millen, R. A. (1985). A methodology for predicting single-stage lot-sizing performance: Analysis and experiments. *Journal of Operations Management, 5*, 433–448.
Boe, W. J., & Yilmaz, C. (1983). The incremental order quantity. *Production and Inventory Management, 24*, 94–100.
Brahimi, N., Dauzere-Prese, S., Najid, N. M., & Nordli, A. (2006). Single item lot sizing problems. *European Journal of Operational Research, 168*, 1–16.
De Bodt, M. A., Gelders, L. F., & Van Wassenhove, L. N. (1984). Lot sizing under dynamic demand conditions: A review. *Engineering Costs and Production Economics, 8*, 165–187.
DeMatteis, J. J. (1968). An economic lot-sizing technique, I: The part-period algorithm. *IBM Systems Journal, 7*, 30–39.
Dobson, G. (1988). Sensitivity of the EOQ model to parameter estimates. *Operations Research, 36*, 570–574.
Gafaar, L. (2006). Applying genetic algorithms to dynamic lot sizing with batch ordering. *Computers & Industrial Engineering, 51*, 433–444.
Ganas, I. S., & Papachristos, S. (1997). Analytical evaluation of heuristics performance for the single-level lot-sizing problem for products with constant demand. *International Journal of Production Economics, 48*, 129–139.
Glock, C. H. (2012). The joint economic lot size problem: A review. *International Journal of Production Economics, 135*, 671–686.
Glock, C. H., Grosse, E. H., & Ries, J. M. (2014). The lot sizing problem: A tertiary study. *International Journal of Production Economics, 155*, 39–51.
Gorham, T. (1968). Dynamic Order Quantities. *Production and Inventory Management, 9*, 75–81.
Groff, G. K. (1979). A lot sizing rule for time phased component demand. *Production and Inventory Management, 20*, 66–74.
Gupta, Y. G., Keung, Y. K., & Gupta, M. C. (1992). Comparative analysis of lot-sizing models for multi-stage systems: A simulation study. *International Journal of Production Research, 30*, 695–716.
Gupta, Y. P., & Keung, Y. (1990). A review of multi-stage lot-sizing models. *International Journal of Operations & Production Management, 10*, 57–73.
Harris, F. W. (1913). How many parts to make at once. *Factory – The Magazine of Management, 10*, 135–136, 152.
Ho, J. C., Solis, A. O., & Chang, Y.-L. (2007). An evaluation of lot-sizing heuristics for deteriorating inventory in material requirements planning systems. *Computers & Operations Research, 34*, 2562–2575.
Jans, R., & Degraeve, Z. (2007). Meta-heuristics for dynamic lot sizing: A review and comparison of solution approaches. *European Journal of Operational Research, 177*, 1855–1875.
Jans, R., & Degraeve, Z. (2008). Modeling industrial lot sizing problems: A review. *International Journal of Production Research, 46*, 1619–1643.
Jeunet, J., & Jonard, N. (2000). Measuring the performance of lot-sizing techniques in uncertain environments. *International Journal of Production Economics, 64*, 197–208.

Karimi, B., Fatemi Ghomi, S. M. T., & Wilson, J. M. (2003). The capacitated lot sizing problem: A review of models and algorithms. *Omega, 31*, 365–378.
Kazan, O., Nagi, R., & Rump, C. M. (2000). New lot-sizing formulations for less nervous production schedules. *Computers & Operations Research, 27*, 1325–1345.
Kian, R., Gürler, Ü., & Berk, E. (2014). The dynamic lot-sizing problem with convex economic production costs and setups. *International Journal of Production Economics, 155*, 361–379.
Leinz, J., Bossert, B., & Habenicht, W. (1997). *Entwicklung eines Verfahrens zur dynamischen einstufigen Einprodukt-Bestellmengenplanung* [Development of a procedure for single-level single-item dynamic lot sizing]. Hohenheim: Arbeitspapier Nr.15 – Universität Hohenheim, Institut für Betriebswirtschaftslehre, Lehrstuhl für Industriebetriebslehre.
Maes, J., & Van Wassenhove, L. (1988). Multi-item single-level capacitated dynamic lot-sizing heuristics: A general review. *Journal of the Operational Research Society, 39*, 991–1004.
Orlicky, J. (1975). *Material requirements planning*. New York, NY: McGraw-Hill.
Pan, C.-H. (1994). Sensitivity analysis of dynamic lot-sizing heuristics. *Omega, 22*, 251–261.
Ritchie, E., & Tsado, A. K. (1986). A review of lot-sizing techniques for deterministic time-varying demand. *Production and Inventory Management, 27*, 65–79.
Robinson, P., Narayanan, A., & Sahin, F. (2009). Coordinated deterministic dynamic demand lot-sizing problem: A review of models and algorithms. *Omega, 37*, 3–15.
Saydam, C., & Evans, J. R. (1990). A comparative performance analysis of the Wagner–Whitin algorithm and lot sizing heuristics. *Computers & Industrial Engineering, 18*, 91–93.
Schulz, T. (2011). A new Silver–Meal based heuristic for the single-item dynamic lot sizing problem with returns and remanufacturing. *International Journal of Production Research, 49*, 2519–2533.
Silver, E. A., & Meal, H. C. (1973). A heuristic for selecting lot size quantities for the case of a deterministic time-varying rate and discrete opportunities for replenishment. *Production and Inventory Management, 12*, 64–74.
Simpson, N. C. (2001). Questioning the relative virtues of dynamic lot sizing rules. *Computers & Operations Research, 28*, 899–914.
Teunter, R. H., Bayindir, Z. P., & Van Den Heuvel, W. (2006). Dynamic lot sizing with product returns and remanufacturing. *International Journal of Production Research, 44*, 4377–4400.
Toy, A. Ö., & Berk, E. (2013). Dynamic lotsizing for a warm/cold process: Heuristics and insights. *International Journal of Production Economics, 145*, 53–66.
Vachani, R. (1992). Performance of heuristics for the uncapacitated lot-size problem. *Naval Research Logistics, 39*, 801–813.
Van Den Heuvel, W., & Wagelmans, A. P. M. (2005). A comparison of methods for lot-sizing in a rolling horizon environment. *Operations Research Letters, 33*, 486–496.
Wagner, H. M., & Whitin, T. M. (1958). Dynamic version of the economic lot size model. *Management Science, 5*, 89–96.
Zoller, K., & Robrade, A. (1988). Dynamic lot sizing techniques: Survey and comparison. *Journal of Operations Management, 7*, 125–148.

Appendix 1

To show: The minimum of the unit costs is equal to the minimum of the total costs in an EOQ model.
Proof:
The total cost function for an EOQ-model is given by

$$C(q) = C_H(q) + C_O(q) = c_H \cdot \frac{q}{2} \cdot T + c_O \cdot \frac{T \cdot d}{q} \rightarrow \text{Min!}$$

The first derivation is set to zero to find the critical point:

$$\frac{dC(q)}{dq} = c_H \cdot \frac{1}{2} \cdot T - c_O \cdot \frac{T \cdot d}{q^2} = 0$$

This leads to the optimal value for the order quantity which can be expressed as follows:

$$q = \sqrt{\frac{2 \cdot d \cdot c_O}{c_H}}$$

The unit costs can be denoted as follows:

$$c(q) = \frac{C(q)}{T \cdot d} = c_H \cdot \frac{q}{2 \cdot d} + c_O \cdot \frac{1}{q} \rightarrow \text{Min!}$$

Here, the first derivation is also needed and has to be set to zero:

$$\frac{dc(q)}{dq} = c_H \cdot \frac{1}{2 \cdot d} - c_O \cdot \frac{1}{q^2} = 0$$

The resulting optimal order quantity is

$$q = \sqrt{\frac{2 \cdot d \cdot c_O}{c_H}}$$

Appendix 2

To show: For the optimal order quantity, the marginal holding costs and the marginal ordering costs are equal according to amount.
Proof:
We need the total cost function of the EOQ-model again which is given by

$$C(q) = C_H(q) + C_O(q) = c_H \cdot \frac{q}{2} \cdot T + c_O \cdot \frac{T \cdot d}{q} \rightarrow \text{Min!}$$

Differentiation and setting equal to zero can be mathematically expressed by

$$\frac{dC(q)}{dq} = c_H \cdot \frac{1}{2} \cdot T - c_O \cdot \frac{T \cdot d}{q^2} = 0$$

And the transposing of the equation leads to

$$c_H \cdot \frac{1}{2} \cdot T = c_O \cdot \frac{T \cdot d}{q^2}$$

$$\left|\frac{dC_H(q)}{dq}\right| = \left|\frac{dC_O(q)}{dq}\right|$$

7

On decoupling points and decoupling zones

Joakim Wikner[a,b]*

[a]*Department of Management and Engineering, Linköping University, Linköping, Sweden;* [b]*School of Engineering, Jönköping University, Jönköping, Sweden*

In a market with demand surplus, it is possible to compete with standard products available in finished goods inventory. Sooner or later, the products will be sold and mass production can prevail. Competition is however increasing and to strike a competitive balance between cost efficiency and market responsiveness, it is becoming ever more important to establish a competitive level of customer-order-based management (COBM). This paper outlines a framework for this management approach based on content, represented by four key decision categories, and an overview of a process for applying the content. The content is based on a generic decision-based decoupling theory that is used for deriving the decision categories; flow driving, flow differentiation and flow delimitation. The derivation of these decision categories is based on analysis of strategic lead-times. Thereafter, the decision category flow transparency is included as the fourth content cornerstone of the framework. A process is then outlined for application of the framework. A basic bill-of-material is used as an illustration of applying the framework for COBM.

Keywords: decoupling points; postponement; decision-making; decision categories; operations strategy

1. Introduction

Decoupling points (see e.g. Blackstone, 2008) have played a crucial role in production and logistics management since the infancy of materials management. The objective of decoupling points, usually associated with stock points, has traditionally been to disconnect the material flow into sub-flows and thus enabling more focused and local flow-management. It is well known that if the flow is not decoupled, it becomes more sensitive to disturbances since the disturbances then easily propagate through large parts of the flow due to the dependencies (Goldratt & Cox, 1984). Managing this combination of uncertainties (referred to as fluctuations by Goldratt and Cox) and dependencies is the key challenge to flow management. In isolation, local uncertainties can be handled by probability theory and dependencies without uncertainties are suitable for tools from optimization theory. However, when uncertainties are combined with dependencies, these tools are challenging to apply.

Henry Ford showed the potential of tightly coupled flows under certainty when he developed the Ford production system (Ford & Crowther, 1988). A major drawback of this type of system is its inability to handle variations in volume or mix, which can be seen as disturbances that affect the continuous one-piece flow. GM and other companies

*Email: joakim.wikner@decouplingpoints.org

introduced frequent changes to the product lines which forced the system developed by Ford beyond its limits (Jones, 2001). Ford realized that one way to handle this challenge was to introduce a functional organization and stock points to decouple and reduce the dependency between different steps of the flow. GM went down a similar path but to a larger extent using arguments from cost accounting for decoupling the flow and this approach was adopted by many western companies for decades (Waddell & Bodek, 2005). This approach was also intrinsically connected to materials management systems such as material requirements planning (MRP). MRP is basically a system that is designed to maintain a target stock level, i.e. safety stock level, at the decoupling points of the materials flow. In general, this approach to decoupling points was indiscriminate to prioritizing different decoupling points even if the introduction of master scheduling and VAX-profiles provided some support (see e.g. Plossl, 1985, p. 177). The letters V, A, and X represent different types of material profiles, where the 'waist' of the profile is associated with a suitable level for performing master scheduling. The waist does not only provide an indication of suitable level for master scheduling but is also associated with the decoupling of customer-order-driven flow from forecast-driven flow (see e.g. Smith, 1989, p. 193), and hence a strategic location for inventory.

The importance of positioning decoupling points was further highlighted by the theory of constraints (TOC) (Goldratt, 1990) that emphasizes the importance of constraints for positioning of decoupling points. This approach proposes a design that can handle disturbances and dependencies by carefully positioning and dimensioning a combination of capacity buffers and material buffers. TOC uses time buffers, which are not in themselves decoupling points but rather an abstraction of flow segments related to non-constraints where the detailed flow analysis is replaced by earliest start time and latest finishing time for the time buffer (Stein, 1996). As a consequence, the time buffer will result in a physical buffer and hence a decoupling point. The most critical decoupling point is related to the bottleneck of the system. This is a different perspective compared to the VAX approach, which is material based. Olhager and Wikner (1998) showed how these two perspectives can be combined in support of master scheduling but provided limited support in terms of how to operationally combine them.

At the same time as MRP was established, Ford's original approach was further developed by Toyota that created Toyota production system (TPS) (see e.g. Ōno, 1988). TPS is the backbone of lean, which is a system that from a flow management perspective uses small decoupling points, referred to as supermarkets and FIFO-lanes, to handle product-mix demand uncertainties, and continuous improvement to reduce supply uncertainties. In all these approaches described above, the focus is on managing operational and tactical decoupling points (Wikner, Johansson, & Persson, 2009) when planning and controlling the materials flow. The lean approach also carries many similarities to the system simplification approach developed by the Cardiff group (see e.g. Wikner, Naim, & Towill, 1992). Both the lean and the Cardiff system approach emphasize simplification through elimination of unnecessary activities, i.e. waste. Lean is mainly concerned about waste at an operational level whereas the system's approach is concerned with waste at a structural level by creating effective control structures. The concept control structures as used here focuses on decision-making for creating effective flows whereas the lean approach has mainly been successful as an approach for creating efficient flows. The control structures of the system's approach can be described in terms of three components related to feedback and feed forward of information, and key decoupling points. These decoupling points are here referred to as strategic decoupling points (Wikner et al., 2009) due to their critical impact on competitiveness. In addition, there are tactical decoupling points related to items and item stock points, and operational decoupling points related to queues and buffers in the flow. Below the focus is exclusively on strategic decoupling points even if they are referred to simply as decoupling points.

The interest in decoupling points has developed in parallel streams of research related to, e.g. lead-time relation (see e.g. Mather, 1984; Shingō, 1989), decoupling point (see e.g. Hoekstra & Romme, 1992), customer order decoupling point (CODP) (see e.g. Bertrand, Wortmann, & Wijngaard, 1990), postponement (see e.g. Schmenner, 2001; van Hoek, 2001), order penetration point (Sharman, 1984), supply chain segmentation (see e.g. van der Veeken & Rutten, 1998), customization (see e.g. Graca, Hendry, & Kingsman, 1999), services (Fließ & Kleinaltenkamp, 2004; Wikner, 2012b) and leagility (Naim & Gosling, 2011; Naylor, Naim, & Berry, 1999). These different streams emphasize slightly different aspects of decoupling points but a fundamental property is the explicit focus on customers from a lead-time perspective. The main focus has been on the customer as the driver of the process but the literature on postponement (see e.g. García-Dastugue & Lambert, 2007), as well as the literature on CODP (see e.g. Hoekstra & Romme, 1992; Olhager & Östlund, 1990) have emphasized product differentiation as a separate aspect. Even if these streams have many similar properties, they have also to a large extent developed in parallel. They do however provide the decision-maker with a similar kind of decision support. In addition, a discussion on 'multiple decoupling points' has emerged in this context (see e.g. Banerjee, Sarkar, & Mukhopadhyay, 2011) where these issues are targeted to some extent but with a disperse foundation. An overall common structure is therefore needed that highlights key decisions related to decoupling points and thus enabling the development of a more cohesive theory related to decoupling points. The intention here is to identify a set of decision categories that outline the key decisions to be made when developing a competitive strategy for customer-order-based management (COBM), which here is defined as:

> *Customer-order based management* (COBM) is a management approach that emphasizes the individual customer's demand as a key input to flow-based decision making in the supply network.

COBM is a management approach and based on decision-making from a generic process perspective in contrast to, e.g. customer-order-based production (Borgström & Hertz, 2011), customer-driven strategy (Wallace, 1992), and customer-driven manufacturing (Berry, Hill, & Klompmaker, 1995; Wortmann, Muntslag, & Timmermans, 1997) that are explicitly targeting manufacturing. In complex decision-making, the sheer number of options makes the situation difficult to embrace for the decision-maker and the decision problem must be organized in a structured manner. One approach is to identify so-called decision categories by disintegrating the decision problem in relatively independent categories. One of the earliest, and probably the most referenced set of decision categories, was defined by Hayes and Wheelwright (1984). They divided decisions into structural and infrastructural decision categories based on a rather resource-oriented approach in that it focuses on the preconditions in terms of invested capital for the value-adding flow rather than the core properties of the value-adding flow per se. With a more process-oriented and lead-time-based approach, the focus shifts from different functional areas of the business to how and where customer value is created by processes in the flow. From a supply perspective, this approach puts the customer in focus and the extent of customer influence on supply management is here referred to as decisions concerning the level of COBM that should be implemented. High level of COBM means that the customer can be offered a product that to a large extent is unique and may consist of services, but at the same time it also means that the customer must wait for delivery, i.e. for demand to be fulfilled. In contrast to this, a low level of COBM implies little customer influence, and that products are standardized goods with the possibility to be delivered with short lead-time.

In some cases, the decision on level of COBM is simplified to a decision about positioning of the CODP. This may be a sufficient description of the decision problem in some cases, but in many others it provides a too simplistic picture of the decision for segmentation of the supply chain from a customer perspective. Competition from low-cost countries has also put increasing pressure on companies in other countries to be more customers oriented. COBM is appropriate in this context as it involves more explicit customer focus both in terms of understanding specific customer requirements and cooperation with customers in different contexts. In this way, COBM rewards closeness to customers from a geographical perspective (due to costs and lead-times) as well as cultural/social perspective (due to the higher level of interaction between the parties) wherefore local suppliers, with high level of COBM, have advantages in this context compared to traditional goods focused providers.

Presently, there is very limited support available in terms of comprehensive frameworks concerning decision support for COBM. One reason is probably that COBM exists in a borderland between different challenges related to traditional manufacturing as well as service, distribution, and engineering activities. Due to these circumstances, it is interesting and important to investigate the challenges facing decision-makers in COBM to identify similarities and differences between these different types of businesses, and how synergies can be exploited to improve our ability to understand and manage these kinds of enterprises. It is also important to avoid adopting assumptions from specific industries. By keeping the approach focused on fundamental and generic concepts, the result can be employed in different kinds of industries. A consequence is however that industry-specific properties of importance may have to be added to the resulting framework.

The purpose of this paper is therefore to outline a framework that captures key aspects of COBM and in particular the role of lead-times and decoupling points from a decision-maker's perspective. As outlined above, the real-world relevance of this purpose is based on that the different widely applied management approaches all would benefit from a more explicit recognition of COBM. The research objectives targeted in this paper are therefore focusing on the foundation of flow-based COBM:

- Define a generic framework for lead-time-based analysis of strategic decoupling points.
- Identify the lead-times of strategic importance for decoupling.
- Outline key decision categories of a flow-based framework for COBM.

To achieve these objectives, the paper initially identifies the so-called logical entities which are defined from a management perspective. Decision domains are then introduced and used to define decoupling points and resource-based and process-based decoupling zones which are positioned using strategic lead-times. Thereafter, more general compounded decision domains are established. Based on this flow-based theory, the three decision categories flow driving, flow differentiation, and flow delimitation are defined based on strategic lead-times. These decision categories are complemented with flow transparency and altogether they constitute the four decision categories of the framework for COBM. Finally, the process for applying the content is outlined.

2. Research approach

The framework for COBM outlined below is the result of many lines of research related to both a deductive theoretical approach to theory development and an inductive empirical approach to both theory development and theory testing.

The theoretical baseline is a set of concepts developed in the literature related to decoupling points such as the CODP, postponement, customization, and leagility. The literature has mainly focused on the application of decoupling points in different areas and not so much on conceptual development. In particular, discussions on positioning of decoupling points (see e.g. Hoekstra & Romme, 1992) have gained much interest but less effort has been spent on investigating the more fundamental aspects of decoupling points and common properties of different kinds of decoupling. This is also the gap targeted here and the main contribution lies in the holistic perspective on decoupling points and the explicit flow-based approach.

The literature provided a set of theoretical concepts that were used in two research projects involving 5-6 companies. As the number of concepts used in the projects increased, a more general theoretical foundation of the decoupling-based concepts was established resulting in the generic framework presented below. The generic framework was then used as a platform for defining decoupling-oriented concepts providing a comprehensive and conceptual identification of COBM. The resulting framework for COBM therefore rests on a combined theoretical and empirical foundation. The actual cases from the research projects are rather complex and still not each covering all aspects of the framework. Hence, a fictitious example is used below that provides an opportunity to illustrate the application of the framework without introducing the complexity of the real cases.

3. Management perspectives

Cost efficiency has for many decades been the most important driver for enterprise development. This approach has turned out to result in focus on local efficiency where customers' needs are of lesser importance. The development of lean thinking (see e.g. Womack & Jones, 1996) as a framework for flow-based enterprise management has however put the customer in focus and as a consequence the enterprise's ability to create customer value becomes decisive. In this context, time is maybe the most important resource and the time that is available should be used to create customer value, otherwise the time is considered as wasted. This puts strain on the management system since time should be used for value creation at the same time as uncertainties must be considered. Before the importance of time in terms of lead-times is discussed, the context of lead-times is defined from a system's perspective.

3.1. Transformation-based system perspective

A lead time analysis is performed in some kind of context in terms of a system's perspective. The context can, for example, be different aggregation levels or hierarchical control structures. The analysis performed here is however more management oriented and based on the transformation process.

Analysis of supply chains is often based on the interaction between different actors (see e.g. Harland, 1996). The actors may however be viewed from different perspectives (Wikner, 2012a). From an overall perspective, the actors are usually companies (legal entities) and the challenges are handled within *business management*. These *legal entities* interact through e.g. customer orders and purchase orders. In this way, the legal perspective handles issues associated with who is ultimately funding different types of transactions, i.e. who is acting as the *sponsor of transformation*, see Figure 1.

The financial and contractual dimension, represented by the legal perspective, is however only a reflection of processes being performed by resources at the physical entities in terms of geographical places or organizational units such as departments, production sites or distribution centres. To define actors from this perspective as *physical entities* is important when analysing localization issues or how different units collaborate from a production perspective. This also involves how different strategies for managing the supply chain are applied related to e.g. leagility and postponement. The management approach is therefore focused on logistics activities in a broad sense, including production, and can be referred to as *supply chain management*. A physical entity is hence a value-adding node in a physical network that performs transformation in terms of form, place or time and is associated with *type of transformation*.

The division into physical entities is not that obvious from an enterprise management perspective since the business information systems, such as ERP systems, now offer the possibility to manage multiple physical units as one integrated network, which may also be referred to as a virtual supply chain (Chandrashekar & Schary, 1999). The idea of virtual supply chains dates back to virtual inventory management and e.g. different types of base stock systems (see e.g. Clark & Scarf, 1960) which is interesting since inventory management in essence is concerned with the management of tactical decoupling points. The concept of virtual supply chains is emphasizing a transient property with short-term orientation as supply chains may change frequently based on market requirements. In contrast, the framework introduced here makes no assumption on short term vs. long term but rather emphasizes a structural management perspective. The key property here is that this type of network, from a planning and control perspective, is a system of multiple geographically or organizationally dispersed units that are managed as one entity and it is here referred to as a *logical entity*. Logical entities are not concerned with the type of transformation that is performed but defines the flow in terms of processes and objects and is thus associated with a generic *transformation* of input to output from a *process management* perspective. The concept logical entity is a fundamental construct of COBM and is hence explicitly defined here:

> A *logical entity* is a network of one or more physical entities that from a management perspective can be considered as one integrated network, offering the same level of controllability in all its parts.

The three perspectives legal, physical, and logical are summarized in Figure 1.

3.2. Legal and logical entities

The three system's perspectives introduced above can be combined in different ways. A company (legal perspective) can be divided into manufacturing and distribution/transportation with different functional belongings (physical perspective) but due to integrated business information systems, they can be managed as one virtual integrated unit

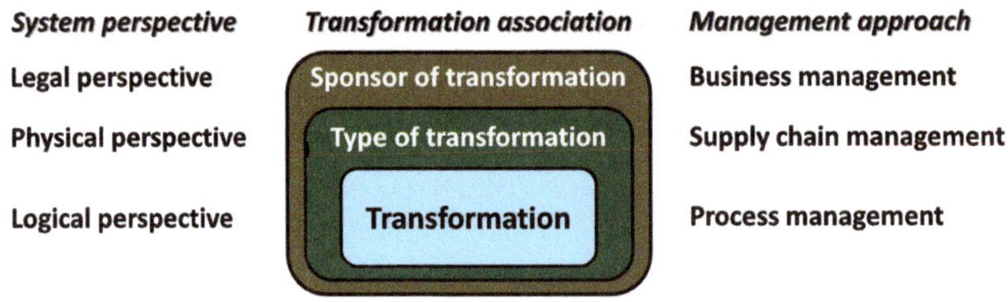

Figure 1. System perspective, transformation association and management approach.

(logical perspective). In particular, it is important to separate between the legal perspective and the logical perspective when performing flow analysis. Even if the economic consequences of the transformation in the flow is related to the legal entities, the mission for effective flow is created by considering the parts of the flow that can be managed as an integrated unit, i.e. as a logical entity. Figure 2 illustrates this with four fundamental configurations that can be identified by combining legal entities with logical entities. More general networks may of course contain multiple legal and logical entities but this would basically be a generalization of the simple configurations of Figure 2.

The configuration with one legal entity and one logical entity corresponds to two legal entities and two logical entities since in both cases there is one logical entity per legal entity, i.e. each company can be managed as one integrated system. This is probably the most common scenario since legal entities with one production unit usually can be managed as one integrated system and hence constitutes one logical entity. When there is one logical entity for two legal entities, it means that the two legal entities are integrated and managed as one unit. If, on the other hand, one legal entity has two logical entities, it corresponds to a situation where one legal entity has not been able to establish one integrated control approach but instead is divided in, e.g. different departments that are managed as separate, non-integrated units.

Below, the point of departure is different logical entities. How these logical entities relate to legal entities is of less importance for this framework since the focus here is on efficient and effective flows rather than who is responsible for different parts of the flow and economic evaluation of the flow. An implicit assumption below is therefore that one logical entity may correspond to a part of a legal entity, a complete legal entity or a multiple legal entities that are managed as one unit. Since the logical entity is defined from a flow and control perspective, it is also important for flow-based decision-making. By defining the specific preconditions that are valid for decision-making

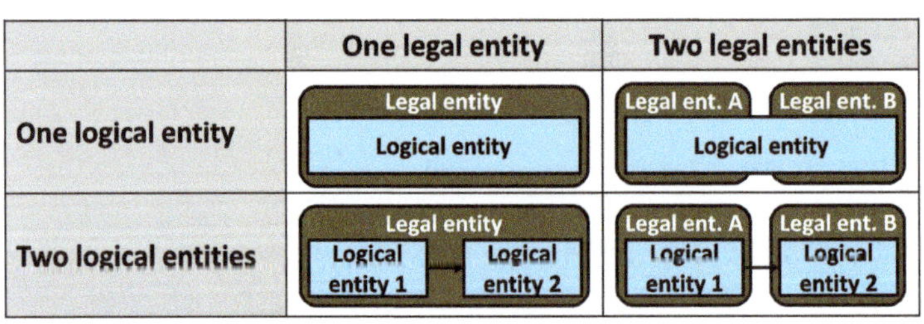

Figure 2. Four fundamental configurations of legal entities (actors) and logical entities.

in different parts of the logical entity, the key decision categories for COBM can be identified.

4. Framework for flow-based decision-making

Flow is a general concept that basically implies a 'change', or in other words a transformation, and a rate of transformation. The transformation can take place in the form, place, or time dimension (see e.g. Bucklin, 1965). Being a rate-based concept means that flow is closely associated to time and understanding implications of time in this context is critical. A detailed flow analysis does however require a more elaborate view in terms of the types of flow involved. Depending on the context of the analysis, different types of flow are significant but in most cases the five types identified by Forrester (1958) (information, materials, money, manpower, and capital equipment) suffice. At a higher level of abstraction, the concept of flow can be defined from a customer perspective as:

> *A flow* represents the value-adding for customers through a set of transformation processes.

In this case, the concept of flow is used to represent that value adding is taking place, where value is added in a number of transformation processes (of course also non-value adding activities may be included but the emphasis is on value adding) which is in the spirit of, e.g. swift and even flow (Schmenner & Swink, 1998) and simplified material flow (Childerhouse & Towill, 2003). In every such process, a number of flow-based decisions are made that are decisive for how value adding is performed. The detailed design of the processes and the value-added flow may of course vary but here a common denominator is that for illustrative purposes, the flow direction is assumed to be from left to right and that the flow rate is directly or indirectly based on customer demand. The flow may hence be associated with materials flow but it is important to also recognize that the suggested framework does not exclude services since service can be defined as a process-based concept (Vargo & Lusch, 2004).

4.1. Decision categories and decision domains

Decision-making in COBM is comprehensive and covers a wide range of aspects to be effective. Providing a comprehensive framework for all types of decisions is both challenging to assemble due to the sheer number of different aspects to cover and would also be of limited value to the decision-maker due to the complexity provided by the comprehensiveness. To reduce the complexity due to a too high level of details, the decisions that should be covered can be classified according to their impact. Hayes and Wheelwright faced a similar challenge when outlining their framework for manufacturing strategy: 'Because of the diversity of manufacturing decisions made in different businesses, an organizing framework that groups them into major categories is a useful tool in both identifying and planning the functional strategy for manufacturing'. (Hayes & Wheelwright, 1984, p. 84). The major categories for COBM are defined later but the decision categories, as a concept that were used by Hayes and Wheelwright, are also suitable for this purpose and is here defined as:

> A *decision category* groups decisions into major categories that are fundamental to decision making for a particular purpose.

The scope of a decision category (DC) is based on some fundamental aspect of decision-making, which is here referred to as a decision criterion. Thus, decision categories are a way of classifying decisions and the decision criterion is what the classification is based on:

> A *decision criterion* is a standard on which a decision may be based.

To manage an enterprise involves making numerous decisions of different character but all related to transformation processes. Some decisions are of more simple character and do not affect the flow to a large extent while others are of decisive importance to the performance of the process, such as initiating a process or not. Effective process management therefore requires that the critical decisions can be isolated and classified to enable the design and application of a management framework with decision categories. These categories should then be applicable on a wide spectrum of processes with similar preconditions. In support of this classification of processes, the concept of decision domain is introduced:

> A *decision domain* is characterized by consistent preconditions for decision making related to a specific decision criterion.

A decision domain thus identifies what is common to a number of processes from a decision perspective and can hence be perceived as sub-categories within a DC (Wheelwright, 1984). A decision domain might, for example, define that the processes related to a decision domain are performed based on that a customer order has been received and hence that the process can be classified as customer-order-driven. This decision domain would then belong to the DC 'flow driving' and be based on the decision criterion 'flow driver' with the property 'driven by customer order' (see e.g. Figure 20 for an overview of decision categories). From this perspective, it would be reasonable to illustrate a decision domain as if the processes are within a decision domain. But for practical reasons, this relation is illustrated with a decision domain whose extension along the flow determines the processes associated with a particular decision domain, as shown in Figure 3. This means that the processes U1 and U2 in Figure 3 are classified as belonging to Upstream decision domain (from a flow perspective) and processes D1 and D2 as belonging to Downstream decision domain.

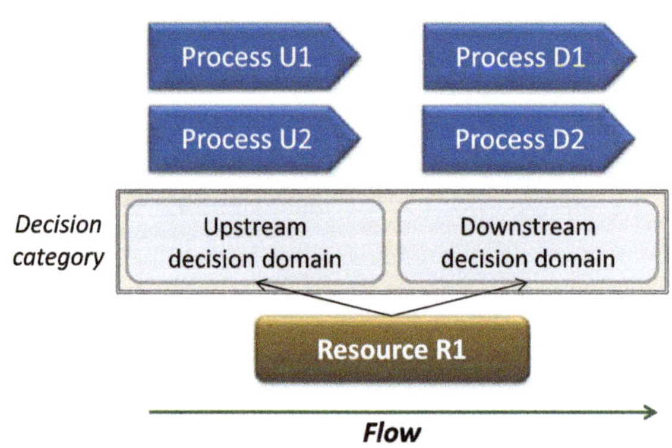

Figure 3. Relation between processes, resources and decision domains.

Resources that perform processes have a relation to decision domains that follow a different pattern compared to processes. Since a resource can perform more than one process, it can also be related to more than one decision domain. Resource R1 in Figure 3 is, for example, related to both process U1 and process D2. Process U1 may be forecast-driven and process D2 customer-order-driven wherefore they are classified differently in terms of the DC flow driving. But, since both these processes are performed by the same resource, the capacity of the resource is required by both forecast-driven and customer-order-driven processes. As Figure 3 shows, a decision domain is here illustrated by a rectangle with rounded corners and with black text on grey background. Processes are depicted with traditional arrows, resources with rectangles with rounded corners and white text on dark background.

4.2. Decision domains and decoupling points

The analysis above defines decision domains as a collecting concept for processes with a specific property in common. The decision domains can be relatively independent of each other in the sense that they can be based on different decision criteria that highlights different process perspectives, such as focusing on credit risk or representing disassembly decisions. From a flow perspective, it is of particular interest to focus on decision criteria that separates the flow into two separate parts. A specific example of this is processes that are customer-order-driven and processes that are forecast-driven. For the DC flow driving, these processes are completely different but still they are connected to the same decision criterion, i.e. what decides that a process is initiated. From a flow perspective these two decision domains can be seen as being in a sequence in the flow and characterized as one upstream decision domain and one downstream decision domain that from a decision perspective are disconnected from each other.

The decision criterion is decisive for the type of property that can be classified as either upstream or downstream. To emphasize this difference, a decoupling point is introduced that indicates that the decoupled decision domains are related to the same decision criterion:

> A *decoupling point* separates decisions that are made under different consistent properties related to a specific decision criterion.

The decision domains are disjunctive in the sense that they represent completely different properties, such as forecast-driven or customer-order-driven, considering the same decision criterion. This relation is depicted in Figure 4 where a general decoupling point is illustrated with an ellipse which symbolically connects the two decision domains of the DC.

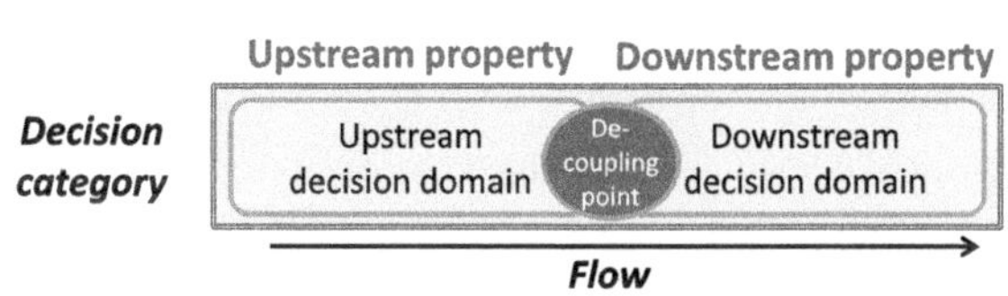

Figure 4. Decision domains with disjunctive properties.

However, in many cases, the decision support is not homogenous and based on only one property at each side of the decoupling point. In some cases, the transition is not instantaneous along the flow, i.e. an upstream property is not changed to a downstream property at one point, but rather can be seen as a gradual transition consisting of a mix of properties from each side of the decoupling point of Figure 4. A process driver may, for example, be a combination of forecast and customer order. This type of mixed properties is here associated with a hybrid decision domain, see Figure 5. It is then logical to model this as three decision domains and consequently two decoupling points for each DC.

Hybrid decision domains represent more complex decision-making where the criterion has mixed properties. In some cases, it can be an advantage to use both decoupling points but this approach also increases the complexity. An alternative is to emphasize only one decoupling point and this is a useful approach when the decision criterion is intrinsically linked to one of the decision domains. A typical example is the flow driver criterion of COBM, which is customer focused and hence the logical place to decouple is upstream of the customer-order-driven decision domain. This domain is customer focused, whereas the other decision domains also have other properties. Since the focus below is on COBM, only decoupling point 1 will be considered and below simply be denoted 'decoupling point'. The mixed properties that the hybrid decision domain relate to represent, from a flow perspective, a transition from the properties valid upstream to the properties valid downstream of the hybrid decision domain. This transition with a hybrid decision domain can be seen as a zone that decouples the two single-property decision domains.

Decoupling points by definition mean, as described above, that different preconditions exist upstream and downstream from the decoupling point. From a material flow perspective, this type of discontinuity corresponds to that the preconditions for the flow changes significantly which leads to disturbances of the flow. Different kinds of buffers are used to reduce the impact of these disturbances. Capacity buffers, related to resources, offer interesting possibilities since they increase the capability to be agile in processes performed to, e.g. customer order or backorders at stock points. A drawback with capacity buffers is however that they do not provide instant reaction but requires a lead-time before a result can be delivered (since the capacity is used to perform some kind of transformation which takes time). The material buffers do not have this drawback since they provide immediate availability of what is demanded even if this type of buffer provides less flexibility since decisions, about what to include in the material buffer, are made before the requirement is identified. In this way, the material buffer is a more lead-time responsive type of buffer that is suitable in relation to decoupling points. The capacity buffer does, however, provide important support downstream from the

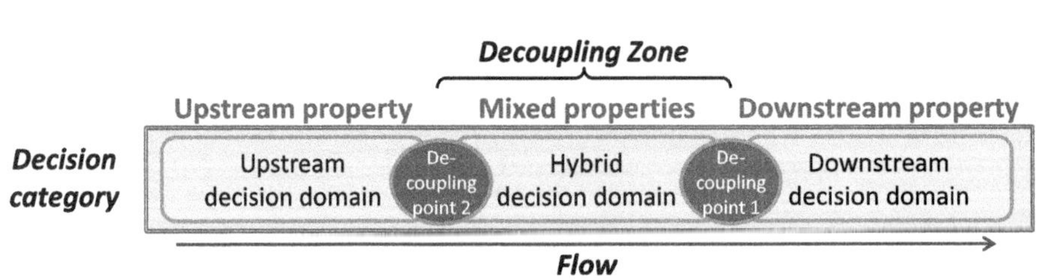

Figure 5. Decision domains and strategic decoupling points.

decoupling point as it enables the resources to be more flexible. The material buffer is here referred as a stock point and later illustrated in the figures by a black triangle pointing downwards representing physical inventory in the flow.

The significance of so-called discontinuities can be reduced, or even eliminated, by introducing decoupling zones that enable a softer transition from the conditions at the upstream decision domain to the conditions at the downstream decision domain. If discontinuities are completely eliminated, there is also no need for a corresponding buffer at a stock point.

4.3. Decision domains and decoupling zones

Decoupling points are positioned at an instantaneous transition along the flow from upstream property to downstream property. This represents a 'black or white' scenario which in some cases is too simplistic. The transition may be gradual and hence represents a zone with different 'shades of grey' related to the hybrid decision domain as described above.

> A *decoupling zone* covers decisions that are made under mixed properties related to one or more decision criteria.

There are mainly two different perspectives that require these 'shades of grey'. The first perspective is the gradual transition from upstream property to downstream property and it is here called *process-based decoupling zone* since it exists along one individual flow and corresponds to one single decision domain. The other perspective exists when multiple flows are considered in parallel, from an aggregate perspective, and is in particular related to when one resource is involved across multiple flows at the same time. This scenario has its roots in how resources are positioned in relation to decision domains in Figure 3 and is therefore here referred to as *resource-based decoupling zone* that corresponds to an aggregation of two or more different decision domains.

4.3.1. Process-based decoupling zones

The concept decoupling zone refers to a mix of different properties related to decision domains. In case there is a gradual transition along a flow, a number of attributes related to the property can be affected. The properties can, for example, be associated with uncertainty or certainty concerning customer orders. The certainty can however be based on different types of attributes related to the customer order such as requested delivery date or requested quantity. When deciding on the number of attributes to use, a balance between the complexity of using many attributes and not challenging the relevance of the attributes, by using too few, must be established. In this framework for COBM, two attributes are used for each DC related to a process-based decoupling zone, but from a more general perspective, one or more than two attributes can also be used.

> A *process-based decoupling zone* is based on the gradual transition between two properties of one decision criterion related to one flow.

The process-based decoupling zone (PB-DZ) is in Figure 6 illustrated by a rectangle placed above the hybrid decision domain. The zone is expanded to a two-dimensional surface that represents two attributes that each can take values on a scale from upstream

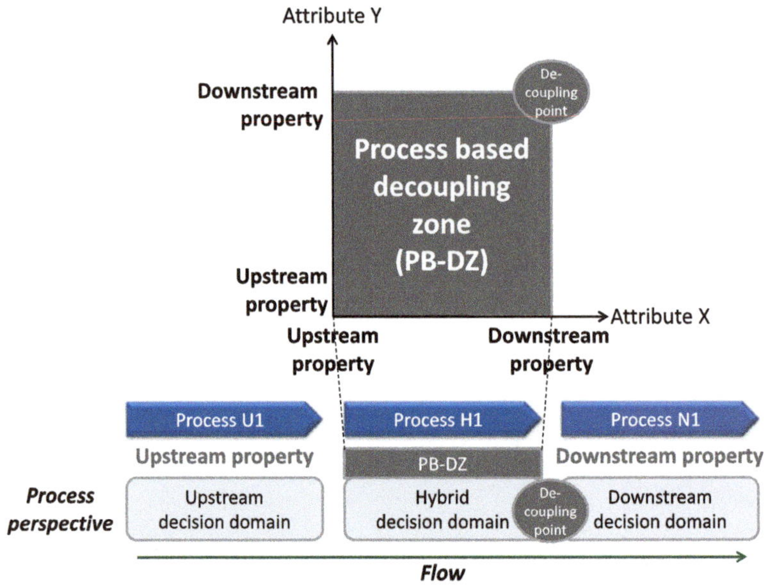

Figure 6. Process-based decoupling zone along one flow.

to downstream property. The end-points (the corners) where the attributes have the same 'value' represent the limit of upstream and downstream decision domain, respectively. Based on the definition of decision domain and how it is related to processes means that there are now, at least, three types of processes along the flow; in Figure 6, this is illustrated with the processes U1, H1 and N1. At the beginning of the flow (on the left in the figure), the upstream property is of relevance to decisions. When the flow reaches the PB-DZ, the decision preconditions correspond to the lower left corner of the two-dimensional surface. The flow through the PB-DZ corresponds to that the preconditions change as the value of the two different attributes changes along a trajectory from the lower left corner to the upper right corner of the surface, i.e. when the decoupling point is reached. Thereafter, the downstream property is relevant for decision-making until the end of the flow is reached.

4.3.2. Resource-based decoupling zones

From a resource perspective, it looks different compared to the process perspective since resources can be connected to multiple decision domains with different properties as shown in Figure 3. This multi-property load on a resource is related to resource-based decoupling zones.

> A *resource-based decoupling zone* is based on the intersection between different properties of one decision criteria across one or more flows.

In Figure 7, this is illustrated with the two flows, Flow A and Flow B, where each consists of two processes related to one upstream and one downstream decision domain, respectively, and thus has one decoupling point each. One individual flow is therefore relatively simple from a process perspective with two basic decision domains both based

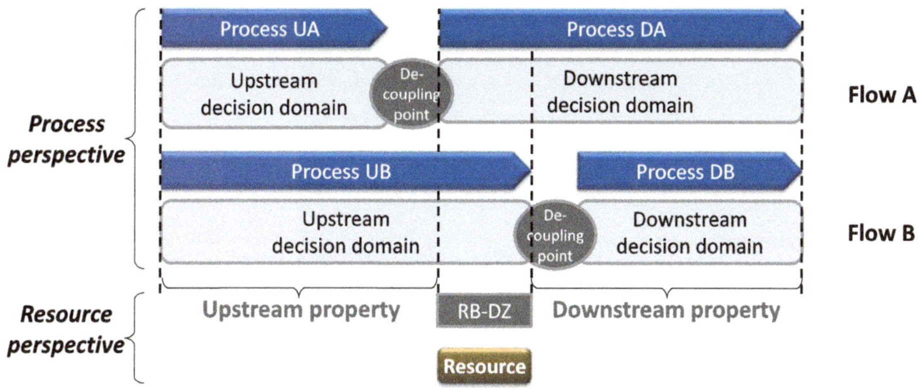

Figure 7. Resource-based decoupling zone across multiple flows.

on the decision criterion in question. In Figure 7, one resource is identified and from the resource's perspective, the situation is more complex since it is involved in process DA and process UB, i.e. two processes that are related to different decision domains with different properties. This type of scenario is here referred to as a resource-based decoupling zone (RB-DZ) and could, for example, mean that some of the load on the resource is forecast-driven (based on UB) and some is customer-order-driven (based on DA). This means that the load analysis becomes considerably more complex compared to when only UA and UB or only DA and DB are involved, where the preconditions would be more homogenous what concerns the actual DC. Note that the assumption is that the same decision criterion is in focus for all processes, i.e. they belong to the same DC. In addition, the individual flows may not contain a process-based decoupling zone but in aggregation a resource-based decoupling zone can be identified. In case the decoupling points of the individual flows are positioned at the same place also, the aggregate would not contain a decoupling zone and the decoupling point could be referred to as a 'resource-based decoupling point'.

The aggregation discussed above is across multiple flows. The corresponding situation can occur in the case of aggregation in time where, for example, weeks are used as time periods rather than days. With a resolution of days, a more detailed analysis can be performed resulting in that resources are either loaded with forecast-driven or customer-order-driven activities. Using the aggregation level of weeks on the other hand might result in a mix of different types of load for a particular resource since the timing of load cannot be determined as exactly as in the case of time periods based on days. Also, resource aggregation may create a similar problem as individual resources are not loaded, but instead groups of resources are loaded. Hence aggregation in terms of flows (i.e. products), time, or resources can generate similar challenges from a capacity requirement perspective.

In summary, the capacity requirement for a resource may be created by seven different combinations of processes related to three different decision domains of one single decision criterion. Note that Figure 7 is based on that each flow consists of two decision domains whereas Table 1 is based on the existence of also a hybrid decision domain, i.e. PB-DZ. In Table 1, these combinations are illustrated and the first three cases (type 1–3) in the table are based on one decision domain each with capacity requirement from

Table 1. The seven different types of capacity requirement.

	Upstream DD	Hybrid DD	Downstream DD
Capacity requirements type 1	X		
Capacity requirements type 2		X	
Capacity requirements type 3			X
Capacity requirements type 4	X	X	
Capacity requirements type 5	X		X
Capacity requirements type 6		X	X
Capacity requirements type 7	X	X	X

the respective decision domain. Three of the combinations originate in the combination of two types of decision domains. Capacity requirement type 7, finally, is based on three different types of decision domains, i.e. one resource would be used in processes of three different decision domains. The example of Figure 7 can, based on this, be classified as a resource with capacity requirement type 5 since the capacity requirements originate from a upstream DD and a downstream DD.

For both the PB-DZ and the RB-DZ, the 'length' of each decision domain is important since it determines the position of both decoupling point and decoupling zone. The decision domain is based on one decision criterion; for example, flow driver, which means that the length of the decision domain and the priority of the chosen decision criterion are different sides of the same coin. The definition of important decision categories for COBM can therefore be based on the identification of important lead-times.

4.4. Positioning of decoupling point and decoupling zone

The extension of decision domains have so far been based on a relative perspective. A decoupling point has been identified that separates the flow into two distinct parts related to one decision domain each. In addition, a hybrid decision domain may be positioned between these two decision domains resulting in a sequence of three decision domains along the flow. Identifying the decision domains is a first step but it is also important to determine the actual extension of each decision domain and as a consequence also the position of the decoupling point and the decoupling zone.

A useful measure for the extension would operationalize the concepts introduced so far. A number of candidates exist for acting as a point of reference such as the physical extension of the flow, organizational properties such as functions involved, or cost aspects related to e.g. resource ownership. It is however important to have a baseline that is absolute in the sense that it is not something that can easily be manipulated to suit different purposes. It should rather be something that can act as a point of reference that is important and transparent to all concerned interest parties. Considering the history of operations and supply chain management, a number of cases have shown that the key resource to manage is time as time lost can never be recovered. In addition, time has also been shown to be a key driver of competitiveness. Henry Ford was one of the first to explicitly highlight time as one of the pillars of his management philosophy (Ford & Crowther, 1988) and time has continued to play a key role in, e.g. Time-based management (Stalk & Hout, 1990), BPR (Hammer & Champy, 1993) and lean thinking (Womack & Jones, 1996). Presently, the main proponent of time as a key resource is probably lean in all its flavours such as lean thinking, lean production, lean services, lean administration, etc. At the core of all these, lean 'flavours' is the continuous improvement to reduce waste, and waste is basically different incarnations of lost time. For example value stream mapping (Rother & Shook, 1998), the most highlighted part of the analysis is the timeline capturing the total lead-time as well as the value adding and non-value adding time. From a time perspective, the timeline positions the different parts of the value stream in relation to the customer, i.e. the end-point of the value stream.

The decision criterion and the related decision domains provide a conceptual division of the flow from a decision perspective. The decision in itself and what it involves are important to define the decision situation, but the timing of the decision is absolutely critical in defining the preconditions for the decision-maker. The timing is defined in relation to when the result of the sub-flow, corresponding to the decision domain, is required. With this as a point of departure, three lead-times can be defined as of Figure 8, related to one decision domain each: upstream (US), hybrid (H), and downstream (DS). Accumulating these lead-times the total lead-time for the flow i (F_i) is obtained, where: $FLT_i = USLT_i + HLT_i + DSLT_i$. The decoupling zone highlighted here is related to a decision domain and hence of type PB-DZ. The RB-DZ can be positioned in a similar way based on that the decoupling zone is related to an intersection of a set of flows. DSLT would in this case correspond to $DSLT = \min_{i \in Flows} \{DSLT_i\}$ where Flows is the set of all flows where the resource is involved. In a corresponding way the lead-time of the RB-DZ can be defined as: $HLT = \max_{i \in Flows} \{HLT_i + DSLT_i\} - DSLT$. USLT can then finally be defined as: $USLT = \max_{i \in Flows} \{FLT_i\} - HLT - DSLT$.

Based on the argument earlier, where the second decoupling point was considered as of less interest, the decoupling point upstream of the hybrid decision domain is not analysed further. The positioning of the decoupling point is hence based on the lead-time of the downstream decision domain, DSLT, in relation to a reference lead-time such as the total flow lead-time, FLT. The position of the decoupling point can hence be expressed in two different ways:

- The lead-time DSLT, which is useful in e.g. analysis based on the time-phased bill-of-material to calculate the position of the decoupling point. This is the 'absolute' lead-time expressed in suitable time units.
- The 'relative' lead-time results in a percentage and expresses how long a lead-time is in relation to a 'benchmark' lead-time and in particular we are interested in the fraction of the benchmark lead-time that is related to downstream property. For positioning of the decoupling point the key lead-time is the relation between the lead-times DSLT and FLT, which is called the DSLT:FLT relation. If DSLT:FLT = 1 then the decoupling point is at the beginning of the flow. If DSLT:FLT << 1

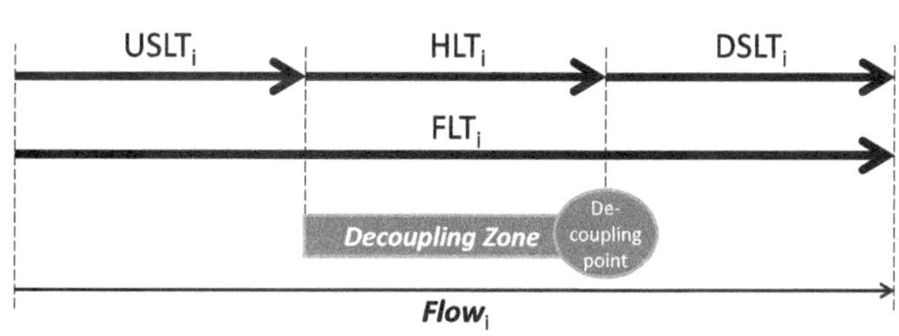

Figure 8. Decision domains and lead-times.

(or DSLT:FLT = 0) then the decoupling point is at the end of the flow and finally if DSLT:FLT < 1 the decoupling point is positioned within the flow.

The 'relative' lead-time is measured in relation to a reference value and is thus of more practical use in COBM. From a general perspective it can be defined as:

A *lead-time relation* is a ratio of two strategic lead-times.

Positioning of the decoupling point also provides information about the extension of the upstream and the downstream decision domains in case there is no hybrid decision domain. If also a hybrid decision domain is present additional lead-time analysis is required to position also this decision domain. This additional analysis can be based on either USLT or HLT. Once the decoupling point and decoupling zones are positioned, the different decision categories can be combined resulting in compounded decision domains.

4.5. Compounded decision domains

In an actual decision situation, multiple decision criteria may be important to consider at the same time. Still each decision criterion can be defined separately as in Figure 9 where this is illustrated on the left side with decision criteria 1 and 2. In most cases, the different decision criteria are independent of each other and capturing different aspects of the challenges facing the decision-maker. Therefore the two decision criteria may be modelled as two separate dimensions as shown on the right of Figure 9. The resulting four decision domains are referred to as compounded decision domains (CDD):

A *compounded decision domain* covers the intersection of two decision domains related to two separate decision criteria.

Each CDD is a combination of upstream (U) and/or downstream (D) decision domains based on decision criteria 1 or 2. For example, D1, D2 is a CDD where the decision criterion is a combination of decision criteria D1 and D2. Depending on the actual decision criteria used some of the combinations might not be possible, or at least not competitive, and hence a strategy for that CDD might not be necessary. However, from a generic point of view all CDDs are assumed to be valid and hence all possible CDDs are included.

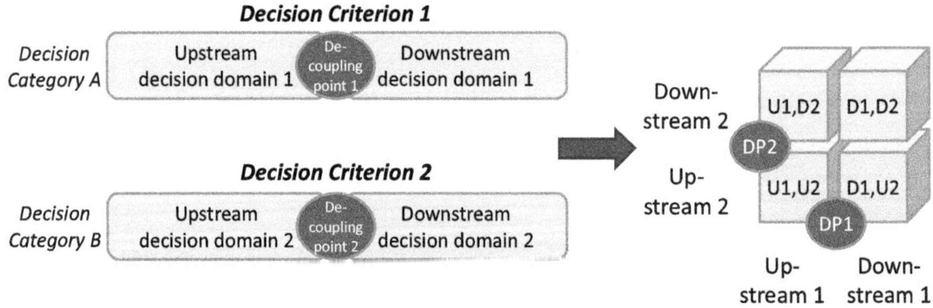

Figure 9. Example of compounded decision domains.

The CDDs add complexity since they introduce a multi-dimensional perspective on decision domains. The example in Figure 9 is based on two flows with decoupling points only. By also including decoupling zones there are some more challenging issues to consider.

4.5.1. Compounded decision domains and process-based decoupling zones

The concept of CDD can easily be expanded to also include hybrid decision domains and PB-DZs. The example of Figure 9 represents the possible combinations when no hybrid decision domains are included. By allowing hybrid decision domains each dimension would be extended to three blocks and in total nine CDDs would be included on the right in Figure 9. This scenario, with nine CDDs, is somewhat similar to the analysis of the resource-based decoupling zone of Table 1 (with seven possible combinations). It is however important to note that in this case the two dimensions are based on different criteria whereas in Table 1 the two dimensions were based on the same criterion. Since the CDD here is based on two criteria the type 7 case cannot occur and the criteria are different for types 4–6 which means that it would result in another three types (for example, U1, D2 does not correspond to D1,U2 when different criteria are used). In summary there would be four CDDs only involving upstream and downstream decision domains and an additional five that include at least one hybrid decision domain. This would result in three decision domains in each dimension of Figure 9 and, consequently, in total nine different CDDs (cf. Figure 16 for an example).

4.5.2. Compounded decision domains and resource-based decoupling zones

The CDDs creates additional complexity in terms of RB-DZs since the number of possible combinations increases significantly. The example of Figure 7 resulted in one additional scenario (totally three) compared to when one flow was analysed. The number of possible combinations of decision domains, $C(\text{Number of flows, Number of decision domains per flow})$ affecting a resource when each flow can be divided into n different decision domains is a combinatorial problem. For the case in Figure 7, which is based on that the decision domains belong to the same DC, the number of combinations would be: $C_{\text{One DC}}(2,2) = 2 + 1 = 3$ different combinations of decision domains. In case each flow would contain a hybrid decision domain the number of combinations would be: $C_{\text{One DC}}(2,3) = 3 + 3 + 1 = 7$ different combinations (as shown in Table 1). When the CDDs are introduced the possible combinations for a resource to be involved in more than one decision domain increases dramatically for RB-DZs. Consider, for example, the case introduced above in Figure 9 with two dimensions and two decision domains in each dimension resulting in a total of four CDDs. One case could be that a resource is acting in the CDD $\text{U1}, \text{U2}$ for one flow and the CDD $\text{D1}, \text{D2}$ in the other flow. By definition the RB-DZ is related to when a resource is involved in two CDDs of different properties which means that if a resource is involved in CDD $\text{U1}, \text{U2}$ of two different flows it is not defined as a RB-DZ. For this case with two flows with four CDDs each there are a total of $C_{\text{Two DCs}}(2,4) = 12$ possible combinations. Introducing PB-DZs in each flow of course increases the complexity further and the number of combinations fast becomes hard to manage. It is therefore important to identify the critical intersections and to exclude the less important from the analysis.

4.6. Summary of framework for flow based decision-making

The generic framework for decoupling points and decoupling zones provides the means for identifying some key decision categories related to supply chain and operations management. In summary, the generic framework for flow based decision-making contains ten key concepts (which have been explicitly defined above):

(1) *Logical entities* are the platform for decisions on flow based management.
(2) *Flow* provides the point of reference for identifying lead-time based decision categories.
(3) *Decision categories* are defined based on decision criteria.
(4) *Decision criteria* are used to define decision domains within a decision category.
(5) *Decision domains* for the same decision criteria are separated by decoupling points.
(6) *Decoupling points* are positioned at the interface between two decision domains.
(7) *Decoupling zones* correspond to mixed properties in a decoupling point scenario.
(8) *Process-based or resource-based* perspective can be applied on decoupling zones.
(9) *Lead-time relations* position decoupling points/zones and decision domains.
(10) *Compounded decision domains* reflect the complexity of an actual decision problem.

The generic framework with these ten key concepts is next used to define three lead-time relations and thereafter four decision categories for COBM.

5. Lead-time based flow analysis

Lead-time analysis is used in different contexts to e.g. reduce lead-times, identify wastes or simply to identify the magnitude of the lead-times. This type of lead-time analysis can be categorized as 'absolute' lead-time analysis since focus is on the length of the lead-times. It is however 'relative' lead-time analysis that is of main interest as an indicator of positioning of decoupling points and decoupling zones since it involves the relation between different lead-times. Before these lead-time relations can be defined it is necessary to identify the key 'strategic' lead-times.

5.1. Lead-time definitions

Balancing the requirements and availability of material and capacity of operations are important challenges in logistics flow. Common to these challenges is that timing is of critical importance. The relationship between available capacity and capacity requirements should be investigated with respect to a timeline and also the balancing of demand and supply of materials requires a time-phased approach (however less significant in a rate-based context). The comprehension of lead-times and how they should be managed is therefore of utmost importance to enterprise management.

The standard item lead-times (L) are fundamental, but even more important from a competitiveness perspective are the five strategic lead-times (Wikner, 2011). From a supply perspective the cumulative lead-time (Blackstone, 2008) is critical since it represents the lead-time of the product including all items. The cumulative lead-time is then split in two parts representing items provided by the logical entity in focus (internal lead-time) and items provided by the upstream logical entities (external lead-time). Another key aspect is the level of possible customization involved and this is here referred to as adapt from a supply perspective. Customization is based on customer requirements and adapt may hence also be seen as based on customer requirements from a demand perspective. Finally, a key aspect from a customer perspective is also the requested delivery lead-time as it also frequently defines a key characteristic of the order winners.

Five strategic lead-times (referred to as SEIAD based on their variable names) are investigated here and further elaborated on below:

- Supply lead-time (*S*) is the cumulative lead-time for the product through the whole (extended) logical entity.
- External lead-time (*E*) can in many cases be seen as related to the (purchased) component and is therefore usually associated with purchase orders.
- Internal lead-time (*I*) corresponds to the controllable part of the bill-of-material and is usually associated with the own provisioning lead-time.
- Adapt lead-time (*A*) is related to the customer order but based on when the supply performed is actually customer order unique. The possible level of customization is related to a supply perspective (A_S) and the requested level of customization is related to a demand perspective (A_D). This lead-time could also be referred to as the customization lead-time but to avoid confusion with different types of customization strategies the more neutral term adapt is used. The adapt lead-time is important in e.g. the context of mass customization but it is also a key component in other cases such as e.g. postponement strategies.
- Delivery lead-time (*D*) is based on market requirements for delivery and directly associated with demand/customer order.

5.2. *Example: lead-times and time-phasing*

The different types of lead-times defined above are related to each other and a simple example is introduced as an illustration of the theory. Some of the concepts introduced here have also been applied on actual cases (Bäckstrand et al., 2013) but in general, actual bill-of-material that can be used as an illustration of several concepts are rather complex. The sheer complexity of the bill-of-material in these cases and their area of application diverge the focus from the illustration of the concepts. The example introduced below is therefore fictitious and designed to be simple but still sufficient to illustrate the key concepts used here. The example is shown in Figure 10 in terms of a traditional material based bill-of-material (product structure) to the left and a time-phased bill-of-material to the right (see e.g. Bäckstrand & Wikner, 2013; Clark, 1979; Wikner & Rudberg, 2005b). The material based bill-of-material includes where-used relations and lead-times. For example, item *Y* consists of item *X* and item *U*, and item *Y* has a lead-time (L_Y) of 3 periods. On the right in Figure 10 this bill-of-material is instead represented by a time-phased bill-of-material where the horizontal distance between two filled circles corresponds to each item's lead-time (*L*). In this case the supply lead-time $S = 12$ periods (the cumulative lead-time for the bill-of-material) and

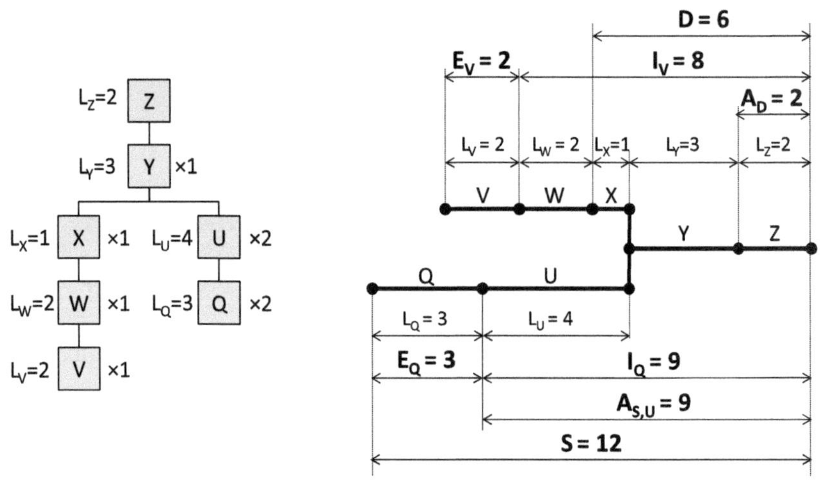

Figure 10. Example: Material-based and lead-time-based bill-of-materials.

the delivery lead-time to the customer is given as $D=6$ periods. (The estimation of the delivery lead-time D will be more elaborated on later, see e.g. Figure 18). Since it is the adaptation of Z that constitutes a customer order unique solution, the demand adapts lead-time $A_D=2$ periods and Y is customer unique. From an engineering and production perspective it would be possible to also make item U customer order unique which means that the supply adapt lead-time $A_{S,U}=9$ periods. The items V and Q, finally, are purchased in a traditional way from suppliers. These two items represent the end of two branches with an external lead-time (E). An internal lead-time (I) can then be calculated for each branch of the bill-of-material. The set Leafs is assumed to contain all N items that are at the lowest level of each branch and would in many cases correspond to purchase items. In the example of Figure 10 it would mean that $N=2$ and $\text{Leaf}=\{V,Q\}$. For the branch of item V the cumulative lead-time is $S_V=E_V+I_V=2+8$ and for item Q the cumulative lead-time is $S_Q=E_Q+I_Q=3+9$. The branch with the longest cumulative lead-time is also equal to the supply lead-time $S=\max_{n\in\text{Leaf}}\{S_n=E_n+I_n\}$. A more elaborate analysis can also be performed where a S is defined for each item in the bill-of-material (Bäckstrand & Wikner, 2013). This approach provides the opportunity to identify for each item if it is within D and can be customer-order-driven, or if it is longer than D and must be forecast-driven.

5.3. Lead-time context

If only one branch of the product structure in Figure 10 is considered, a network without different branches is obtained and this can be labelled as a 'linear' chain. Using the terminology introduced above the chain may be referred to as consisting of two logical entities, one focal entity ('focal' refers to the unit that is in focus for the analysis) and one supplying entity, i.e. two logical entities in sequence and with an implicit entity represented by a customer. The context of this analysis is therefore a triad consisting of a customer entity and two logical entities representing the supply. Each one of the logical entities can be illustrated with the entity's strategic lead-times SEIAD resulting in

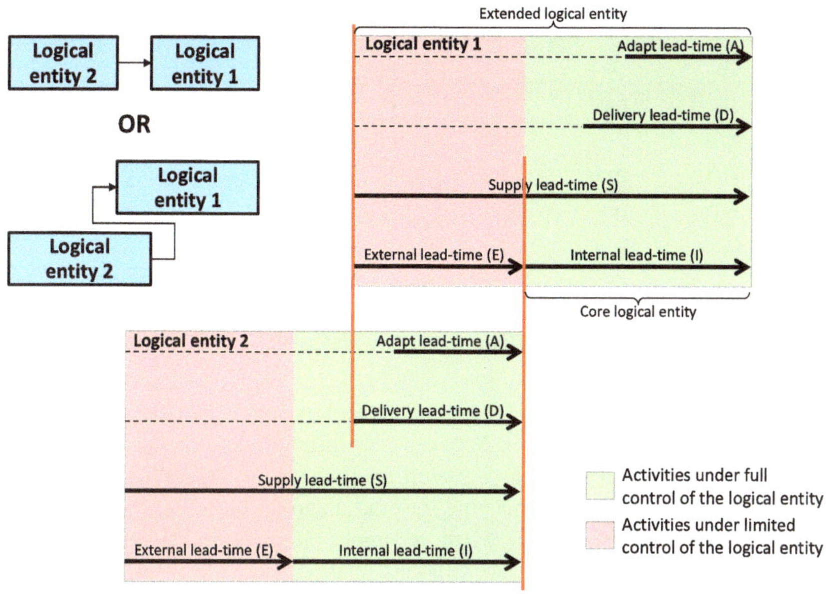

Figure 11. Dyad of two logical entities with no integration.

Figure 11. Logical entity 1 would then correspond to all activities required to supply with item *Z* in the example of Figure 10, i.e. the focal entity. Logical entity 2 is related to the supply of item *V* or *Q* (since $E+I=S$ in logical entity 1). These items may in turn consist of other items but that is not shown in the example but covered by other structures within the supplying logical entity (logical entity 2). The green part on the right of each logical entity represents the part of the flow that is controllable for the logical entity and the red part on the left represents the part that is not controllable for the logical entity. As shown in Figure 11 there is an overlap between the entities but in the overlap only one logical entity has the ability to control the activities (logical entity 2 in the overlap in Figure 11). This assumption is relaxed later in section 6.3.2. This type of overlap is illustrated in Figure 11 on the top left of the figure where two logical entities are in sequence with some overlap. The top entity-dyad illustrates in this way the core logical entities (green segment) while the lower entity-dyad illustrates the extended logical entity (which corresponds what is referred as a logical entity in the text unless otherwise stated).

The external lead-time (*E*) for logical entity 1 corresponds to the lead-time for a purchase order, e.g. for item V or item Q in Figure 10. From the supplying entity's perspective (logical entity 2) it corresponds to the delivery lead-time (*D*) for a customer order. Hence, this is also an illustration of how a purchase order of the focal company is connected to a customer order at the supplier.

The strategic lead-times thus represent a link between different logical entities as exemplified by *E* for logical entity 1 and *D* for logical entity 2. Within each logical entity a number of important lead-time relations can be identified. In particular three lead-time relations are of key important to identify decision categories for COBM and related strategic choices in terms of positioning of decoupling points and decoupling zones.

Table 2. Lead-time relations.

	S	E	I	A	D
S	-	n/a	×	×	×
E	n/a	-	n/a	n/a	n/a
I	***I:S***	n/a	-	×	×
A	A:S	n/a	A:I	-	***A:D***
D	***D:S***	n/a	D:I	×	-

5.4. Lead-time relations

Five different strategic lead-times were identified above and referred to as SEIAD. In total 20 different lead-time relations, between two different lead-times, can be identified based on SEIAD as shown in Table 2 (based on Bäckstrand, 2012) (excluding the relation between lead-times of the same type, i.e. the diagonal of Table 2). There are some overlaps between these lead-times when a supply network is investigated since the delivery lead-time *D* for a supplier corresponds to the external lead-time *E* for that item at the customer. Note, however, that the analysis performed here is based on one logical entity. Since the focus here is on COBM the emphasis is on customer facing lead-times (see Figure 11), i.e. *A*, *D*, *S*, and *I* (lead-time relations related to *E* are therefore indicated with *n/a* in Table 2 and since $S=E+I$, the *E*-based relations may be derived based on *I*-based relations if necessary). The lead-time relations above the diagonal are the inverse of the lead-time relations below the diagonal and only one relation of each 'pair' is included. As a general rule the relation where we normally would expect the smaller value in the numerator is included and the other is indicated by a '×' in Table 2. In total there are hence six potential lead-time relations left to investigate: *D*:*S*, *A*:*D*, *I*:*S*, *A*:*I*, *A*:*S*, and *D*:*I*.

In a lead-time relation one lead-time is used as a point of reference and the relation then shows how long the other lead-time is in relation to the reference lead-time. The lead-time relations further investigated are mainly related to *D* since the delivery lead-time *D* is the fundamental lead-time from customers' perspective. Note that the *I*:*S* relation and the *I*:*A*-relation are multiple relations since there are *N* pieces of *I* lead-times (one for each branch of the bill-of-material) and hence also *N* pieces of *I*:*S* relations. In case of the *A*:*I* relations it is even more complex since there may also be multiple *A* lead-times, and in addition *A* may be supply based or demand based. The *A*:*I* relation could be of some interest as it can be used to analyse if customization affects more than the focal entity. This information is, however, implicitly provided by the *I*:*S*-relation in combination with the *A*:*D*-relation when the compounded decision domains are introduced. In a similar fashion the *A*:*S* relation represents how customization affects supply but it is less exact compared to the *A*:*I*-relation and is hence not included. The *D*:*I* relation, finally, could provide some interesting information on if customer-order-driven flow impacts suppliers but this information can also be obtained from compounded decision domains which are described later. The three remaining lead-time relations are of key importance to identify important decision categories for COBM and they are highlighted in Table 2 as bold and italic. As a result there are three lead-time relations that connects the two aspects of supply (*I* and *S*), the two aspects of demand (*A* and *D*, in particular if A_D is assumed), and finally the key link between supply and demand (*D* and *S*).

5.4.1. D:S-relation

The fundamental lead-time relation is based on the delivery lead-time (*D*) in relation to the supply lead-time (*S*). If the customers can accept to wait longer than it takes the supplier to provide the product it becomes possible to perform all provisioning activities at the request of a customer. If, on the other hand, the customer cannot accept to wait the time it takes to perform the activities at least some of these activities must be performed on speculation about future customer orders. This lead-time relation has received relatively large attention in both practical application and literature (Shingō, 1989) (originally published in Japanese in 1981) and is usually seen as the earliest reference on this even if Shingō used the denotation 'relation of *D:P*'. This relation was introduced to a broader audience by Mather (1984) who observed the potential of this concept. Mather did however note that *D:P* could be mixed up with DP, which at that time was widely used as an acronym for 'data processing'. To reduce the risk for this confusion he suggested that it would be better to call it the *P:D* relation (or *P:D* ratio as Mather called it). Unfortunately the numerical result of the *P:D* relation does not provide an intuitive sense of how large part of *P* that is customer-order-driven. Due to this the original relation *D:P*, as defined by Shingō, is used here as a point of departure.

Initially (Shingō, 1989) referred to *P* as the product lead-time but it has also been associated with production lead-time. Today supply is used as a terminology for all types of activities related to the provisioning of goods and hence *S* is here used instead of *P*. Both Shingō (1989) and Mather (1984) used *D* as an abbreviation for delivery lead-time and this is also used here. Note that *D* implies that activities within *D* may be performed to customer order. It is, however, not a requirement but should be seen as an 'option' to provide to customer order. If the products are standardized and delivered frequently the activities may also be performed on speculation with reasonable risk taking. This option is sometimes not fully used by for instance production engineering reasons where it is suitable to produce a quantity that deviates from an individual customer order. Some standard examples of the *D:S*-relation are:

- $D << S$: Basically all activities must be performed on speculation (corresponds to the strategy Make-to-Stock, MTS).
- $D < S$: Some activities are performed on speculation (corresponds to the strategy Assemble-to-Order, ATO).
- $D \approx S$: All activities can be performed on commitment (corresponds to the strategy Make-to-Order, MTO, Purchase-and-Make-to-Order, PMTO, or Engineer-to-Order, ETO, depending on if engineering activities are included or not).

5.4.2. A:D-relation

The relation between adapt lead-time (*A*) and delivery lead-time (*D*) is relating two aspects of demand to each other. The distinction made here is based on how large part of *D* that is actually related to a part of the lead-time that is customer order unique, *A*. This relation highlights that even if the flow is customer-order-driven it is not necessarily also tailor made for that particular customer. The *D:S* relation shows what is customer-order-driven and therefore also that there is an 'option' for providing something customer order unique. A more detailed approach could also, as mentioned earlier, differentiate if *A* is what is actually possible from a supply perspective (A_S) or if *A* is related to what is required by the customer (A_D) (Bäckstrand & Wikner, 2013). From a supply perspective the different branches of the bill-of-material may provide different opportunities for customization and hence A_s should also be defined for each and every branch of the bill-of-material. In summary A_D represent the market requirements and A_s the opportunities available from a supply perspective. Hence the difference corresponds to the options available for responding to new market requirements due to e.g. increasing competition. In this context A is however assumed to represent the customers' requirements. The A:D relation shows to what extent the provider has decided to exercise the 'option' to provide something customer unique:

- $A << D$: Basically all activities are standardized.
- $A = D$: Customer-order-driven activities are also customized.
- $A > D$: Some forecast-driven activities are customer order adapted (a high risk strategy that usually should be avoided).

The connection between customer adaptation and customer-order-driven flow has many facets covering e.g. that it is not necessary for a product to first be completely engineered and then produced. It may also involve, for example, engineering adaptations where some semi-finished goods are in stock when engineering adaptations are made. This can be described as a two dimensional problem where engineering and production activities can be combined in different ways (Wikner & Rudberg, 2005a).

5.4.3. I:S-relation

In contrast to the D:S and A:D relation, which are based on demand, the I:S relation is completely based on the supply side. The internal lead-time (I) represents the lead-time performed within control of the focal logical entity and the supply lead-time (S) represents the cumulative lead-time of the system in question. In the case of traditional relations to the suppliers, where integration is low and purchase order and customer order are the foundation for cooperation, this means that if I is much smaller than S, basically all activities within S are performed by suppliers upstream in the flow. But, the supplier plays a less important role in the supply chain in case the total lead-time S is short and I therefore relatively long. Also here three cases are of particular interest and note that in this case there is a lead-time relation for each 'branch', cf. Figure 10:

- $I_k = S_k$: Basically all activities of the k-branch are performed by the core logical system.
- $I_k < S_k$: Some activities of the k-branch are outsourced to a supplier
- $I_k << S_k$: Basically all activities of the k-branch are outsourced to a supplier (the extended logical entity).

Based on the three lead-time relations defined above it is possible to define and position three decoupling points and three process-based decoupling zones that constitute three decision categories and thus also the core of the flow based framework for COBM.

5.5. Example: Lead-time relations

Going back to the example of Figure 10 a set of lead-time relations can be calculated and used to classify both the whole product and the individual items. In this case the

Table 3. Example: Item-based *D*:*S* relations.

Item (k)	Make/buy	L_k	S_k	D:S_k
Z	Make	2	2	6/2 = 3
Y	Make	3	5	1.2
X	Make	1	6	1
W	Make	2	8	0.75
V	Buy	2	10	0.6
U	Make	3	9	0.67
Q	Buy	4	12	0.5

Table 4. Example: classification using item-based D:S_k relations

Category	Item
Requires speculation	W, V, U, Q
Possible to speculate on	X
Not suitable for speculation	Y, Z

D:*S* relation is analysed and in particular the item based *D*:*S* relations are analysed. The indented bill-of-material of Table 3, based on Figure 10, includes the item based *D*:*S* relations in the rightmost column. S_k is calculated for each item based on the cumulative lead-time for that particular item. For item X the cumulative lead-time $S_X = 2 + 3 + 1 = 6$ periods which is the same as the delivery lead-time according to Figure 10 (D=6). Consequently the item based D:S_k relation is one which means that item X can be produced based on customer order.

According to the introduction of the example, item *Z* is customer order unique and *Y* is customer unique. Both of these items are within the delivery lead-time which means that they can be produced based on customer order. Since they are unique for the customer they are not suitable for speculation even if recurring customer orders would make it possible to speculate on *Y* since it is unique for the customer and is hence the same for different customer orders from the customer, The items *W*, *V*, *U*, and *Q* all have a D:S_k relation < 1 and must be made to forecast and thus requires speculation. Item *X* finally is interesting since it has a D:S_k relations $= 1$ indicating that is should be customer order driven but at the same time it is a standard item used for all types of customer. In this case the D:S_k relation should be interpreted as a standard item with an option to produce either to customer order or to forecast, i.e. on speculation. This discussion is summarized in Table 4.

6. Customer-order based management

The different lead-time relations introduced above each represents a way of dividing the flow into two parts with different preconditions. The interface between these two types of flow involves a discontinuity along the flow in the sense that the preconditions change significantly when the flow crosses that point, which also corresponds to the interface between two decision domains. Each one of the lead-time relations identified as important in Table 2 corresponds to such an interface and is of critical importance as decision support for COBM. The three decision categories covered by this analysis are flow driving, flow differentiation and flow delimitation. In addition the DC flow transparency, with a slightly different approach, is introduced below.

6.1. Decision category: flow driving

Flow driving refers to what it is that decides if a process should be initiated or not. In the case of *flow driving*, it is assumed that a high *level of certainty* of the driver is defined from a customer perspective. The decoupling point that separates certain from uncertain flow driving is referred to as the CODP (see e.g. Giesberts & van der Tang, 1992) and based on the *D*:*S*-relation, see Table 2. Downstream from the CODP demand is certain in terms of a 'perfect' customer order where what (in terms of form/place) the customer wants and when the customer wants delivery is determined. The hybrid decision domain here represents that there is some knowledge of what will be demanded but that some information is missing for complete certainty. This domain corresponds to the concept customer order decoupling zone (CODZ), introduced by Wikner & Rudberg (2005b). The starting point of the flow in Figure 12 is Source, which represents what is beyond the extended logical entity, and Sink, which is the end point (usually also corresponding to a customer) of the logical entity.

6.1.1. Customer order decoupling point

The interest in the CODP gained ground when different so called hybrid strategies (see e.g. Sun, Ji, Sun, & Wang, 2008) came into focus. The pure make-to-stock (MTS) and make-to-order (MTO) strategies were employed in different companies, but as customer requested short delivery lead-times in combination with customization the assemble-to-order (ATO) strategy was highlighted (Wemmerlöv, 1984). The ATO strategy can be described as a combination of MTS with MTO where MTS is applied upstream from the CODP and MTO downstream. In this context, the CODP is usually seen as identical to a physical stock point. This material-based approach to decoupling points has a long tradition in the literature. There are a number of 'local' publications in e.g. industry magazines in different countries but the two publications usually quoted as the origins of this concept are Hoekstra & Romme (1992), originally published in 1985 in Dutch and Sharman (1984). Sharman referred to the concept as order penetration point (OPP) since it is a measure of how far deep into the supply process (from a lead-time perspective) that the actual customer order penetrates. The CODP has been used in many contexts and is also known as e.g. decoupling point (Hoekstra & Romme, 1992), customer order point (COP) (Mason-Jones & Towill, 1999; Olhager & Östlund, 1990), OPP (Sharman, 1984), supply stream decision point (Hines & Rich, 1997), material

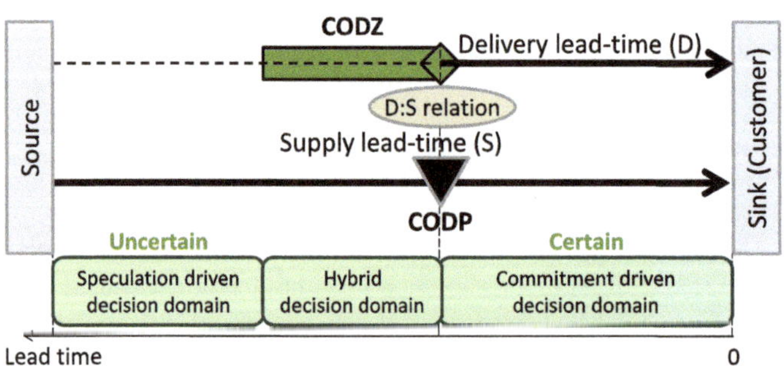

Figure 12. DC for customer-order based management – flow driving.

decoupling point (Mason-Jones & Towill, 1999) and push-pull boundary (Chopra & Meindl, 2004). In the postponement literature, the CODP corresponds to time postponement relating to the time when the customer order actually drives the activities (Pagh & Cooper, 1998). This explicit focus on the physical stock point (see e.g. Banerjee et al., 2011; Sun et al., 2008) as the CODP is the dominant perspective in the literature but here the CODP is first and foremost seen as something related to decision-making in line with Wikner and Rudberg (2005b) and the CODP is therefore defined as:

> The *customer order decoupling point* (*CODP*) separates decisions about initiating flow based on speculation on future customer orders from commitment against actual customer orders.

The CODP is based on the *D*:*S*-relation, where *D* indicates the part of the flow where the (future) customer's demand is known. In Figure 12, the decision-based aspect of the CODP is indicated with a diamond (quadrangle). The stock point (black triangle in Figure 12) corresponding to the CODP constitutes a buffer between the speculation-driven decision domain and the commitment-driven decision domain. The CODP buffer is thus a safety stock for uncertainty in demand. The material in the CODP buffer is replenished based on speculation (forecast), i.e. an estimation of future demand for materials in the stock point has been done. The commitment-driven withdrawals represent real demand and the dimensioning of the CODP buffer is therefore based on the deviation between forecasted withdrawals and actual withdrawals which is in line with how safety stocks usually are calculated. It is however important to observe that the 'real' demand used in dimensioning should reflect the demand that the commitment-driven process can deliver. If there is capacity to handle variations in demand downstream, the pressure on the material availability in the CODP buffer increases, but if the capacity downstream is not that flexible it is also a constraint for how much the withdrawals can vary. The CODP hence play a critical role in order promising and this is further discussed in e.g. Fleischmann and Meyr (2004); Rudberg and Wikner (2004). Dimensioning of the CODP buffer should therefore be based on the flexibility that the flow downstream from the CODP can offer. Product mix variations can, however, increase demand on the buffer even if the volume is stable. Another interpretation of this scenario is that flow upstream from the CODP is goods based but downstream also can involve services (Fließ & Kleinaltenkamp, 2004). In this context, the CODP has also been called the service decoupling point (Wikner, 2012b).

6.1.2. Customer order decoupling zone

The CODZ involves a mix of uncertainty and certainty about customer orders' timing. Downstream from CODZ the flow is driven by customer orders and upstream from CODZ the flow is driven by a market-based forecast. Within the CODZ, the flow driver is uncertain to some extent depending on either partial customer order information (process-based CODZ) or aggregate capacity requirements for resources where some of the materials are forecast-driven and some are customer-order-driven (resource-based CODZ).

> The *customer order decoupling zone* (*CODZ*) covers decisions about initiating flow based on a compromise of speculation and commitment on customer orders.

For process-based CODZ, the certainty increases along the flow (Wikner & Rudberg, 2005b) and the level of certainty can then be defined by two dimensions based on igure 6:

- Spatial dimension (what?): On a scale from uncertain to certain about what the customer wants (in terms of form and place).
- Temporal dimension (when?): On a scale from uncertainty to certainty of when the customer wants delivery.

Resource-based CODZ is based on that the level of certainty can vary depending on to what extent resources are involved in a combination of speculation-driven and commitment-driven decision domains. This type of CODZ has been noticed in different contexts such as the assortment hybridity (Giesberts & Tang, 1992). The mixed load on resources can be created both as a consequence of that customers with different demand in terms of *D* purchase the same product and that different products are loading the same resource but with different flow drivers. If the customers request different delivery lead-times, a backlog profile can be created where the resource-based CODZ represents the time horizon where some customer orders are known, but not all, and hence the resources performing the corresponding activities operate under mixed load. This could also be referred to as customer-based CODZ. Correspondingly, different products may create a mixed load on resources which means that the aggregate could be referred to as product-based CODZ. But, for simplicity, both these types of mixed load are associated with the concept resource-based CODZ. In both these cases, there is a challenge in terms of how to schedule the flow on the resources. For example, the forecast-driven flow can generate a levelled first loading and then being complemented with customer-order-driven load, or the customer-order-driven flow may be prioritized and first loaded on the resources and thereafter the forecast-driven flow can be used to level out the load on the resources.

6.2. Decision category: flow differentiation

Flow driving concerns the driver of the process but provides no information about the level of customization involved, which is a separate issue (see e.g. García-Dastugue & Lambert, 2007; Hoekstra & Romme, 1992; Mason-Jones, Naylor, & Towill, 2000; Olhager & Östlund, 1990; Wikner & Wong, 2007). For decisions about customization, the point of departure is *flow differentiation*, which means the *level of uniqueness* in terms of form and place. A completely unique process is adapted for a specific customer order and is limited, from a flow perspective, by the customer adaptation decoupling point (CADP) related to the *A*:*D*-relation. Generic processes are upstream from CADP and provide standard products and therefore lack connection to specific customers in terms of form and place at this stage. The customer adaptation decoupling zone (CADZ) corresponds to some degree of adaptation, such as when the flow is customer unique, i.e. the product is customer unique, but not customer order unique, in terms of design (Wikner & Bäckstrand, 2012). The CADZ may hence extend upstream from the CODZ since speculation may be performed even on customer unique materials if the customer orders are recurring. A more comprehensive discussion on these issues, from a customization aspect, can be found in Graca et al. (1999). These concepts are illustrated in Figure 13.

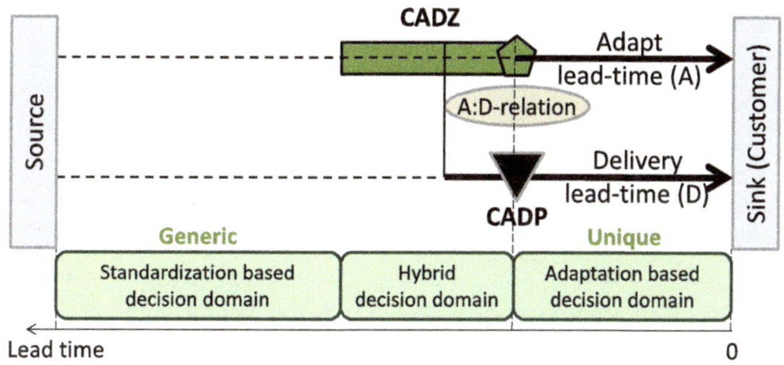

Figure 13. DC for customer-order based management – flow differentiation.

6.2.1. *Customer adaptation decoupling point*

The CADP is based on the *A*:*D* relation (Bäckstrand & Wikner, 2013; Wikner, 2012a) and represents the interface between flow of standard products and customer order adapted products (indicated with a pentagon in Figure 13). As mentioned above, the *A*:*D*-relation can be defined from a demand perspective or a supply perspective. When the decoupling point and decoupling zone for customer adaptation are defined, the demand perspective is assumed but in most cases the discussion is equally valid for the supply perspective.

> The *customer adaptation decoupling point* (*CADP*) separates decisions about differentiating flow based on standardization for a market of different customers from adaptation against actual customer orders.

The requirement for a stock point positioned at the CADP is in this case not as obvious as in the case of the CODP. It is, however, important to note that upstream from CADP there is recurring need for the same product which enables some speculation (it is here assumed that CADP is positioned downstream from CODP since it otherwise would result in speculation on customer order unique products, which is not recommended). If it is possible to perform standard activities between the CODP and the CADP to customer order there might be a lot-sizing buffer of standard products positioned at the CADP as a stock point. This can be useful when it is preferable to supply with larger quantities (due to e.g. long set-up time) than the customer order that initiated the batch requires (a lot-sizing method other than lot-for-lot is used). This type of lot-sizing inventory replenishement can also be created in other places upstream from CADP but CADP indicates the last point in the flow where this type of inventory should occur.

6.2.2. *Customer adaptation decoupling zone*

The CADZ (Wikner & Bäckstrand, 2012) is a hybrid between what is generic from a customer perspective and what is unique from a customer order perspective. This compromise means that the flow is customer unique but not unique for a single customer order. This distinction between what is customer unique and what is customer order unique is in many cases of decisive importance when it is combined with the decision to speculate since it can be a reasonable decision for customer unique products but basically not for customer order unique products. In general, the CADZ is based on the *A*:*D*-relation and can be defined as:

> The *customer adaptation decoupling zone* (CADZ) covers decisions about differentiating flow based on a compromise of standardization and adaptation for customer orders.

As indicated in the general decoupling theory, the decoupling zones can be of two types: process-based or resource-based. Adaptation of a product can be performed in many different aspects related to both form properties of product, including the service performed, and the actual location, place, of the product. For process-based CADZ, two key dimensions for flow differentiation can hence be identified in line with Figure 6:

- Form dimension (product): On a scale from generic form (standard product) to unique form (customer order unique product).
- Place dimension (location): On a scale from generic place (e.g. central distribution site) to a unique place (e.g. the customer's site or a specific delivery point/site).

Within the process-based CADZ the product is not completely standard but has some characteristics that only is of interest to some customers which means that the product can be adapted in the meaning customer unique but not customer order unique (Wikner & Bäckstrand, 2012). The CADZ hence provides a more detailed view of customization where one level of customization is product related in the sense that it is unique for a particular customer, whereas customization for a particular customer order is unique for that particular customer order. In some cases, only the form dimension is applied since it could be argued that place is only unique once the product is in the hands of the customer. Unless form and place uniqueness are handled separately, the form uniqueness may then be dominated by the place uniqueness.

For resource-based CADZ, it means that one resource is involved in multiple processes that have different characters in terms of differentiation. This can create a number of challenges from a capability perspective since standard products usually have cost as an order winner, and customer order adapted products usually have order winners related to delivery speed or delivery precision. This of course also affects the loading challenge indirectly since products with focus on delivery precision have other requirements on the process compared to products of standard character which can be kept in inventory.

6.2.3. Product differentiation

Customer adaptation (or customization) as a concept is part of the wider concept of product differentiation since the latter by definition is not explicitly connected to customer demand but covers products in general, including customer generic products. Also for the flow upstream from the CODZ, a number of important decisions may be made related to what in the TOC (see e.g. Blackstone, 2008) are known as control points related to diverging flows which means that one particular material or component may be used for different purposes due to supply-based differentiation. A set of standard components can be assembled to a wide range of products based on forecast and available inventory for each product. Hence, the decision to use the components for a particular product is in this case also resulting in differentiation since different variants could be created by the standard components (see e.g. Hoekstra & Romme, 1992). Note however that the differentiation in this case is based on supply aspects and not related to a particular customer. The subject can therefore be divided into demand-based and supply-based differentiation:

Demand-based differentiation is based on customer requirements as above, which means that the product (the result from a process) in some sense is customer adapted. The customer unique flow is related to the CADZ in Figure 13 and the customer order unique flow is related to the flow downstream from the CADP.

- Customer adaptation type A: Customer order unique product: Properties are related to requirements that are unique for a specific customer order of one-time character, wherefore the product is difficult, and maybe even impossible, to sell at a later time.
- Customer adaptation type B: Customer unique product: Properties that are related to requirements that are unique for a particular customer but covered by recurring customer orders. This type of products provide some opportunity for speculation since it is a recurring customer.

Supply-based differentiation is independent of specific customer requirements and is therefore only related to customer generic product (standard products). In Figure 13, this corresponds to the flow upstream from the CADZ but may in some instances also be an additional cause of differentiation in the CADZ due to the design of the product in terms of the process-based CADZ or due to mixing different flows in terms of resource-based CADZ. Note that CADP_S represents potential customer-based differentiation, whereas supply-based differentiation is independent of customer requirements. Supply-based differentiation is therefore an important concept, but not in focus for COBM as outlined here.

- Standard products can be sold to a market consisting of multiple customers. Product differentiation in this case is related to that material changes in terms of form or place to be part of different types of customer generic products.

6.3. Decision category: flow delimitation

The third flow-based DC complements the other two in the sense that flow driving is related to why and when a flow is initiated and flow differentiation is related to decisions about the uniqueness of the flow. The remaining issue, according to the analysis of Table 2, is how the flow is managed and this DC is here referred to as *flow delimitation*. This DC is based on the extension of the logical entities and is related to the $I{:}S$-relation and the *level of finiteness* that can be applied in managing the flow. The decoupling point is defined from a control perspective but is despite this referred to as the purchase order decoupling point (PODP), (Wikner et al., 2009). Since the PODP is related to the $I{:}S$-relation it is important to remember that since there might be multiple $I{:}S$-relations, there might correspondingly also be multiple PODPs related to one product (see e.g. Figure 10 with I_V and I_Q). The background to this name is that what is referred to as 'external' and 'uncontrollable' frequently is also related to suppliers and hence related to purchase orders (note that this is however not necessary). The analogy with CODP and customer order also supports the naming of the decoupling point in the framework. Somewhat simplified, processes performed by the focal company (same as focal logical entity in this case) can be considered as controllable as only own resources are involved. Processes performed by suppliers can in a corresponding fashion be considered as uncontrollable. In some cases, the customer can control parts of the supplier's resources (for example, when capacity is 'purchased') or depending on divided responsibility so that all the own resources can be controlled as one entity. Correspondingly, a supplier may have influence on some of the customers flow through vendor-managed inventory (VMI), which would move the PODP downstream compared to the situation before VMI was implemented. Wikner and Bäckstrand (2011) provide an overview of different configurations. Since controllability cannot unambiguously be related to customer/supplier interface the controllable decision domain is referred to as 'Internal' and the uncontrollable decision domain as 'External'.

Also in this case, a hybrid decision domain can be identified. Supply of material and capacity can vary between finite and infinite. If material and capacity can be analysed as finite resources there is full controllability but this is not always possible to achieve. The hybrid decision domain, where resources to some extent can be managed as finite, are in Figure 14 represented by the purchase order decoupling zone (PODZ).

6.3.1. Purchase order decoupling point

The PODP is indicated with a hexagon in Figure 14 and is based on the *I*:*S*-relation. The interface between the decision domains is in this case based on the controllability of the process which means that the PODP can be defined in line with CODP and CADP:

> The *purchase order decoupling point* (*PODP*) separates decisions about delimiting flow based on what is external to the logical entity from what is internal and hence controllable.

The PODP is based on the *I*:*S* relation and the interface is in this case based on the flow delimitation and therefore also the controllability. Upstream from PODP is another logical entity that 'owns' the controllability but downstream the flow is controllable from the focal logical entity's perspective. Since the flow upstream from the PODP is not controllable for the focal entity, it can also involve some uncertainty considering the supply to the stock point indicated in Figure 14 at the PODP. To handle this uncertainty, either safety lead-time is used, which results in some stock since materials are delivered

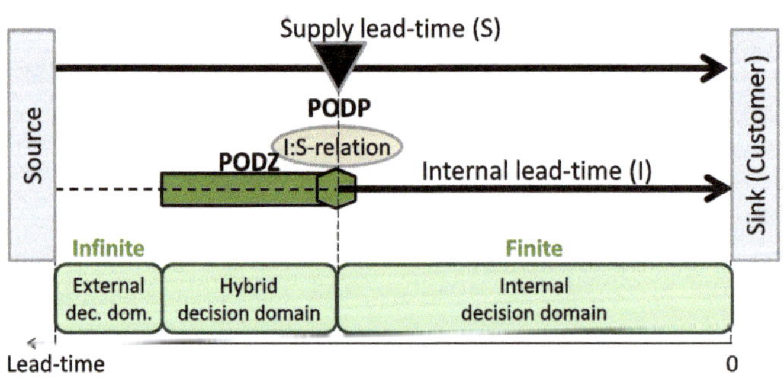

Figure 14. DC for customer-order based management – flow delimitation.

before there is a requirement, or a traditional safety stock is used with a physical quantity. In combination with CODP, the PODP creates different preconditions for supplier interaction (Wikner & Bäckstrand, 2011).

6.3.2. Purchase order decoupling zone

Planning and control of materials flow has traditionally been divided into planning and control for manufacturing and for materials supply (basically purchased materials). This delimitation along the flow is becoming less significant and focus is more and more on creating a holistic and integrated management approach to the whole flow independent of, if it covers one department or multiple integrated actors in a supply network. In this context, the logical entities play a crucial role since they are defined from a management and control perspective. The activities that are included in the core logical entity, see Figure 11, are controllable and finite principles considering resource constraints can be applied. The upstream activities are considered as non-controllable which means that infinite control principles can be applied where a lead-time is given but the availability of resources in terms of materials and capacity is unknown to the core logical entity. The delimitation between these domains is not always that obvious but different degrees of finite can exist, which is also what defines the PODZ.

> The *purchase order decoupling zone* (*PODZ*) covers decisions about delimiting flow based on a compromise on what is external and what is internal to the logical entity.

Two key dimensions for flow control can be identified and they are applied in many contexts related to planning and control. For example, in MRPII, materials are planned using a time-phased approach and then a capacity evaluation is performed based on these plans (Wight, 1984). In rate-based approaches, capacity is usually planned first through e.g. line balancing and then the product mix and required materials are scheduled as needed and replenished using pull systems (see e.g. Duggan, 2012). At an overall level, most techniques for planning and control are emphasizing these two dimensions:

- Material dimension (transformed resources): On a scale from infinite to finite materials and with focus on availability of materials at stock points.
- Capacity dimension (transforming resources): On a scale from infinite to finite capacity and with focus on availability of capacity at resources.

For CODZ and CADZ, it is obvious that there is a process-based and a resource-based variant of the decoupling zone. For PODZ, it is however not as intuitive to identify these two perspectives since resources play such a key part in this DC on controllability. But, also in this case the two types of decupling zones can be identified.

Process-based PODZ represents a scenario where the possibility to apply finite techniques gradually increases along the flow. In general, it is assumed that the more downstream the more information about the resources is available to the logical entity. It is, for example, possible to apply finite scheduling in the own production (at least in theory, in practice all information might not be available due to e.g. technical reasons) but in the parts that involve suppliers, the information about availability might be limited and the only given information might be the planned lead-times. For the receiving logical entity, it means that the supplier is upstream from PODP and hence classified as infinite. It could however be that the availability of some of the capacity at the supplier is known and hence it would be possible to consider as finite, which would correspond to being inside a process-based PODZ, where some of the capacity can be finite scheduled or that the capacity can be finite scheduled to some degree.

The resource-based PODZ is an aggregate of multiple flows that consumes the same resource. In most cases, all load would be treated in the same way meaning for instance that if a resource is scheduled using finite capacity all load would have to compete for this capacity using some kind of priority as a base line but when the maximum capacity is reached more capacity is only added as an exception. An example of this scenario is when different customers have different priority. Normally, the resources are finite scheduled but in some cases fast delivery is highest priority and by using, e.g. overtime the extra capacity can be provided to secure delivery reliability. In this way, the resource is managed as finite for some flows/customers and as infinite for other flows/customers.

6.4. Decision domains for customer-order-based management

With flow-based decision-making as a point of departure, three different decision categories have been identified above with support from analysis of lead-time relations. Three different sets of decision domains have been identified together with three decoupling points and related decoupling zones. The four strategic lead-times *S*, *I*, *A*, and *D* have been used to position the decoupling points and therefore also decoupling zones from a flow perspective. To be able to decide the starting point of the zones require further information which either can be based on the first decision domain in the flow or simply the length of the decoupling zones, which is included in Figure 15 as CODZ lead-time, CADZ lead-time or PODZ lead-time. From a planning and control perspective, it can also be observed that the supply lead-time *S* corresponds to the planning time fence (PTF) if master scheduling (Blackstone, 2008) is applied at the level of the delivery object and defines how far into the future that flow is a constraint according to planned lead-times. Note however that sometimes a master schedule is used at lower levels in the flow such as at the CODP, and in this case the PTF would cover the time

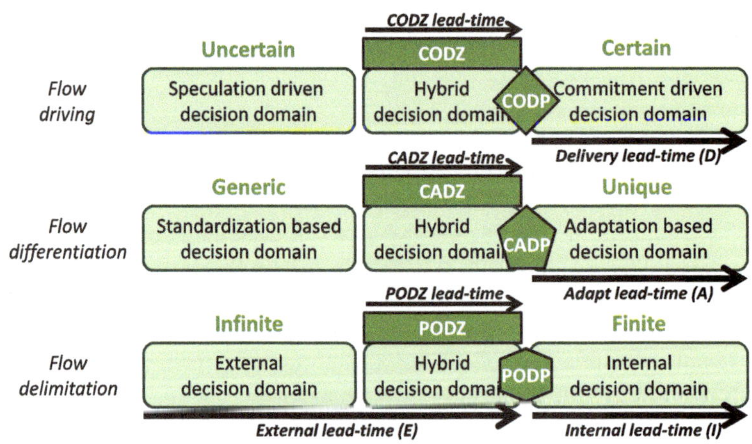

Figure 15. Decision domains for three decoupling points and three decoupling zones.

span from the CODP to the beginning of the flow, i.e. the PTF would at the CODP be set to PTF = *S–D*. If there are several branches in the product structure, such as in the example of Figure 10, PTF is related to the longest branch in terms of lead-time. PTF is also referred to as frozen time fence, since firm planned order is used within PTF. Correspondingly, the demand time fence (DTF) that also can be used for planning and control of delivery objects (Blackstone, 2008) is associated with the CODP since DTF indicates the part of the flow that, from a lead-time perspective, is customer-order-driven with no respect to forecast. DTF is also used in master scheduling. All deliveries within DTF must be related to customer order to make it possible for all customer-order-driven activities to be performed within the delivery lead-time. In case master scheduling is applied at the CODP, the DTF would be positioned at time zero since the master schedule then would be applied in a MTS-scenario.

The three decoupling points and the three decoupling zones can therefore be summarized as:

- CODP: *D*:*S* relation, safety stock for uncertain demand, DTF.
- CODZ: Process-based: Different levels of certainty between uncertain and certain for the attributes what and when. Resource-based: Mixed load for resources based on forecast-driven and customer-order-driven flows.
- CADP: *A*:*D* relation, last point in flow with lot-sizing inventory.
- CADZ: Process-based: Different levels of uniqueness between customer generic and customer order unique for the attributes form and place. Resource-based: The same resource is involved with different levels of customization.
- PODP: *I*:*S* relation, safety stock for uncertain supply.
- PODZ: Process-based: Different levels of controllability between infinite and finite for the attributes material and capacity. Resource-based: Different levels of finiteness are applied to different load-generating activities for the same resource.

Decoupling points thus mean that the preconditions for decision-making can vary depending on which part of the logical entity that is considered. This more sophisticated description of the decision problem is modelled in Figure 15 in terms of three decision categories with three decision domains each. The resource-based decoupling zones are however not explicitly illustrated in Figure 15.

So far the three decision categories have been analysed individually, but in practice a combination of multiple decision categories are usually relevant and the compounded decision domains are suitable to use for this purpose in COBM.

6.5. Compounded decision domains for customer-order-based management

The three decision perspectives related to flow driving, flow differentiation and flow delimitation are important but do not exist in isolation from each other. In some decision situations, one of these could dominate the others which may then be less important and therefore neglected. From a more general perspective, the context of the decision in a specific situation is based on a combination of different decision domains. A flow decision may, for example, be based on a certain requirement (flow driver is a customer order) for a standard product (flow uniqueness is standard in terms of standard product from stock) and own resources are used (flow control can be performed in terms of finite material and finite capacity for delivery from finished goods inventory).

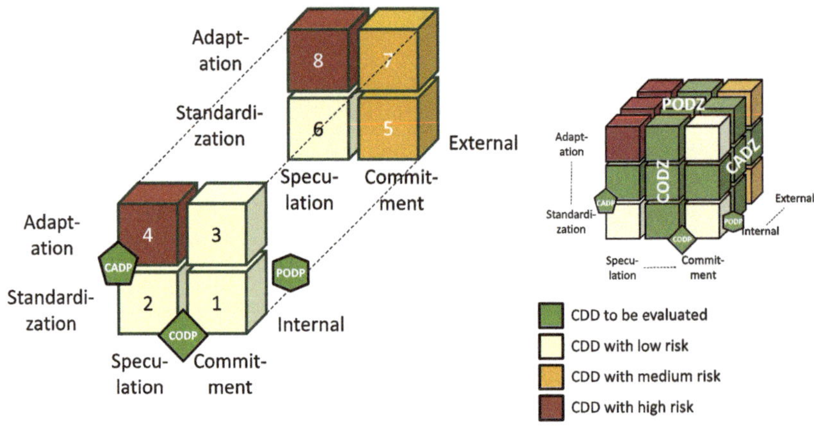

Figure 16. Compounded decision domains.

Since the three different types of decision domains, based on lead-time relations, are not mutually dependent, in terms of what they represent, they can be considered as three separate dimensions as shown in Figure 15. By combining these three dimensions of decision domains, eight compounded decision domains (CDD) can be identified if only decoupling points are included (on the left of Figure 16) or a total of 27 CDD if also decoupling zones are included (on the right of Figure 16). This approach is based on Figure 9. A CDD can be used to identify risk levels for different process configurations. In the case with eight CDD, the CDDs 1, 2, 3 and 6 (as indicated in Figure 16), represent scenarios that are well-established scenarios with relatively low risk. CDDs 5 and 7 represent a slightly higher level of risk when uncontrollable parts of the flow (e.g. a traditional supplier) are involved in the customer-order-driven part of the flow (which also can be customer order unique). CDDs 4 and 8, finally, have high risk since a customer order unique flow is performed on speculation.

To make it easier to reference to individual CDDs in the more complex model with 27 CDDs, to the right in Figure 16, a typology is introduced to be able to refer to one individual CDD where the first dimension (x-axis, to the right) represents flow driving, the second dimension (y-axis, upwards) flow differentiation and the third dimension (z-axis, outwards) flow delimitation. The example above with a certain demand for standard products supplied by own resources would then be represented by CDD (3,1,3), i.e. the cube to the right, furthest down and at the front, which is a CDD with low risk.

The three light yellow CDDs correspond to the standard cases with own production of standard products to forecast (1,1,3) and customer order (3,1,3), respectively, and to make something customized to customer order (3,3,3). The slightly darker CDDs at the back on the right (3,1,1) and (3,3,1) are CDDs with medium risk in the sense that this is about standard products but the supply is performed by resources that are outside the control of the logical entity (e.g. being performed by a supplier). CDDs (1,3,1), (1,3,2) and (1,3,3) are CDDs with high risk since they involve customer order unique production under speculation, which obviously should be avoided. The remaining CDDs include one or more zones and require a more thorough analysis to be evaluated. The present knowledge about these CDDs is simply not so developed in neither the literature nor in practical application. A further complicating factor is that the decoupling zones can be of two types, i.e. process-based or resource-based. Above the decoupling zones included were process-based. As discussed in the section on CDDs and RB-DZs, the complexity would increase even further if RB-DZs were included since intersections of the 27 different CDDs would then be involved.

6.6. Decision category: flow transparency

The lead-time-based decoupling points define different decision categories in terms of decision domains as outlined above. A complementary type of DC is based on the sharing of information. This type of decoupling point does not define the decision situation or how the information may be used but indicates if a specific type of information is available as decision support. If there is a need for information that the owner of the information cannot, or does not want to, share this is indicated in the flow by an information-based decoupling point. From a flow perspective, there are in particular two types of information that are in focus: information about demand and information about supply.

Sharing of demand information upstream in the flow is sometimes associated with sharing sales information such as point-of-sales data. From a more general perspective, all types of sales information are involved related to, e.g. customer orders, direct sales, delivery schedules and call-offs. The sharing of demand information upstream is limited by the demand information decoupling point (DIDP), which corresponds to information decoupling point according to Mason-Jones and Towill (1999). The DIDP should be positioned upstream from the CODZ, or upstream from the CODP if no CODZ is present (Olhager, Selldin, & Wikner, 2006), as shown in Figure 17.

> The *demand information decoupling point* (*DIDP*) constrains the transparency upstream of demand information.

If DIDP was positioned downstream from the CODP, it would mean that parts of the customer-order-driven flow would not have access to information about the customer order which would obviously be problematic. This could, however, occur if the CODP is positioned at a supplier who is not allowed to take part of the customer order information directly but only after it has been 'filtered' by the logical entity in direct contact with the customer. Considering also the CODZ, the DIDP should be positioned upstream of CODZ, if it is included in the analysis, since the CODZ indicate that some

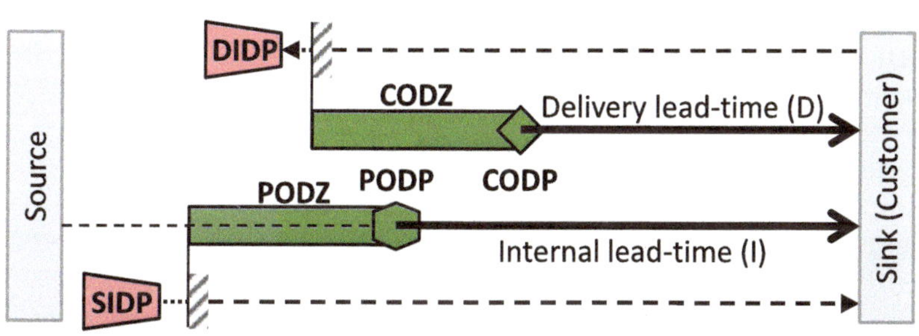

Figure 17. DC for customer-order-based management – flow transparency.

of the process driving is based on customer order. DIDP therefore highlights the interface between the part of the flow where all requested demand information is available and the part where different types of filter have been applied to the demand information.

Sharing of supply information downstream, i.e. downstream units having access to information of an upstream state, in the flow, enables for decision-makers to have information available regarding supply resources, such as available capacity and/or the usage of capacity. The will to share information may also in this case be limited. The user requesting the information is sometimes not allowed to take part of it and this constraint in information sharing is indicated by the supply information decoupling point (SIDP), see Figure 17.

> The *supply information decoupling point* (*SIDP*) constrains the transparency downstream of supply information

Since PODP marks a limit in the flow what concerns the possibility to control the flow, it also means that SIDP cannot be positioned downstream from PODP as indicated in Figure 17. If PODZ is also considered, the SIDP should be positioned upstream of PODZ since the PODZ indicates that some information on availability is required. Also note that since the general case can consist of multiple PODPs there would also be corresponding multiple SIDPs.

In both the case with demand information and the case with supply information, it is about the availability of information. The actual use of information is instead related to the decision domains or CDDs in combination with resource-based decoupling zones.

6.7. Example: decision categories

A fictitious product *Z* was introduced in Figure 10 in terms of a time-phased bill-of-material. Strategic lead-times were identified for the product and later some of the lead-time relations were analysed, see e.g. Table 3. Continuing on this example, some aspects of COBM can be illustrated. As a first step, a representative order book is assembled. In this example, the representative order book consists of 15 customer orders. The order book is sorted according to the requested delivery lead-time and illustrated as a time-phased order book in Figure 18. An alternative would be to use the promised delivery lead-time but to actually reflect the market requirements the requested delivery lead-time is preferable. The customer orders are positioned horizontally, where the requested delivery lead-time corresponds to the orange part at the right of each bar. The delivery lead-times vary between 6 periods and 10 periods. Consequently, all activities during the last 6 periods of the supply lead-time can be customer-order-driven and the CODP is positioned at $D=6$ periods. During the time from 6 periods to 10 periods before delivery, the activities can be based on a compromise of forecast and actual customer orders. This corresponds to a resource-based CODZ which extends from the CODP and $10–6=4$ periods upstream. Production of *V*, *W* and *U* may hence be produced/purchased based on a combination of forecast and customer orders.

Product *Z* is customized for individual customer orders at the last stage which means that $CADP_D$ is positioned two periods upstream. In addition, *Y* is unique for different customers which means that a $CADZ_D$ is positioned across the item *Y*. Items *V* and *Q* are purchased items, which here corresponds to the respective PODP since the product is assumed to be produced in one logical entity. The information about the activities

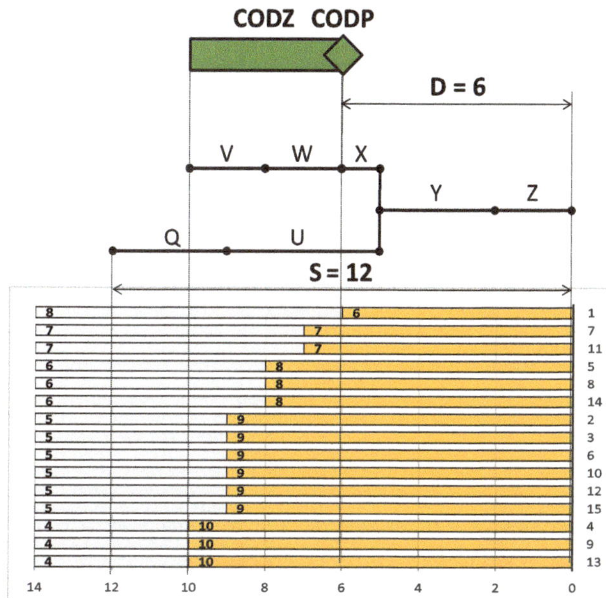

Figure 18. Example: the time-phased bill-of-material in relation to the time-phased order book.

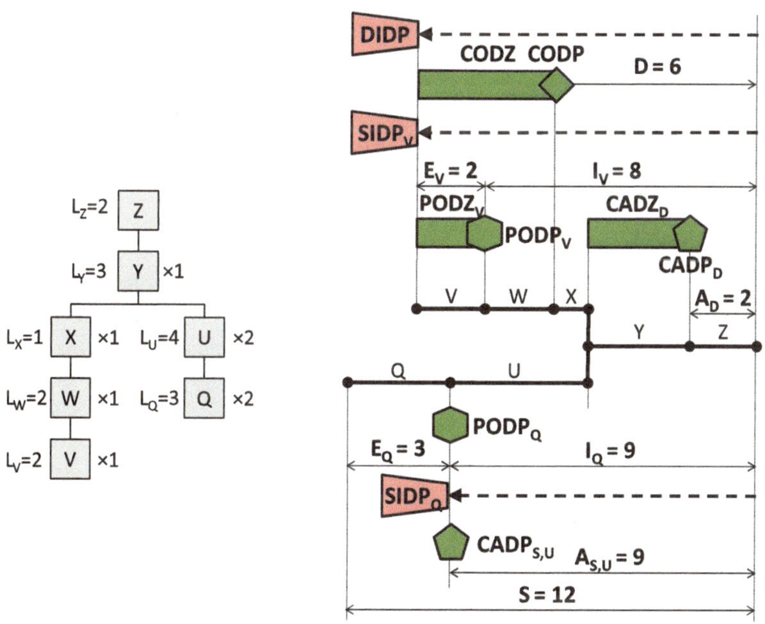

Figure 19. Example: the time-phased bill-of-material with decoupling points and decoupling zones.

and resource load at the supplier of Q is limited, wherefore the $SIDP_Q$ is positioned at the $PODP_Q$. The supplier of V provides load information but still is independent, wherefore this supplier is positioned within $PODZ_V$. Demand information (i.e. point of sales

Table 5. Example: CDD classification of the items (columns ordered in line with Figure 20).

Item (*i*)	Driving	Differentiat.	Delimitation	CDD	SIDP	DIDP
Z	Customer order	Customer order	Internal	(3,3,3)	DS	DS
Y	Customer order	CADZ	Internal	(3,2,3)	DS	DS
X	Customer order	Standard	Internal	(3,1,3)	DS	DS
W	CODZ	Standard	Internal	(2,1,3)	DS	DS
V	CODZ	Standard	PODZ	(2,1,2)	DS	DS
U	CODZ	Standard	Internal	(2,1,3)	DS	DS
Q	Forecast	Standard	External	(1,1,1)	US	US

data) is provided to the supplier of V which is indicated by the position of DIDP which is upstream from *V*. The supplier of *Q* does however not receive demand information as the DIDP is positioned downstream from the supplier. In this case, only one DIDP is used but in some cases it might be necessary to have multiple DIDP depending on how the provision of demand information is differentiated between different suppliers. Using the notation introduced above, the example could be illustrated as in Figure 19.

At this stage, it is also possible to categorize all items according to CDDs. Based on Figure 19, each item can be categorized in relation to the decoupling points as shown in Table 5. Item *Y* for instance is positioned downstream from the CODP and is hence customer order driven. It is also customer unique which corresponds to a position within the CADZ and finally it is manufactured and downstream of the PODPs which correspond to internal supply. In summary, this corresponds to CDD (3,2,3). It is important to note that each item is classified based on the most upstream position of the item and how it is related to the decoupling points. Finally, the information decoupling points are included in the two columns on the right. Each item is then categorized as being upstream (US) or downstream (DS) in relation to SIDPs and DIDP.

7. Framework for customer-order-based management

Decision categories with decision domains related to lead-time relations and also decoupling points and decoupling zones are the core elements of the framework for COBM. This type of components of a strategy framework is usually referred to as the *content* (Leong, Snyder, & Ward, 1990). To operationalize the content, it is necessary to identify how the content is supposed to be applied and this is usually referred to as the work process or simply *process*.

7.1. COBM framework: content

The framework's different decision categories, i.e. its content, have been defined and exemplified above. Figure 20 summarizes the most important elements of the four decision categories. Three of the decision categories in Figure 20 are decision based and of similar structure. The DC Flow transparency is of information character and has in this way a more supportive function, but nevertheless it is covering decisions about level of transparency and hence also about positioning of two decoupling points. Flow transparency is also absolute in the sense that it is based on lead-times, whereas the other decision categories are relative since they are based on lead-time relations.

The four decision categories of Figure 20 represent the core aspects of the COBM framework as it is defined here. The purpose of the approach is to highlight that by

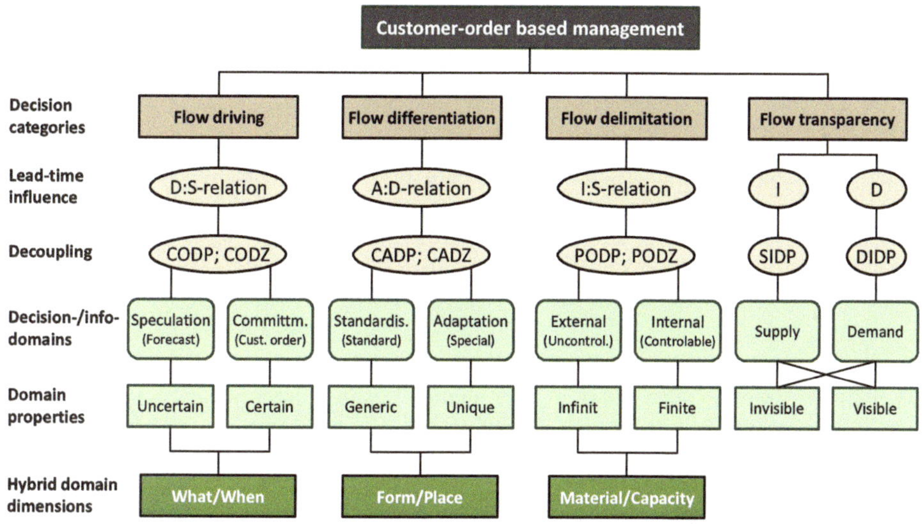

Figure 20. Key content in the framework for customer-order-based management.

identifying the different decision domains, positioning decoupling points and decoupling zones, and consequently identifying the CDDs, an appropriate approach to flow management can be outlined. The procedure for applying this type of approach is then referred to as the process of the framework.

7.2. *COBM framework: process*

A complete process for analysing COBM is relatively comprehensive with a large number of alternative activities depending on the specific circumstance in the particular case. The processes below are only outlined to provide the general idea and are based on Wikner (2012a). A more comprehensive process for supplier interaction applied in industrial cases is presented in (Bäckstrand, 2012; Bäckstrand et al., 2013). There are mainly two scenarios that are of interest and they are based on the analysis of a present state and the design of a future state, respectively. In both cases, the point of departure is usually the analysis of one product family, i.e. one value stream, at a time.

7.2.1. *Analysing a present state*

When analysing a present state, the objective is usually to investigate if the used control model is suitable for the present preconditions. The process can be summarized in seven steps:

(1) Identify the strategic lead-times S, A_s, D and the relevant I_s.
(2) Analyse lead-time relations.
(3) Position the strategic decoupling points and use decoupling zones when suitable.
(4) Analyse the positioning of the information decoupling points SIDP and DIDP.
(5) Identify which CDDs that are applicable.
(6) Evaluate the concerned supplier/customer relations from a CDD perspective.

(7) Decide/evaluate the appropriate approach for managing the flow based on identified CDDs and resource-based decoupling zones.

When this process is finished, a set of CDDs have been identified for the logical entity. This output supports the decision-maker in applying the appropriate approach for managing the enterprise. Each CDD has its particular characteristics and hence require a particular management approach. In this sense, the output can be seen as a foundation for 'plan for every CDD' in a similar fashion as 'plan for every part' that is frequently applied to select the appropriate approach for planning and control.

7.2.2. *Example: analysing a present state*

The fictitious example used above can be seen as representing a present state and hence also matching the outlined process for analysing a present state.

(1) The strategic lead-times were identified and illustrated in Figure 10.
(2) Lead-time relations were analysed in Tables 3 and 4.
(3) Position of the strategic decoupling points and decoupling zones was illustrated in Figures 18 and 19.
(4) SIDPs and DIDP were positioned in Figure 19.
(5) CDDs were identified in Table 5.
(6) The relation with the two suppliers were briefly discussed based on Figure 19 and Table 5.
(7) The CDDs were identified in Table 5 and provides the foundation for managing flow since items are categorized according to driving, differentiation and delimitation of flow. The details of flow management are however not within the scope of this paper.

7.2.3. *Designing a future state*

Design of a future state involves a completely different challenge compared to analysis of a present state. Instead of focusing on a present state this process is about innovatively designing a future flow in the best possible way. This can of course involve many types of scenarios, such as designing new flows when new products are introduced, which involves the largest degrees of freedom, to designing a desired future state for an existing value stream with existing products and resources. At an overall level, the process has similar properties for most alternative scenarios. Also, in this case, the process can be summarized in seven steps:

(1) Identify customer value and requirements on the delivery lead-time D and customizations (related to A_D).
(2) Position the CODP and decide the length of CODZ to handle variations in the requirement(s) for D.
(3) Position CADPs and decide the length of CADZs to highlight potential differences between standard, customer unique and customer order unique.
(4) Position the PODPs that are required to identify where the control possibilities are limited, e.g. supplier interfaces.
(5) When required, decide on the length of the respective PODZs.

(6) Position the information decoupling points SIDP and DIDP.
(7) Decide on the appropriate approach for managing the flow based on identified CDDs and resource-based decoupling zones.

Designing a future state obviously provides more degrees of freedom. It is a great opportunity to balance efficiency and effectiveness and consequently to establish competitive advantage by carefully positioning decoupling points and decoupling zones.

8. Discussion

Producing standard products to a mass market is still the main business for many companies. In these cases, the main competitive advantage is through cost efficiency and this has accelerated the trend for outsourcing based on finding the supplier that can provide the lowest possible cost. An alternative avenue is to differentiate the offering to the market. Differentiation can be based on increasing the variety of standard products but unless the variety is large the possibility of fulfilling individual customer needs is limited. Instead, different approaches for customization and servitization have developed over time to support this type of agility. Some activities are performed in the time span between the customer order is received and the delivery is performed and this is the target scenario for the framework developed here. Strategic decoupling points are an important concept for understanding the implications of these trends but so far the different perspectives have mainly been covered separately in the literature. In most cases the focus has been on the process driver which also can be seen as the origin of decoupling point analysis. The interest in a more integrated approach has however increased and in particular 'multiple decoupling points' (see e.g. Banerjee et al., 2011; Sun et al., 2008) has been used as a keyword. In addition, a number of CODP-based extensions have evolved over time. The most extensively explored is the different aspects of customization. Even in the early literature on CODP, such as Hoekstra and Romme (1992), this aspect was thoroughly investigated and many have followed. In parallel, the postponement literature provided an integrated approach to these two perspectives (see e.g. Pagh & Cooper, 1998). The CODP itself has been extended with the introduction of CODZ (Wikner & Rudberg, 2005b). Engineering has mainly been seen as occurring upstream from manufacturing but later these were treated as two separate dimensions (Giesberts & Tang, 1992; Wikner & Rudberg, 2005a). In addition, purchasing has been suggested as an additional dimension (Wallin, Rungtusanatham, & Rabinovich, 2006; Wikner & Bäckstrand, 2011). The resources and planning & control and order promising perspective has also been used in some cases (Giesberts & Tang, 1992; Rudberg & Wikner, 2004), and also implications for reverse logistics (Wikner & Tang, 2008) and services (Fließ & Kleinaltenkamp, 2004) have been included. The list above is not all-inclusive but still all these aspects are introduced more or less separately and not based on an integrative approach and on a common platform.

The frameworks outlined here provide both a common platform, in terms of the framework for flow-based decision-making, which is a general framework that can be used for many purposes, and the COBM framework, which provides an integrated approach to all these aspects of extending the CODP in terms of, e.g., multiple decoupling points. The generic framework for flow-based decision-making is related to the first research objective and outlines the fundamental aspects of COBM in terms of 10 key concepts. The concepts provide the building blocks of the decision categories for COBM. The second research objective resulted in the identification of SEIAD, the five

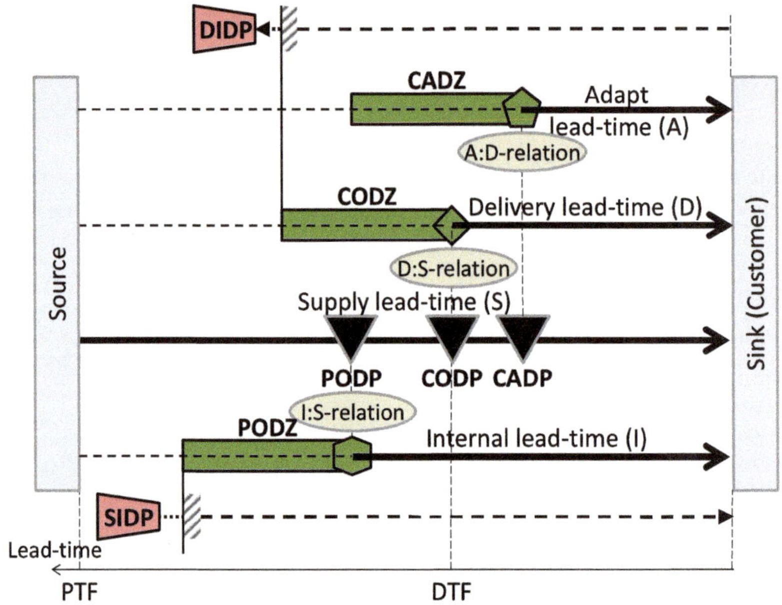

Figure 21. Lead-time-based summary of content for customer-order-based management.

strategic lead-times. Three of the decision categories are defined based on these strategic lead-times and each involves a lead-time relation, a decoupling point, decoupling zones and a strategic stock point. In addition, a DC related to transparency is included which resulted in a total of four decision categories which correspond to the third research objective. Figure 21 provides a lead-time-based overview of these four decision categories. Note however that the compounded decision domains and the resource-based decoupling zones are not explicitly represented in Figure 21. These two aspects are related to the intersection between multiple flows and different decision categories, respectively.

In addition to supporting the design of the flow, the COBM framework has also proved itself to be very useful for communication between different functions within a company and for communication with suppliers. Financial information is readily expressed using cost accounting terminology, engineering information in terms of product design and material selection, quality information in six sigma-based terminology, etc. Logistics have however had more difficulties in providing a comprehensive picture of the implications of different types of decisions on logistics-related issues (also including manufacturing). The COBM framework highlights strategic lead-times and strategic decoupling points and decoupling zones based on a time-phased bill-of-material and a time-phased order book. In combination they provide a comprehensive picture that highlights implications of lead-times. For example, a lead-time reduction that moves the CODP upstream in the bill-of-material has more important implications than a lead-time reduction that does not affect the position of the CODP. In a similar way, the procurement of focusing on low cost might extend the lead-time and thus changes a purchase-to-order to a purchase-to-forecast scenario. If this is also within the CADZ, it will be even more critical since customer unique material should preferably not be purchased to forecast.

The COBM framework rests on a generic foundation independent of, e.g., industry and product portfolio. Some level of customer order influence on the flow obviously makes the approach more relevant. The application can, however, be demanding during some circumstances. An important challenge is if the context is very dynamic, the strategic lead-times and consequently the position of decoupling points and decoupling zones may be dynamic, i.e. their position may change over time. In this type of situation, it would be necessary to either re-evaluate the positioning according to the decision categories, or to prepare for a number of scenarios with predefined configuration of COBM for each scenario. Another challenge is when the demand is not easily classified as customer order or forecast. In some cases, the contracts may stipulate some flexibility for the customer depending on the time horizon. The PB-CODZ might be suitable to handle this but still it can be difficult to actually position the CODP. This is even more accentuated in rate-based environments where the rate is set for an extensive time period.

9. Conclusions

The COBM framework has been derived using a three-layer approach, where the generic framework consisting of 10 key concepts provides the foundational constructs. The first level of instantiation is represented by the COBM framework consisting of content (see Figure 20) and process. The second level of instantiation is represented by the example used throughout the text, which is an instantiation of the COBM framework. The example highlights some detailed aspects of the framework but the implications of applying the framework are important also at a more general management perspective. Enterprise development with emphasis on operational efficiency has played a central role for many years in improving industrial productivity. Lean production has been a key enabler in this context and also introduced more emphasis on the customer. A cost efficient internal flow is however only the first step towards improved competitiveness. To reach a higher level of competitiveness, it is important to also develop the effectiveness, which is here referred to as the strategic efficiency, to provide the right preconditions for the operational flow through improved interaction. The flow-based COBM framework introduced here can hopefully provide increased understanding of, and support for, how strategic efficiency may be established.

The suggested COBM framework provides numerous opportunities for further research related to theory development, as well as industrial applications. In particular, three areas are being developed further:

- *Strategic decoupling and supply chain strategies* connects the logical entities to the physical entities and focus on how common supply chain strategies can be interpreted in terms of the COBM framework.
- *Strategic decoupling points and supplier/customer interactions* connects the logical entities to the legal entities and focuses on how supplier/customer interactions can be interpreted in terms of the COBM framework.
- *Strategic decoupling points and implementation* establishes a method for how to implement competitive COBM and thus also how to best position decoupling points and decoupling zones.

Acknowledgements

The development of the COBM framework has been partly supported by six companies based on the project KOPeration, covering the alignment of key aspects of purchasing strategy with operations strategy, and the project KOPtimera, focusing on the evaluation of positioning of decoupling points and decoupling zones. The projects are funded by the Swedish Knowledge foundation (KKS) and by the participating companies.

References

Bäckstrand, J. (2012). *A method for customer-driven purchasing: Aligning supplier interaction and customer-driven manufacturing*. Jönköping: School of Engineering, Jönköping University.

Bäckstrand, J., Johansson, E., Wikner, J., Andersson, R., Carlsson, B., Kornebäck, F., ... Spaak, B. (2013). *A method for customer-driven purchasing*. Paper presented at the 20th EurOMA Conference, Dublin, Ireland.

Bäckstrand, J., & Wikner, J. (2013). *Time-phasing and decoupling points as analytical tools for purchasing*. Paper presented at the IEEE International Conference on Industrial Engineering and Engineering Management, Bangkok, Thailand.

Banerjee, A., Sarkar, B., & Mukhopadhyay, S. K. (2011). Multiple decoupling point paradigms in a global supply chain syndrome: A relational analysis. *International Journal of Production Research, 50*, 3051–3065.

Berry, W. L., Hill, T. J., & Klompmaker, J. E. (1995). Customer-driven manufacturing. *International Journal of Operations & Production Management, 15*, 4–15.

Bertrand, J. W. M., Wortmann, J. C., & Wijngaard, J. (1990). *Production control: A structural and design oriented approach*. Amsterdam: Elsevier.

Blackstone, J. H. (Ed.). (2008). *APICS dictionary* (12th ed.). Athens, GA: APICS.

Borgström, B., & Hertz, S. (2011). Supply chain strategies: Changes in customer order-based production. *Journal of Business Logistics, 32*, 361–373.

Bucklin, L. P. (1965). Postponement, speculation and the structure of distribution channels. *Journal of Marketing Research, 2*, 26–31.

Chandrashekar, A., & Schary, P. B. (1999). Toward the virtual supply chain: The convergence of IT and organization. *International Journal of Logistics Management, 10*, 27–39.

Childerhouse, P., & Towill, D. R. (2003). Simplified material flow holds the key to supply chain integration. *Omega, 31*, 17–27.

Chopra, S., & Meindl, P. (2004). *Supply chain management: Strategy, planning, and operation* (2nd ed.). Upper Saddle River, NJ: Prentice Hall.

Clark, J. T. (1979). *Capacity management*. Proceedings of the 22nd APICS Annual International Conference.

Clark, A. J., & Scarf, H. (1960). Optimal policies for a multi-echelon inventory problem. *Management Science, 6*, 475–490.

Duggan, K. J. (2012). *Creating mixed model value streams: Practical lean techniques for building to demand*. New York, NY: CRC Press.

Fleischmann, B., & Meyr, H. (2004). Customer orientation in advanced planning. In H. Dyckhoff, R. Lackes, J. Reese, & G. Fandel (Eds.), *Supply chain management and reverse logistics* (pp. 297–321). Berlin: Springer.

Fließ, S., & Kleinaltenkamp, M. (2004). Blueprinting the service company: Managing service processes efficiently. *Journal of Business Research, 57*, 392–404.

Ford, H., & Crowther, S. (1988). *Today and tomorrow*. Cambridge, MA: Productivity Press.

Forrester, J. W. (1958). Industrial dynamics. *Harvard Business Review, 36*, 37–66.

García-Dastugue, S. J., & Lambert, D. M. (2007). Interorganizational time-based postponement in the supply chain. *Journal of Business Logistics, 28*, 57–81.

Giesberts, P. M. J., & van der Tang, L. (1992). Dynamics of the customer order decoupling point: Impact on information systems for production control. *Production Planning & Control, 3*, 300–313.

Goldratt, E. M. (1990). *What is this thing called theory of constraints and how should it be implemented?* Croton-on-Hudson, NY: North River Press.

Goldratt, E. M., & Cox, J. (1984). *The goal: Excellence in manufacturing*. Croton-on-Hudson, NY: North River Press.

Graca, A., Hendry, L., & Kingsman, B. (1999). Competitive advantage, customisation and a new taxonomy for non make-to-stock companies. *International Journal of Operations & Production Management, 19*, 349–371.
Hammer, M., & Champy, J. (1993). *Reengineering the corporation: A manifesto for business revolution*. New York, NY: HarperBusiness.
Harland, C. M. (1996). Supply chain management: Relationships, chains and networks. *British Journal of Management, 7*, S63–S80.
Hayes, R. H., & Wheelwright, S. C. (1984). *Restoring our competitive edge: Competing through manufacturing*. New York, NY: Wiley.
Hines, P., & Rich, N. (1997). The seven value stream mapping tools. *International Journal of Operations & Production Management, 17*, 46–64.
Hoekstra, S., & Romme, J. (Eds.) (1992). *Integral logistic structures: Developing customer-oriented goods flow*. New York, NY: Industrial Press.
Jones, D. T. (2001). The lean transformation. In D. H. Taylor & D. Brunt (Eds.), *Manufacturing operations and supply chain management: The Lean approach* (pp. 17–24). London: Thomson Learning.
Leong, G. K., Snyder, D. L., & Ward, P. T. (1990). Research in the process and content of manufacturing strategy. *Omega, 18*, 109–122.
Mason-Jones, R., Naylor, B., & Towill, D. R. (2000). Engineering the leagile supply chain. *International Journal of Agile Management Systems, 2*, 54–61.
Mason-Jones, R., & Towill, D. R. (1999). Using the information decoupling point to improve supply chain performance. *International Journal of Logistics Management, 10*, 13–26.
Mather, H. (1984). *Attack your P:D ratio*. Paper presented at the APICS Conference.
Naylor, B. J., Naim, M. M., & Berry, D. (1999). Leagility: Integrating the lean and agile manufacturing paradigms in the total supply chain. *International Journal of Production Economics, 62*, 107–118.
Naim, M. M., & Gosling, J. (2011). On leanness, agility and leagile supply chains. *International Journal of Production Economics, 131*, 342–354.
Olhager, J., & Östlund, B. (1990). An integrated push-pull manufacturing strategy. *European Journal of Operational Research, 45*, 135–142.
Olhager, J., Selldin, E., & Wikner, J. (2006). Decoupling the value chain. *International Journal of Value Chain Management, 1*, 19–32.
Olhager, J., & Wikner, J. (1998). A framework for integrated material and capacity based master scheduling. In D. Andreas & A. Kimms (Eds.), *Beyond manufacturing resource planning (MRP II): Advanced models and methods for production planning* (pp. 3–20). Berlin: Springer-Verlag.
Ōno, T. (1988). *Toyota production system: Beyond large-scale production*. Cambridge, MA: Productivity Press.
Pagh, J. D., & Cooper, M. C. (1998). Supply chain postponement and speculation strategies: How to choose the right strategy. *Journal of Business Logistics, 19*, 13–33.
Plossl, G. W. (1985). *Production and inventory control: Principles and techniques* (2nd ed.). Englewood Cliffs, NJ: Prentice-Hall.
Rother, M., & Shook, J. (1998). *Learning to see: Value stream mapping to add value and eliminate muda*. Brookline, MA: Lean Enterprise Institute.
Rudberg, M., & Wikner, J. (2004). Mass customization in terms of the customer order decoupling point. *Production Planning & Control, 15*, 445–458.
Schmenner, R. W. (2001). History of technology, manufacturing, and the industrial revolution: A rejoinder. *Production and Operations Management, 10*, 103–106.
Schmenner, R. W., & Swink, M. L. (1998). On theory in operations management. *Journal of Operations Management, 17*, 97–113.
Sharman, G. (1984). The rediscovery of logistics. *Harvard Business Review, 62*, 71–79.
Shingō, S.(1989). *A study of the Toyota production system from an industrial engineering viewpoint* (rev. ed.). Cambridge, MA: Productivity Press.
Smith, S. B. (1989). *Computer-based production and inventory control*. Englewood Cliffs, NJ: Prentice-Hall.
Stalk, G., & Hout, T. M. (1990). *Competing against time: How time-based competition is reshaping global markets*. New York, NY: Free Press.

Stein, R. E. (1996). *Re-engineering the manufacturing system: Applying the theory of constraints*. New York, NY: M. Dekker.

Sun, X. Y., Ji, P., Sun, L. Y., & Wang, Y. L. (2008). Positioning multiple decoupling points in a supply network. *International Journal of Production Economics, 113*, 943–956.

Vargo, S. L., & Lusch, R. F. (2004). The four service marketing myths. *Journal of Service Research, 6*, 324–335.

van der Veeken, D. J. M, & Rutten, W. G. M. M. (1998). Logistics service management: Opportunities for differentiation. *International Journal of Logistics Management, 9*, 91–98.

Waddell, W., & Bodek, N. (2005). *Rebirth of American industry: A study of lean management*. Vancouver, WA: PCS Press.

Wallace, T. F. (1992). *Customer-driven strategy: Winning through operational excellence*. Essex Junction, VT: Oliver Wight.

Wallin, C., Rungtusanatham, M. J., & Rabinovich, E. (2006). What is the "right" inventory management approach for a purchased item? *International Journal of Operations & Production Management, 26*, 50–68.

van Hoek, R. I. (2001). The rediscovery of postponement a literature review and directions for research. *Journal of Operations Management, 19*, 161–184.

Wemmerlöv, U. (1984). Assemble-to-order manufacturing: Implications for materials management. *Journal of Operations Management, 4*, 347–368.

Wheelwright, S. C. (1984). Manufacturing strategy: Defining the missing link. *Strategic Management Journal, 5*, 77–91.

Wight, O. W. (1984). *Manufacturing resource plannig: MRP II* (rev. ed.). Essex Junction, VT: Wiley.

Wikner, J. (2011). *Ledtidsbaserat ramverk för kundorderstyrning* [in Swedish] [Lead-time based framework for customer-order based management]. Paper presented at the 13th Plan Conference on research and application, Norrköping, Sweden.

Wikner, J. (2012a). Grad av kundorderstyrning [Level of customer-order based management]. In G. Hultén (Ed.), *Inköp & Logistik* (pp. 5.1/1–5.1/25). Stockholm: Bonnier.

Wikner, J. (2012b). A service decoupling point framework for logistics, manufacturing, and service operations. *International Journal of Services Sciences, 4*, 330–357.

Wikner, J., & Bäckstrand, J. (2011). *Aligning operations strategy and purchasing strategy*. Paper presented at the EurOMA Conference, Cambridge, UK. Retrieved from: http://urn.kb.se/resolve?urn=urn:nbn:se:hj:diva-17224

Wikner, J., & Bäckstrand, J. (2012). *Decoupling points and product uniqueness impact on supplier relations*. Paper presented at the 4th World Conference P&OM, Amsterdam.

Wikner, J., Johansson, E., & Persson, T. (2009). *Process based inventory classification*. Paper presented at the The 21st Annual Nofoma Conference, Jönköping, Sweden.

Wikner, J., Naim, M. M., & Towill, D. R. (1992). The system simplification approach in understanding the dynamic behaviour of a manufacturing supply chain. *Journal of Systems Engineering, 2*, 164–178.

Wikner, J., & Rudberg, M. (2005a). Integrating production and engineering perspectives on the customer order decoupling point. *International Journal of Operations & Production Management, 25*, 623–641.

Wikner, J., & Rudberg, M. (2005b). Introducing a customer order decoupling zone in logistics decision-making. *International Journal of Logistics Research and Applications, 8*, 211–224.

Wikner, J., & Tang, O. (2008). A structural framework for closed-loop supply chains. *International Journal of Logistics Management, 19*, 344–366.

Wikner, J., & Wong, H. (2007, September 17–19). Postponement based on the positioning of the differentiation and decoupling points. In J. Olhager & P. Fredrik (Eds.), *International IFIP TC 5, WG 5.7 Conference on Advances in Production Management Systems (APMS 2007)* (Vol. 246, pp. 143–150). Linköping: Springer-Verlag.

Womack, J. P., & Jones, D. T. (1996). *Lean thinking: Banish waste and create wealth in your corporation*. New York, NY: Simon & Schuster.

Wortmann, J. C., Muntslag, D. R., & Timmermans, P. J. M. (1997). *Customer-driven manufacturing* (1st ed.). London: Chapman & Hall.

8

Inclusive design drivers and barriers – a manufacturing perspective from Pakistan

Amjad Hussain[a], Aftab Ahmad[a] and Keith Case[b]*

[a]Department of Industrial and Manufacturing Engineering, University of Engineering and Technology, Lahore 54890, Pakistan; [b]Mechanical and Manufacturing Engineering, Loughborough University, Loughborough, Leicestershire LE11 3TU, UK

The demographics of older people and people with disabilities in developing countries are discussed in the context of inclusive design and the drivers and barriers to inclusive design have been identified. Data were collected from 50 individuals from various industrial sectors in Pakistan. Corporate social responsibility (CSR) relates to inclusive aspects of products, but most respondents either did not know about CSR or did not have a CSR post in their organizations, but 64% had awareness of inclusive design terminology. The study concluded that motivation through social responsibility; innovation and differentiation; demographics and consumer trends; brand enhancement; customer satisfaction; new market opportunity; and legislation were the perceived drivers for manufacturers in Pakistan. Most respondents felt that lack of resources and guidance, lack of awareness about inclusive design, difficulty in changing the business culture, lack of government regulations, and the perception that inclusive design is expensive were the most significant barriers.

Keywords: inclusive design; drivers and barriers; manufacturing; corporate social responsibility (CSR); developing countries

1. Introduction

Inclusive design aims to address the design needs of the largest proportion of the population where special consideration is given to accommodating older people and people with disabilities along with the rest of the population in a single design solution. The percentage of the older population is increasing throughout the world; however, this trend is quite prominent in developed countries such as the USA, UK, Australia, Japan, Canada, and Germany because of better living facilities, medical treatment, and healthier working environments (U.N.O., 2009). On the other hand, in developing countries, while the average age of the population is lower than that of developed countries, these countries constitute a significant proportion of the overall world population. Attention has previously been given to assessing the level of awareness and exploring drivers and barriers for the promotion of the inclusive design approach in product, process, environment, and service design in developed countries like the UK, USA, and Japan. In spite of the challenging demographics of developing countries considered as the main contributors to the world population, no effort has been made to increase the level of

*Corresponding author. Email: k.case@lboro.ac.uk

awareness of the inclusive design approach, or to explore and highlight the drivers and barriers to the implementation of this design approach. This research is an effort to contribute in this respect by assessing the perspectives of manufacturers on the implementation of the inclusive design method.

2. Literature review

2.1. Demographics

The total population of many countries is increasing. However, this trend is quite significant in developing countries like China, India, Pakistan, and Bangladesh, where the population of these four countries is about 42% of the world population. In the same way, the populations of these countries contain a significant proportion of older people (37% of the world population,) and people with disabilities (6% of the world population) (CIA World Fact Book, 2014). These statistics clearly highlight the need for the implementation of an inclusive design approach in these countries as their contribution to the world population is quite significant.

In general, it is clear from the global demographics that older people and people with disabilities are a considerable proportion of the world population, and this justifies financial as well as legislative incentives for including these groups in the design of products, environments, and services (Coleman, 2001). However, the trends in this demographic change are different in different parts of the world. For example, the overall percentage of older workers is increasing in majority of the developed countries and this justifies their accommodation in the design of products, processes, environments, and services. However, in the developing countries, the same increase is prominent as the overall population of these countries is growing very quickly, and the proportion of older people and people with disabilities is increasing accordingly. These facts draw the attention of designers, ergonomists, engineers, psychologists, planners, and entrepreneurs to seriously consider the provision of healthy living conditions for this significant proportion of the world population. The following sections describe how these challenges can be met by design scenarios that are equally acceptable for the broadest range of the population and also identify the drivers and barriers in the promotion of these ideas.

2.2. Inclusive design – drivers and barriers

'Design is the process of converting an idea or market need into the detailed information from which a product or system can be made' (Royal Academy of Engineering, 2005).

The British Standards Institute (2005) defined inclusive design as 'The design of *mainstream* products and/or services that are accessible to, and *usable* by, *as many people* as reasonably possible … without the need for special adaptation or specialized design'. Subsequently, the inclusive design term has also been related to providing quality of life and independent living for the aging population (Waller & Clarkson, 2009). It can be said that approaches like 'Inclusive Design,' 'Universal Design,' and 'Design for All' have been developed to help designers in developing design strategies that can promote design scenarios that are equally acceptable for all, including older people and people with disabilities (European Institute for Design & Disability, 2006; Keates & Clarkson, 2004; Preiser & Ostroff, 2001).

Published literature shows that there have been previous efforts in different parts of the world to explore the drivers and barriers for inclusive design. Vanderheiden and Tobias (2000) conducted telephone interviews of 26 manufacturers of consumer products in the USA, and identified a range of barriers and motives including government regulation, market data, training, consumer demands, technical complexity and unavailability of highly relevant knowledge, data and techniques. A similar survey was conducted in Japan where 307 companies from different industrial categories were surveyed. Interestingly, Japanese companies provided results that were similar to the USA companies (Unpublished report, 2000).

In the UK, Keates & Clarkson (2004) found that few industries knew about inclusive design and that there were misconceptions in the fundamental understanding of this design method. Companies believed that inclusive design meant designing only for older people and people with disabilities. In another survey conducted by Sims (2003), 32 design professionals working with different types and sizes of companies were surveyed, and it was concluded that 'design for all' is widely known but unfortunately not practiced within the design community. The majority of designers were aware of the philosophy of 'design for all' but rarely considered the approach because of the perceived time and financial costs. Underwood and Metz (2003) and Bellerby and Davis (2003) also discussed how inclusive design methods can be promoted and design-related issues could be addressed. They suggested that the provision of guidelines and standards could be important drivers, as currently these are not presented appropriately. Moreover, legislation and brand imaging can also play an effective role as generic business drivers.

Dong, Keates, and Clarkson (2004) and Dong, Clarkson, Ahmed, and Keates (2004) conducted a more comprehensive study with SMEs, where a survey was carried out with 38 manufacturing and retailing companies, along with 35 design consultancies. It was concluded that different companies perceive different factors as major barriers. However, drivers within these groups were found to be the same. For example, manufacturers and retailers mentioned key barriers because of the assumptions that inclusive design is more expensive, difficult to practice and learn and time consuming. In 2006 Goodman, Dong, Langdon, and Clarkson (2006) unlike Dong, Keates, et al. (2004) and Dong, Clarkson, et al. (2004), targeted large organizations along with SMEs and used a survey method for obtaining a more detailed insight into the drivers and barriers for inclusive design, and used the same questionnaire for comparison purposes. Complete responses were collected from 101 UK companies and organizations, and a detailed analysis was carried out. Barriers most frequently identified were a lack of time and budget for supporting inclusive design, lack of knowledge and tools to practice it, and it not being perceived as a need of the end users. Moreover, the perception that there was no justifiable business case for inclusive design was considered extremely important by most of the respondents.

As mentioned earlier, the population of developing countries is growing rapidly and the proportion of older and disabled people is also increasing accordingly. There are important questions as to how the living standards and quality of life of these people can be maintained and improved. Currently, in these countries the needs of older people and people with disabilities are not considered sufficiently by designers, planners, architects, and ergonomists. It is critically important to create awareness for policy-makers, planners, manufacturers, and designers about the importance of the inclusive design approach for creating environments, products, and services that meet the diverse and changing needs of the whole population.

3. Research focus

The main focus of the research was to assess the level of awareness of inclusive design in developing countries like China, India, Pakistan, and Bangladesh as a significant proportion of the world population lives in these countries. Until now, no research has been conducted in these countries to explore the main drivers that can motivate manufacturers and barriers that are potentially the reasons for lack of interest and resistance to its promotion. This is potentially a huge market where manufacturers can gain benefits by implementing design solutions that are equally acceptable for a wide range of the population. An underlying purpose of the research was to provide awareness to the manufacturers about the inclusive design approach, so that manufacturing industries in this part of the world can create business opportunities by attracting older customers and customers with special needs (disabilities) along with rest of the population.

4. Research method

The main objective of the survey was to identify barriers and drivers for inclusive design from the perspective of manufacturers in Pakistan. Initially, a list of manufacturing companies was prepared from which some were recruited after telephone inquiries about their willingness to participate in the survey. Manufacturing companies from textile, beverage, sports, automotive, design consultancy, and automotive sectors were included in this research. The survey questionnaire was designed with reference to the method adopted by the Engineering Design Centre's Inclusive Design survey at the University of Cambridge (Coleman, 2001) and is shown in abbreviated form as Appendix A. However, additional drivers and barriers found in research studies from the USA and Japan were also included in the study. A pilot survey was carried out to assess the reliability of the method. During the pilot study, it was found that some people had no understanding of any of the terms, 'inclusive design,' 'universal design,' and 'design for all.' So, prior to final distribution, a revised questionnaire was developed including a two-page summary for basic understanding of these terms, elaborated with the help of practical examples, so that the underlying objective of increasing the level of awareness about inclusive design could be achieved. (the illustrations of the practical examples have been removed from Appendix A, but they covered a wide range of activities of daily living that people with disabilities might have difficulty with such as accessing cash machines and dealing with escalators). It was further concluded that questionnaires should be completed in the presence of a research project team member, so that more reliable results could be collected. Representatives from 8 manufacturing companies (linked with the design process within their organizations) were selected, and a total of 50 responses were collected. Pakistan has 6417 manufacturing companies, 3590 of them in Punjab state where this study was conducted (Pakistan Bureau of Statistics, 2006).

5. Results and discussion

Responses to the distributed questionnaires were analyzed using descriptive statistics, where frequencies, average scores (means), and standard deviations (SD) were calculated, keeping in view the type of question and which output parameters could be used as indicators. The overall analysis was divided into three main areas: awareness, drivers, and barriers.

5.1. Awareness

The first portion of the questionnaire was about the assessment of the level of awareness and current understanding of the terms 'Inclusive Design,' 'Universal Design,' and 'Design for All'. As mentioned, the data were collected from 50 respondents, from companies whose profiles with respect to the number of employees are shown in Table 1. Table 2 shows the categories of manufacturing organizations included in the study and their proportional relevance in the sample population. Table 1 shows that about 70% of the sample population belonged to organizations having between 251 and 5000 employees. Moreover, 26% were from large organizations having over 5000 employees. The respondents were asked if their organization had a post for corporate social responsibility (CSR), and all claimed to understand the term CSR. However, 42% admitted that they did not know whether or not their organization had such a post. This clearly indicates that many of those involved in design and manufacturing activities have some preliminary understanding of CSR, but that there is no clear focus on this in their organizations. On the other hand, 30% responded that their organizations had a CSR post and 28% clearly claimed that their organizations had no CSR post. Overall, it can be concluded that in nearly 70% of organizations those with responsibility for design and manufacturing either did not know of or had no clear focus on CSR (shown in Figure 1).

Current understanding or awareness of different terms like 'Inclusive Design,' 'Universal Design,' and 'Design for All' were explored by asking respondents if they had heard of the terms and if they understood their meaning. By their responses, 36% showed no understanding of these terms. However, 64% did show understanding and the most popular term was 'Design for All' with 30% having previously heard of the term. Awareness or understanding of 'Inclusive Design,' and 'Universal Design' were recorded as 14 and 20%, respectively (Figure 2).

The above statistics clearly indicate a lack of awareness about the terminology and consequently of design inclusivity. Similar results were found in the pilot study and on the basis of this a two-page comprehensive summary of these terms along with useful and relevant examples were prepared, and made a necessary part of the questionnaire so that those who had no knowledge of the concepts could at least understand and give appropriate feedback about drivers and barriers to inclusive design. During final data collection, respondents were not allowed to see this information until they had recorded their responses about current understanding and awareness. As mentioned earlier, all the responses were collected in the presence of a member of the project team, so that the respondents were able to ask questions on the scope of drivers and barriers. A large majority of the respondents actively participated in discussion sessions by asking interesting questions and gained a clear understanding about these initiatives. Interestingly, 100% of the respondents requested our team members to send them final conclusions and findings of the research study, and this clearly shows their interest in this area. The following sections discuss findings related to drivers and barriers.

Table 1. Company profiles: number of employees.

No of employees	Frequency	Percent
Up to 250	3	6.0
251–1000	14	28.0
1001–5000	20	40.0
Over 5000	13	26.0
Total	50	100.0

Table 2. Types of manufacturing industries and number of responses.

Type	Responses
Textile	7
Beverages	10
Electronics	10
Design consultancies	12
Sports	3
Automotive	8
Total	50

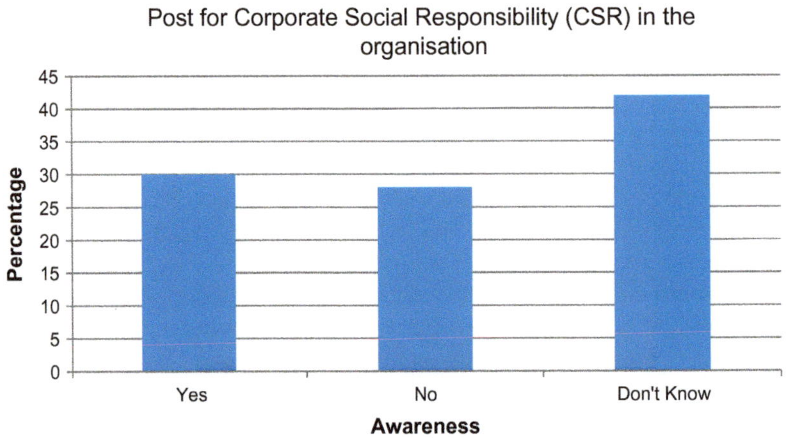

Figure 1. Corporate social responsibility (CSR) awareness.

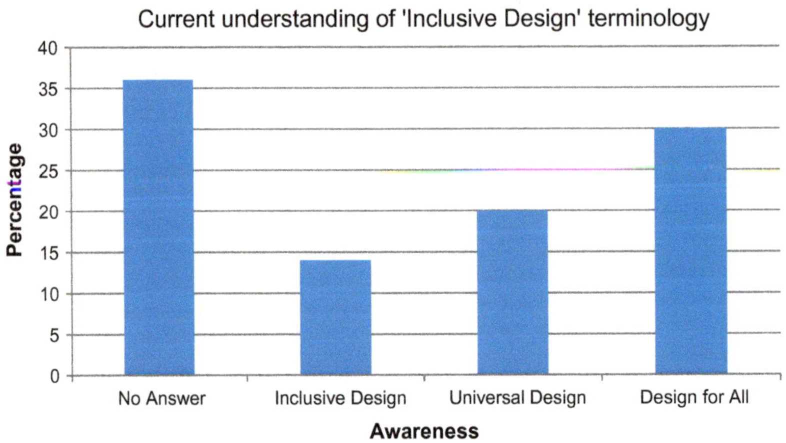

Figure 2. Current understanding of 'Inclusive Design' terminology.

5.2. *Drivers*

The perceptions of drivers of inclusive design from manufacturers have been computed as average scores (mean values) and SD. Responses were collected on a 1–4 Likert-type scale, where 22 drivers were listed and divided into two categories. Drivers, such as legislation, The Employment and Rehabilitation Act, social responsibility, demographic and consumer trends and brand enhancement were included in category one, where responses were collected in the form of agreement or disagreement ('strongly disagree' to 'strongly agree'). In the second category, all other drivers mentioned in Table 2 were included and responses were again collected on a four-point Likert-type scale, where effectiveness in terms of commercial benefits was measured ('not effective' to 'very effective'). The responses indicated that the vast majority of individuals in manufacturing felt that motivation through social responsibility awareness would be the most significant driver for the promotion of inclusive design in their organizations. As discussed in the preceding section, there is a need to promote CSR initiatives in organizations so that the level of awareness about inclusive aspects of products, environment, or service design might be increased. Motivation through social responsibility awareness was perceived as the most effective driver (mean 3.28 and SD .970) in the implementation of design inclusivity. Similarly, other drivers that were perceived as significant were: source of innovation and differentiation (mean 3.28); demographic and consumer trends (3.26); brand enhancement (3.22); increase customer satisfaction (3.16); and new market opportunity (3.12). However, the least significant drivers were: consumer dissatisfaction with current products (mean 2.40); assessment of how many people are excluded (2.58); champion for inclusive design on company board (2.66); major competitors adoption of inclusive design (2.68); and increase the large share of current market (2.70). Full details of means and SD values for all 22 drivers are shown in Table 3.

5.3. *Barriers*

This section discusses the results of data analysis concerning perceived barriers to inclusive design. The perspectives of respondents on 27 potential barriers to inclusive design promotion among manufacturers in Pakistan were sought. Table 4 shows the results of the data analysis, and highlights the percentages of respondents in agreement with the barrier statements. It was concluded that lack of resources/guidance and awareness of inclusive design are the most prominent perceived barriers as about 84% (42 respondents out of 50) and 80% (40 out of 50) respondents showed agreement, respectively. Moreover, change in culture of business (76%), lack of government regulations (76%), perception that inclusive design is more expensive (68%), and that it increased time to market (68%) were perceived as significant barriers to inclusive design. Interestingly, about 80% of the respondents perceived that implementation of inclusive design was not an unachievable goal, and a significant number of people felt that inclusive design was not a passing trend and did not compromise the esthetics of design. Similarly, more than 50% of respondents from manufacturing organizations did not feel that inclusive design was a too difficult thing to practice. These findings highlight that the majority of manufacturing-related people were of the opinion that inclusive design was not too difficult to implement. However, inclusive design can be promoted by launching an organized campaign about the associated usefulness and benefits, by providing proper resources and guidance, along with changes in organizational culture and enforcement of focused regulations regarding inclusive design. The percentage agreements of the respondents with other barriers are shown in Table 4.

Table 3. Perceptions of manufacturers regarding 'drivers' for inclusive design.

Drivers	Mean	SD
Motivation through social responsibility	3.28	.970
Source of innovation and differentiation	3.28	.858
Demographic and consumer trends	3.26	1.121
Brand enhancement	3.22	1.166
Increase customer satisfaction	3.16	.934
Employment and Rehabilitation Act 1981 will help in practicing inclusive design	3.16	1.131
New market opportunity	3.12	.872
Chances of innovation by practicing inclusive design	3.10	.931
Entrance to new market	3.08	.944
Legislation	3.08	1.209
Increase revenue through increased usage	3.06	.913
Increase customer loyalty	3.04	.903
Increase the size of potential market	2.94	.793
Availability of tools/methods to help practicing	2.88	.872
Public awareness of inclusive design	2.88	.982
Availability of expert consultation on inclusive design	2.82	.941
Availability of training opportunities on inclusive design	2.80	.948
Increase the large share of current market	2.70	.863
Major competitors adoption of inclusive design	2.68	.935
Champion for inclusive design on company board	2.66	.823
Assessment of how many people are excluded	2.58	.950
Consumer dissatisfaction with current products	2.40	1.030

From the analysis of the data presented above, many of the drivers and barriers found important in this study were the same as those highlighted by other studies conducted in the UK, USA, and Japan. For example, drivers such as new market opportunity, brand enhancement, and source of innovation were commonly perceived drivers in all these studies. Similarly, barriers like awareness of inclusive design, lack of resources and guidance, and lack of time and budget were common barriers presented in all of these studies. Moreover, barriers such as inclusive design was too difficult and an unachievable goal that compromised the esthetics of design were perceived as relatively less significant barriers. This shows that almost all studies have concluded that people who have seen the relevance of the promotion and implementation of inclusive design felt that there was a great need to increase levels of awareness within companies especially amongst designers. At the same time, they felt that design for inclusivity is not an unachievable goal; it can be achieved by increasing awareness about its importance, usefulness, and relevance with our daily life and by providing technical and financial assistance at the organizational and country level.

Tables 5 and 6 present a comparison of results on perceived drivers and barriers for the different manufacturing sectors (textile, beverages, electronics, sports, automotive and design consultancy). It is quite clear that there were variations in the perceptions of inclusive design drivers and barriers depending upon the type of manufacturing sector. For example, in the electronics manufacturing sector, legislation was perceived as the most prominent driver with a mean score of 3.9, not only highest in this sector but overall as well (Table 5). Similarly, in the beverage sector, it was commonly perceived that putting effort into an increase in customer satisfaction by applying inclusive design methodologies was the most important driver. In the automotive sector, it was felt that new market opportunity and increase in customer loyalty were the most prominent drivers, whereas the same had relatively less importance (in the middle of the list) in

Table 4. Perceptions of manufacturers regarding 'barriers' for inclusive design.

Barriers	No. of responses	Percentage agreed
Lack of resources/guidance on inclusive design	42	84
Lack of awareness of inclusive design	40	80
Difficulty in changing the culture of business	38	76
Lack of government regulations	38	76
Implementing inclusive design could require significant cultural change	35	70
Perception that inclusive design is more expensive	34	68
Perceived longer development time to market	34	68
Perception – more complex to design inclusively	34	68
Lack of interest in inclusive design	34	68
Stigma associated with inclusive design	33	66
Lack of time and budget	33	66
Lack of company policy on inclusive design	32	64
Inclusive design is a perceived need of our end user	32	64
Lack of methods/tools for practicing inclusive design	32	64
Lack of motivation for tackling inclusive design	32	64
Perception of brand association with disabled/older people	32	64
Lack of availability of good design examples	31	62
Perception – there is no need to practice inclusive design	31	62
Working for short term financial objectives	30	60
Perception that inclusive design represents a niche market	29	58
Unavailability of internal support for inclusive design	29	58
No justifiable business case to support inclusive design	27	54
Lack of knowledge and tools to practice inclusive design	26	52
Perception – inclusive design is too difficult	23	46
Perception – inclusive design is a passing trend	20	40
Inclusive design compromises the aesthetics of design	18	36
Perceived as an unachievable goal	10	20

the overall priority list. Generally, it can be said that the textile, beverages, design consultancy, and electronics manufacturing sectors provided similar kinds of results to those of the overall analysis. As shown, only three respondents from sports industries recorded their responses, however, those were not of high significance because of the smaller sample size. The sports sector is one of the leading export-based manufacturing sectors in Pakistan, and so there is a need to explore this sector in a detailed way.

A comparison of different perceived barriers among a variety of manufacturing sectors is shown in Table 6, where differences in perceptions can be observed. For example, in the automotive sector, the vast majority of respondents agreed that there was no need to practice inclusive design, that it reduces short-term financial advantage and was associated with stigma. In comparison with the overall perception, these factors showed relatively less significance. A detailed comparison is shown in Table 6.

6. Future work

Future work will focus initially on capturing the perceptions about inclusive design drivers and barriers from designers working in different areas like architecture, civil engineering, town planning and design, transportation design and management in Pakistan, so as to widen the scope of the investigations. Subsequently, the same study will be carried out in China, India, and Bangladesh, and efforts have been started to

Table 5. Comparison of different manufacturing sectors on perceived inclusive design drivers.

	Overall		Textile		Sports		Beverages		Design consultancies		Electronics		Automotive	
Drivers	Mean	SD	Mean	SD	Mean	SD	Mean	SD	Mean	SD	Mean	SD	Mean	SD
Motivation through social responsibility	3.28	.970	3.57	.787	3.00	1.000	2.90	.876	3.67	.651	3.40	1.265	2.88	1.126
Source of innovation and differentiation	3.28	.858	3.71	.488	3.33	1.155	3.20	.919	3.33	.985	2.90	.994	3.38	.518
Demographic and consumer trends	3.26	1.121	3.00	.816	2.33	1.528	3.10	1.287	3.67	.985	3.50	1.179	3.13	1.126
Brand enhancement	3.22	1.166	3.00	1.155	2.33	1.528	2.90	1.287	3.75	1.215	3.60	.966	2.88	.835
Increase customer satisfaction	3.16	.934	3.57	.787	3.33	1.155	3.30	1.059	3.17	.937	2.60	.843	3.25	.886
Employment and Rehabilitation Act 1981will help in practicing inclusive design	3.16	1.131	3.29	1.113	4.00	1.732	3.10	.994	3.42	1.240	3.00	1.054	2.63	1.061
New market opportunity	3.12	.872	3.71	.488	2.67	1.528	2.80	.919	2.92	1.084	3.20	.632	3.38	.518
Chances of innovation by practicing inclusive design	3.10	.931	3.86	.378	3.00	1.732	2.70	.949	3.25	.965	2.70	.823	3.25	.707
Entrance to new market	3.08	.944	3.29	1.113	3.00	1.732	3.10	.876	3.17	1.030	3.10	.738	2.75	.886
Legislation	3.08	1.209	2.86	1.069	4.33	.577	3.20	1.229	2.58	1.311	3.90	.876	2.38	.916
Increase revenue through increased usage	3.06	.913	3.43	.787	2.67	1.155	3.20	.919	2.83	1.030	3.10	.876	3.00	.926
Increase customer loyalty	3.04	.903	3.00	1.000	3.00	1.000	3.20	1.033	2.58	.996	3.20	.632	3.38	.744
Increase the size of potential market	2.94	.793	3.29	.951	3.00	1.000	2.90	.876	2.83	.718	2.80	.632	3.00	.926
Availability of tools/methods to help practicing	2.88	.872	2.86	.900	2.33	1.528	3.00	1.054	2.67	.888	3.00	.471	3.13	.835
Public awareness of inclusive design	2.88	.982	3.29	.756	3.00	1.732	2.80	1.033	3.00	1.044	2.70	.823	2.63	1.061
Availability of expert consultation on inclusive design	2.82	.941	3.14	.900	2.67	1.528	2.90	.994	3.00	.953	2.50	.707	2.63	1.061
Availability of training opportunities on inclusive design	2.80	.948	3.00	1.000	2.33	1.528	2.40	1.174	2.92	.793	2.70	.823	3.25	.707
Increase the large share of current market	2.70	.863	3.00	1.000	2.67	1.528	2.70	.949	2.50	.798	2.60	.516	2.88	.991
Major competitors adoption of inclusive design	2.68	.935	2.71	1.113	2.33	1.155	2.80	.919	2.83	1.030	2.20	.632	3.00	.926
Champion for inclusive design on company board	2.66	.823	2.71	.951	2.33	.577	2.50	1.080	2.75	.622	2.60	.699	2.88	.991
Assessment of how many people are excluded	2.58	.950	3.00	.816	2.33	1.528	2.40	.843	2.75	1.215	2.50	.850	2.38	.744
Consumer dissatisfaction with current products	2.40	1.030	2.43	1.272	1.33	.577	2.10	.994	2.25	.965	2.80	.789	2.88	1.126

Table 6. Comparison of different manufacturing sectors on perceived barriers to inclusive design.

Barriers	Overall	Textile	Sports	Beverages	Design consultancies	Electronics	Automotive
Lack of resources/guidance on inclusive design	84	85.7	100.0	90	91.7	80	62.5
Lack of awareness of inclusive design	80	85.7	66.7	80	91.7	70	75.0
Difficulty in changing the culture of business	76	71.4	66.7	70	83.3	70	87.5
Lack of government regulations	76	71.4	100.0	90	91.7	60	50.0
Implementing inclusive design could require significant cultural change	70	85.7	66.7	80	66.7	50	75.0
Perception that inclusive design is more expensive	68	71.4	33.3	70	91.7	40	75.0
Perceived longer development time to market	68	42.9	66.7	50	75.0	90	75.0
Perception – more complex to design inclusively	68	57.1	66.7	70	91.7	40	75.0
Lack of interest in inclusive design	68	57.1	66.7	80	83.3	50	62.5
Stigma associated with inclusive design	66	85.7	33.3	30	83.3	60	87.5
Lack of time and budget	66	57.1	66.7	50	75.0	70	75.0
Lack of company policy on inclusive design	64	71.4	66.7	60	75.0	40	75.0
Inclusive design is a perceived need of our end user	64	85.7	66.7	40	66.7	80	50.0
Lack of methods/tools for practicing inclusive design	64	85.7	66.7	70	75.0	40	50.0
Lack of motivation for tackling inclusive design	64	85.7	100.0	70	83.3	20	50.0
Lack of availability of good design examples	62	71.4	66.7	50	75.0	50	62.5
Perception – there is no need to practice inclusive design	62	71.4	33.3	60	58.3	50	87.5
Working for short term financial objectives	60	42.9	100.0	40	66.7	50	87.5
Perception that inclusive design represents a niche market	58	42.9	66.7	40	75.0	50	75.0
Unavailability of internal support for inclusive design	58	57.1	66.7	60	58.3	40	75.0
Perception of brand association with disabled/older people	54	57.1	100.0	70	66.7	60	62.5
No justifiable business case to support inclusive design	54	57.1	33.3	40	50.0	70	62.5
Lack of knowledge and tools to practice inclusive design	52	57.1	66.7	40	58.3	40	62.5

engage relevant people in these countries. Moreover, during this research exercise, efforts will be made to educate the relevant people about the importance of inclusive design and its impact on our daily life in terms of work productivity and better living. In addition to the social aspects of inclusive design, another important associated factor is the potential business opportunities, and efforts will be made to develop useful business cases for companies to motivate them in the promotion of inclusive design.

7. Conclusion

Changing demographics demand the promotion of the inclusive design method throughout the world. Developing countries like China, India, Pakistan, and Bangladesh make a major contribution to the world population, and the overall population in these countries is increasing and the proportion of older people and people with disabilities is increasing accordingly. This article reveals the need to explore the main barriers and drivers for inclusive design from the perspective of manufacturers, so that drivers can be promoted and barriers can be removed. It has been concluded that people from manufacturing organizations feel that motivation through social responsibility, source of innovation, brand enhancement, and new market opportunities are the main drivers for inclusive design in Pakistan. On the other hand, lack of awareness about inclusive design, lack of resources, guidance, change in culture, government regulations, time and budget are perceived as barriers toward the implementation of design inclusivity in manufacturing organizations. The conclusions on drivers are in close agreement with those found by similar studies in the UK, USA, and Japan as are those on barriers except that this study found that inclusive design was not considered too difficult, less achievable, or likely to compromise esthetics.

However, a difference in the perceptions on drivers and barriers has been observed in different manufacturing sectors that shows that inclusive design promotion can be carried out by focusing on the prioritized relevant factors in each sector. Hence, for example, the electronics industry showed considerable concern for legislation, whereas for the automotive industry customer loyalty and market opportunities were prominent. This supports the idea that efforts are needed to make people aware of the usefulness of inclusive design, not only in the social responsibility context, but also as a potential business opportunity.

The promotion of inclusive design practices will not only be useful in improving living conditions of the people of the developing world, but also will open greater business opportunities for companies. It is not only the responsibility of manufacturing companies, but also of governments to provide adequate resources/guidance that can help in increasing awareness about the usefulness of inclusive design. Development and implementation of proper legislation is also an important area on which to focus where regulatory agencies would have to make sure that organizations are properly implementing inclusive design strategies in their true spirit. These investigations can help in shaping different strategies for the promotion of inclusive design awareness in manufacturing industries in general; and textile, automotive, electronics, beverages, and design consultancies in particular.

This research has contributed in terms of promoting a better understanding of the inclusive design approach and this will ultimately help manufacturers and designers in designing products, services, processes, and environments that are equally safe, healthy, productive, and acceptable for a wide range of population. The aim is to achieve an improvement in living conditions of older people and people with disabilities along with the rest of the population, while developing a business case for manufacturers as these countries constitute a huge potential market of product consumers.

Acknowledgements

This research was carried out in collaboration with the Department of Industrial and Manufacturing Engineering, University of Engineering and Technology, Lahore, Pakistan and the Wolfson School of Mechanical and Manufacturing Engineering, Loughborough University, UK.

Disclosure statement

No potential conflict of interest was reported by the authors.

References

Bellerby, F., & Davis, G. (2003). Defining the limits of inclusive design. *Include 2003* (pp. 1:00–1:17). London: Royal College of Art.
British Standards Institute. (2005). British Standards 7000-6:2005. Design Management Systems – Managing Inclusive Design – Guide.
CIA World Fact Book. (2014). www.cia.gov/library/publications/the-world-factbook/
Coleman, R. (2001). Designing for our future selves. In W. F. E. Preiser & E. Ostroff (Eds.), *Universal design handbook* (pp. 4.1–4.25). New York, NY: McGraw-Hill.
Dong, H., Clarkson, P. J., Ahmed, S., & Keates, S. (2004). Investigating perceptions of manufacturers and retailers to inclusive design. *The Design Journal, 7*, 3–15.
Dong, H., Keates, S., & Clarkson, P. J. (2004). Inclusive design in industry: Barriers, drivers and the business case. In 8th ERCIM Workshop "User Interface for All", Vienna, Austria.
European Institute for Design and Disability. (2006). www.design-for-all.org.
Goodman, J., Dong, H., Langdon, P., & Clarkson, P. J. (2006). Increasing the uptake of inclusive design. In industry. *Gerontechnology, 5*, 140–149.
Keates, S., & Clarkson, P. J. (2004). *Countering design exclusion*. London: Springer.
Pakistan Bureau of Statistics. (2006). *Census of Manufacturing Industries 2005–06, District-wise Report*. Islamabad: Government of Pakistan.
Preiser, W., & Ostroff, E. (2001). *Universal design handbook*. New York, NY: McGraw-Hill.
Royal Academy of Engineering. (2005). Educating engineers in design. Retrieved from www.raeng.org.uk
Sims, R. E. (2003). *'Design for All': Methods and data to support designers* (PhD thesis). Loughborough University, Loughborough, UK.
Underwood, M., & Metz, D. (2003). Seven business drivers of inclusive design. In *Include 2003*. 1, 1:39–1:44.
U.N.O. (2009). *Ageing*. Retrieved from http://social.un.org/index/Ageing.aspx
Unpublished report. (2000). Kyoyo-hin (Universal Design) in Japan. Available from the i~ design collection, Helen Hamlyn Research Centre, Royal College of Art, UK.
Vanderheiden G., & Tobias, J. (2000). Universal design of consumer products: Current industry practice and perceptions; 2000. Retrieved from http://trace.wisc.edu/docs/ud_consumer_products_hfes2000/index.htm
Waller, S., & Clarkson, P. J. (2009). Tools for inclusive design. In C. Stephanidis (Ed.), *The universal access handbook*. Boca Raton, FL: Taylor and Francis.

Appendix A. SURVEY Inclusive Design Survey for assessing the level of awareness, drivers and barriers among Designers

This survey consists of six parts and will only take a few minutes to complete.
Please answer all Questions:

Contact Name
Company name
Position in Company
Telephone Number
E-mail Address
Company Web Site:

Please Note: We will not share this information with any third party or use these detail to contact you unless you ask us to share the results from this study where these results will be presented in an anonymized form.

- Please provide an email address and tick this box if you would like us to send you the anonymized result from this study.
- Please tick this box if you would like us to contact you to discuss how inclusive Design may assist your organization.

1.Please describe what your organization does:
2.Please list down the names of the products/Services your organization provides

Part-1: Company Profile

1. Number of Employees approximately	☐ Up to 250 ☐ 251-1000 ☐ 1001-5000 ☐ Over 5000
2. How many people use your products/services?	☐ Up to 1000 ☐ 1001-10,000 ☐ 10,001-100,000 ☐ 100,001-1 Million ☐ Over 1 Million
3. How many companies represent most of your sales?	☐ Up to 5 ☐ 6-50 ☐ 51-500 ☐ Over 500 ☐ Direct to customer
4. Global Profile (tick all if apply)	☐ Pakistan ☐ Asia ☐ Europe ☐ USA ☐ Global
5. Does your organization have a post for corporate social responsibility (CSR)?	☐ Yes ☐ No ☐ Don't know ☐ Don't understand

Part-2: Current Understanding

1. Have you heard of the following terms? (tick all if apply)	☐ Inclusive Design ☐ Universal Design ☐ Design for all
2. What do you think they mean?	

Part-3: Current Company Position

In answering the following. Please think of the definition of inclusive design.“A process whereby designers, manufacturers, and service providers ensure that their products and services address the needs of the widest possible audience.”

(Please select the appropriate number)

1. What level of awareness of Inclusive Design is there in your organization?	☐ 1 (No awareness) ☐ 2 ☐ 3 ☐ 4 (Extremely aware)
2. How inclusive do you think your products/services are?	☐ (Not inclusive) ☐ 2 ☐ 3 ☐ 4 (very inclusive))
3. What level of effort is utilized to make your products/ services inclusive?	☐ 1 (Little or no effort) ☐ 2 ☐ 3 ☐ 4 (Large amount of effort)
4. Is your company interested in making your products/ services more inclusive?	☐ 1 (Not interested) ☐ 2 ☐ 3 ☐ 4 (Extremely interested)
5. How aware is your company or organization?	**A.** The (Employment and Rehabilitation) Act,1981 of Pakistan ☐ 1 (No awareness) ☐ 2 ☐ 3 ☐ 4 (Extremely aware) **B.** Other legislation, Policies and, or codes of practice? ☐ 1 (No awareness) ☐ 2 ☐ 3 ☐ 4 (Extremely aware)
6. Have you ever heard about the (Employment and Rehabilitation) Act,1981	☐ Yes ☐ No
7. Please specify and particular products/services that are designed to be inclusive.	
8. What level of effort is currently utilized to make your products/services inclusive?	☐ 1 (Little or no effort) ☐ 2 ☐ 3 ☐ 4 (Large amount of effort)
Please elaborate on any specific initiatives that you are able to discuss and comment on their success.	

Part-4: Drivers for inclusive Design

1. How much do you agree with the following five statements?	**1**. Legislation is a major driver for inclusive design within our organization. ☐ 1 (Strongly disagree) ☐ 2 ☐ 3 ☐ 4 (Strongly agree) ☐ Don't know **2.** The (Employment and Rehabilitation) Act, 1981 will help us to practice and manage inclusive design. ☐ 1 (Strongly disagree) ☐ 2 ☐ 3 ☐ 4 (Strongly agree) ☐ Don't know **3**. Social responsibility motivates our organization to consider inclusive design. ☐ 1 (Strongly disagree) ☐ 2 ☐ 3 ☐ 4 (Strongly agree) ☐ Don't know **4**. Demographic and consumer trends are driving our organization commitment to inclusive design. ☐ 1 (Strongly disagree) ☐ 2 ☐ 3 ☐ 4 (Strongly agree) ☐ Don't know **5**. Brand enhancement is a key driver for inclusive design within our organization. ☐ 1 (Strongly disagree) ☐ 2 ☐ 3 ☐ 4 (Strongly agree) ☐ Don't know
2. How effective do you think inclusive design could be in helping you to achieve the following commercial benefits?	Entrance to a new market. ☐ 1 (Not effective) ☐ 2 ☐ 3 ☐ 4 (very effective) Increase the size of your potential market. ☐ 1 (Not effective) ☐ 2 ☐ 3 ☐ 4 (very effective) Increase the size of your potential market. ☐ 1 (Not effective) ☐ 2 ☐ 3 ☐ 4 (very effective)

(*Continued*)

Achieve a large share of your current market.
- ☐ 1 (Not effective)
- ☐ 2
- ☐ 3
- ☐ 4 (very effective)

Increase customer loyalty
- ☐ 1 (Not effective)
- ☐ 2
- ☐ 3
- ☐ 4 (very effective)

Increase revenue through increased usage
- ☐ 1 (Not effective)
- ☐ 2
- ☐ 3
- ☐ 4 (very effective)

Increase customer satisfaction
- ☐ 1 (Not effective)
- ☐ 2
- ☐ 3
- ☐ 4 (very effective)

A source of innovation and differentiation
- ☐ 1 (Not effective)
- ☐ 2
- ☐ 3
- ☐ 4 (very effective)

New market opportunities by practicing inclusive design
- ☐ 1 (Not effective)
- ☐ 2
- ☐ 3
- ☐ 4 (very effective)

Assessment of how many people are excluded
- ☐ 1 (Not effective)
- ☐ 2
- ☐ 3
- ☐ 4 (very effective)

Consumer dissatisfaction with current products
- ☐ 1 (Not effective)
- ☐ 2
- ☐ 3
- ☐ 4 (very effective)

Chances of innovation by practicing inclusive design
- ☐ 1 (Not effective)
- ☐ 2
- ☐ 3
- ☐ 4 (very effective)

Public/consumer awareness of inclusive design
- ☐ 1 (Not effective)
- ☐ 2
- ☐ 3
- ☐ 4 (very effective)

Availability of expert consultation on inclusive design
- ☐ 1 (Not effective)
- ☐ 2
- ☐ 3
- ☐ 4 (very effective)

(Continued)

	Availability of tools/methods to help the practice of inclusive design ☐ 1 (Not effective) ☐ 2 ☐ 3 ☐ 4 (very effective) Availability of training opportunities on inclusive design for staff/designers ☐ 1 (Not effective) ☐ 2 ☐ 3 ☐ 4 (very effective) Champion for inclusive design on company boards ☐ 1 (Not effective) ☐ 2 ☐ 3 ☐ 4 (very effective) Major competitor's adoption of inclusive design ☐ 1 (Not effective) ☐ 2 ☐ 3 ☐ 4 (very effective)
If there are other key Drivers for your organization, please specify?	

Part-5: Barriers to inclusive Design

1. Please respond to the following statements on barriers to inclusive design with respect to your organization.	There is little or no internal support for inclusive design. ☐ Yes ☐ No Implementing inclusive design could require significant cultural change. ☐ Yes ☐ No We lack the knowledge and tools to practice inclusive design. ☐ Yes ☐ No There is no justifiable business case to support inclusive design. ☐ Yes ☐ No There is a lack of time and budget to support inclusive design. ☐ Yes ☐ No There is a perception that inclusive design is too difficult. ☐ Yes ☐ No Inclusive design compromises the aesthetics of the design. ☐ Yes ☐ No Inclusive design is a perceived need of our end users. ☐ Yes ☐ No There is a stigma associated with inclusive design. ☐ Yes ☐ No Inclusive Design is seen as an unachievable goal. ☐ Yes ☐ No Lack of awareness of inclusive design ☐ Yes ☐ No Lack of interest in inclusive design ☐ Yes ☐ No Lack of motivation for tackling inclusive design ☐ Yes ☐ No Perception that inclusive design is more expensive ☐ Yes ☐ No Perception that it can be more complex to design inclusively ☐ Yes ☐ No

(*Continued*)

	Perception that inclusive design represents a niche market ☐ Yes ☐ No Perceived problems of brand association with disabled/order people ☐ Yes ☐ No Perceived longer development time to market ☐ Yes ☐ No Perception that there is no need to practice inclusive design ☐ Yes ☐ No Perception that inclusive design is a passing trend ☐ Yes ☐ No Lack of resources/guidance on inclusive design ☐ Yes ☐ No Lack of availability of good design examples ☐ Yes ☐ No Lack of government regulations ☐ Yes ☐ No Lack of methods/tools for practicing inclusive design ☐ Yes ☐ No Difficulty in changing the culture of business ☐ Yes ☐ No Lack of company policy on inclusive design ☐ Yes ☐ No Working for short-term financial objectives ☐ Yes ☐ No
2. Which of the following barriers do you perceive as most important.	☐ Most important. ☐ Second most important. ☐ Third most important.
3. Which of the above barriers do you perceive as least important and.	☐ Least important. ☐ Second least important. ☐ Third least important.
4. Are there any other significant barriers to the adoption of inclusive Design within your organization?	

Part-6: Tools to increase the Usage of Inclusive Design

1. How important do you perceive the following tools to encourage inclusive Design within your organization?	(A) Convincing arguments (business case) for top level management. ☐ 1 (not important) ☐ 2 ☐ 3 ☐ 4 (extremely important) (B) An increased understanding of inclusive design. ☐ 1 (not important) ☐ 2 ☐ 3 ☐ 4 (extremely important) (C) Skills or tools to assist with designing inclusively ☐ 1 (not important) ☐ 2 ☐ 3 ☐ 4 (extremely important) (D) Effective tools to market inclusive Design ☐ 1 (not important) ☐ 2 ☐ 3 ☐ 4 (extremely important)
Which of these (A-D) is the most important?	Most important Second Most important Third most important
2. Which of the following would be useful to your business (tick at that apply)	**A.** A national awareness campaign. **B.** A national centre of excellence for inclusive Design. **C.** Attending an Inclusive Design event. **D.** Receiving an exclusion assessment on your product/service. **E.** Consultancy and/or training on inclusive design for your organization. **F.** Other, please specify:
Which of these (A-F) is most important, and why?	Most important Second Most important Third most important

9

Validating quality productivity improvement framework for sponge iron industry in India

Vinod S. Gorantiwar* and R.L. Shrivastava

Department of Mechanical Engineering, Yeshwantrao Chavan College of Engineering, Nagpur 441 110, India

In this work, an attempt has been made to validate the quality–productivity improvement framework developed for sponge iron (SI) industry, through its implementation in two different SI plants. The study proved that the developed framework is valid and reliable for coal-based SI industry in India. The same framework may be implemented elsewhere in the world with little modification according to the prevailing environment in that country. The concept can also be expanded to other fields.

Keywords: quality; productivity; sponge iron; critical success factors; performance measurement factors

1. Introduction

Identification of critical success factors (CSFs), performance measurement factors (PMFs), and its relation with each other (called as framework) for particular industry helps improving the desired parameters through its implementation. However, developed framework must be validated by implementing it in representative unit. Through the literature review, it was noticed that the quality–productivity improvement framework has not yet been developed for sponge iron (SI) industry anywhere in the world. The authors have developed a framework empirically for coal-based SI industry in India and validated it through its implementation in two different SI plants.

2. Literature review

India accounted for major share of global SI production and has occupied the top slot across the globe. With the country's strong economic environment, the industry prospects are certainly bright. There has been an inspiring track record of growth of SI industry in India during the last decade.

Table 1 shows the data from the ministry of steel about production of gas and coal-based SI in India from 2006–2007 to 2012–2013; while Table 2 shows segment-wise production and capacity utilization for year 2010–2012. Reduced growth rate in the year 2012 demands something to be done towards the improvement in quality and productivity in SI industry. Figure 1 shows the graph of production of SI from 2000 to 2010.

The year 2011–2012 was the year of political turmoil in India. A lot of policy changes have occurred which affected the SI industry, directly or indirectly. Availability

*Corresponding author. Email: principal@ycce.edu

Table 1. Production of SI in India.

Year	Production of SI (in MT)		
	Gas based	Coal based	Total (gas + coal) based (in MT)
2006–2007	5.26	13.08	18.34
2007–2008	5.84	14.53	20.37
2008–2009	5.52	15.57	21.09
2009–2010	6.15	18.18	24.33
2010–2011	5.79	20.92	26.71
2011–2012	5.17	15.20	20.37
2012–2013(Apr–Dec)	3.23*	11.66*	14.89*

Source: Annual Report (2011–2012), and Annual Report (2012–2013), Ministry of Steel.
*Indicates the production for three quarers.

Table 2. Segment-wise production and capacity utilization in India.

Particulars	2010–2011	2011–2012	% Growth
(A) Gas-based units – production (tons)	6,189,917	5,150,018	(−) .83
Annual installed capacity (lakh tons)	96	96	
Capacity utilization (%)	64.47	53.64	
(B) Coal-based units – production (tons)	17,065,432	15,407,645	(−) .90
Annual installed capacity (lakh tons)	253.35	257.9	
Capacity utilization (%)	67.35	59.93	
Grand total (A + B)	23,255,349	20,557,663	(−) .88
Total capacity (lakh tons)	349.35	353.09	
Overall capacity utilization	66.56	58.22	

Source: www.spongeironindia.in/prodfig0607.html, The official website of SIMA India, Retrieved from 10 January 2013.

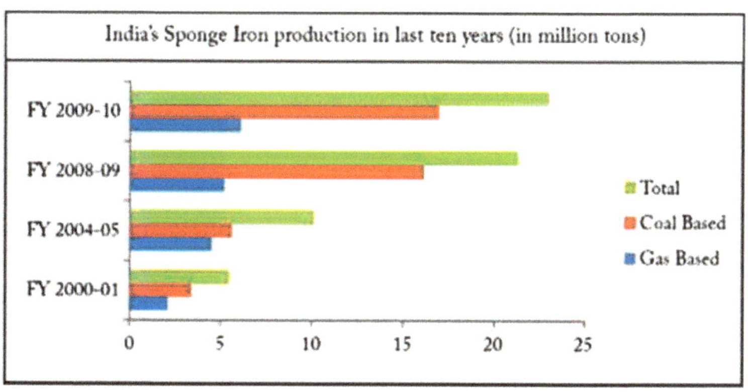

Figure 1. India's SI production.
Source: Ore Team Research (2011).

of iron ore and coal reduced drastically. Many units could not operate to its full capacity. Efforts for quality and productivity improvement were totally at bay. However, few companies (e.g. Tata Sponge Iron Ltd) still managed to earn the profit, resulting from their efforts for quality and productivity improvement through total quality management (TQM) proving the importance and necessity of implementation of TQM in SI industry in India. (www.spongeironindia.in/presentations/PCTATASPONGESIMA Presentation100912.pdf, Retrieved from 22 December 2012; www.tatasponge.com/investor/annualreports/annualreport2010-11.pdf, Retrieved from 10 November 2012; www.tatasponge.com/investor/annualreports/annualreport2011-12.pdf, Retrieved from 10 January 2013.)

During 80s, there was a great demand for SI due to shortage of scrap. Moreover, many new SI plants were established after 1985. Thus, during 80s, quality of SI was not an important factor as there was a huge demand due to the shortage of scrap. Very little attention was paid to the quality and productivity of SI. But as the competition grew, there was a need to focus on the quality and the productivity improvement to survive. After the formation of SIMA i.e. Sponge Iron Manufacturers Association in 1992, the manufacturers had to set standards of quality of SI. Thus, quality of SI gradually began a point of concern in the past.

2.1. *Targeted vs. achieved SI production in India*

Table 3 shows the data for targeted production and achieved production for gas-based and coal-based SI manufacturing in India for the year 2004–2005 to 2011–2012.

The graph drawn from the data from Table 3 is shown in Figure 2. It is clear from the figure that the achieved production is always less than the targeted production in coal-based SI plants. This fact demands the need for identification and validation of critical factors affecting quality and productivity in coal-based SI industry.

2.2. *Problems of SI industry in India*

A panoramic view through the review of Sengupta (2000), Sengupta (2001), Pandit, Sarangi, Babu, and Sheshadri (2003), Sengupta (2005), Raju (2005), Raja and Pal (2006), Kalluraya, Ramchandra, and Sriniwasan (2006), Battacharjee (2007), Equitymaster.com (2009), Sengupta (2009), Khattoi (2010), Reddy (2010), Borwankar (2011), Infocus (2011), Sengupta (2011), Steel Sector Analysis Report (2011), Dhopatkar (2012), Invest India (2012), Kashiva (2012), Scribe.com (2012), and Sengupta (2012a, 2012b), etc. indicates the SI industry with investments of over Rs. 5000 crore contributing more than Rs. 450 crore per annum by way of taxes to the national exchequer, saving sub-

Table 3. Targeted vs. achieved production for gas- and coal based SI in India.

Year	Targeted production (MT)			Achieved production (MT)		
	Gas-based	Coal-based	Total	Gas-based	Coal-based	Total
2004–2005	6.1	6.0	12.1	4.6	5.5	10.1
2005–2006	6.1	8.5	14.6	5.7	6.5	12.2
2006–2007	7.1	11.0	18.1	7.0	8.5	15.5
2007–2008	7.1	13.0	20.1	7.0	10.0	17.0
2008–2009	7.1	17.0	24.1	7.0	15.3	22.3
2009–2010	7.1	18.0	25.1	7.0	14.0	21.0
2010–2011	9.6	25.34	34.94	6.19	17.07	23.26
2011–2012	9.6	25.34	34.94	5.15	15.41	20.56

Source: Chatterjee (2010) and www.steelguru.com, Retrieved from 10 June 2014.

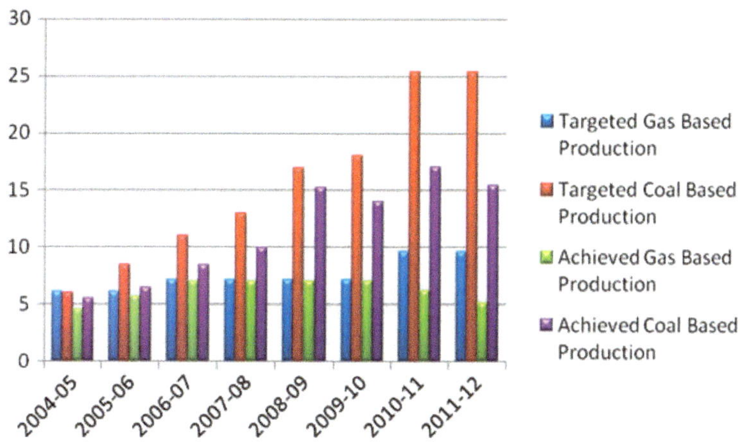

Figure 2. Gas- and coal-based, targeted vs. achieved production of SI in India.

stantial foreign exchange and employing directly and indirectly nearly 1,25,000 people. On the other hand, it is also an industry which is under tremendous viability pressure due to raw material availability and pricing constraints. Selling prices are determined by market forces, whereas all input costs are controlled by the government. The problems of the SI industry are:

- The cost of basic inputs like iron ore, coal, and gas have steeply gone up.
- All input costs (administered pricing) are higher in India than overseas.
- Inputs/services for SI production are mainly in government hands.
- High cost of capital.
- Slowdown of economy, resulting in demand recession.
- Availability of right grade and quality of non-coking coal is a must to optimize cost of production of SI. Indian coal having high-ash content (30–35%) will necessarily have to be blended with imported low-ash content coal (below 12%). Therefore, reduction of custom duty on non-coking coal (below 12% ash content) for metallurgical use is necessary.
- High prices of natural gas in comparison to international prices and cost of production resulting from high input cost and inadequate export incentives.

Table 4. List of CSFs (input factors) and PMFs (output factors).

Critical success factors (CSFs)/input factors	Performance measurement factors (PMFs)/output factors
IF 1: Process management	*OF 1*: Productivity improvement
IF 2: Top management commitment	*OF 2*: Business result
IF 3: Employee involvement	*OF 3*: Employee satisfaction
IF 4: Organization culture	*OF 4*: Time dimension
IF 5: Effective networking	*OF 5*: Customer satisfaction
IF 6: Quality assurance	*OF 6*: Quality performance
IF 7: Competitive assessment	
IF 8: Training	

- The iron ore used as a raw material for SI needs to be of higher quality with more than 62% Fe content. Currently, out of India's estimated reserves of 25 billion tons of iron ore, higher grade constituents are only about 8.7%.

Thus, the various gaps those have been identified are as follows:

(1) Large numbers of SI plants have come up with small capacity of 50–100 TPD due to government's de-licensing policy. They are relatively new and facing many problems. They are coal based due to large availability of non-coking coal and availability of technology. However, for the sake of higher and higher production, Quality and productivity is highly neglected. In the present crisis of closure of many plants, an urgent need has been generated to take care of quality and productivity in SI industry.

(2) Most of the plants are run in traditional ways. Thumb rules and conventional knowledge is employed for getting the work done. Except very few plants, awareness of quality management system is extremely poor and its implementation is the matter of concern.

(3) Exhaustive literature review indicated that CSFs and PMFs for SI industry have not yet been derived. An attempt was made by Gorantiwar, Kshirsagar, and

Table 5. Multiple regression analysis.

	OF					
IF	OF – 1: Productivity improvement	OF – 2: Business result	OF – 3: Employee satisfaction	OF – 4: Time dimension	OF – 5: Customer satisfaction	OF – 6: Quality performance
IF – 1: Process management	.223 (2.641)**	.472 (6.694)**	.270 (3.161)**	.227 (2.967)**	−.039 (−.380)	.305 (3.602)**
IF – 2: Top management commitment	.209 (2.993)**	.147 (2.223)*	.112 (1.403)	−.060 (−.785)	.283 (3.470)**	.184 (2.421)*
IF – 3: Employee involvement	.179 (2.418)*	.138 (1.890)	.272 (3.333)**	.298 (3.824)**	.185 (2.270)*	.099 (1.165)
IF – 4: Organization culture	.136 (2.186)*	.254 (4.196)**	.150 (2.277)*	.054 (.800)	.052 (.710)	.105 (1.561)
IF – 5: Effective networking	−.272 (−5.676)**	−.137 (2.931)**	−.042 (−.767)	.305 (6.198)**	.015 (.247)	.003 (.042)
IF – 6: Quality assurance	.186 (3.391)**	−.038 (−.693)	−.002 (−.036)	−.061 (−1.018)	.067 (.982)	.274 (4.075)**
IF – 7: Competitive assessment	.044 (.746)	.086 (1.624)	.001 (.019)	−.004 (−.061)	−.020 (−.337)	−.110 (−1.975)*
IF – 8: Training	.029 (.514)	.083 (1.481)	.082 (1.308)	.062 (.999)	−.026 (−.385)	.049 (.738)
R	.7463	.7483	.6253	.6656	.4370	.6285
R^2	.557	.560	.391	.443	.191	.395
F-value	53.768**	82,439**	55.577**	69.049**	30.772**	42.230**
Intercept	2.210	2.127	2.392	2.345	3.051	2.632

*Indicates one tailed significance.
**Indicates two tailed significance.

Shrivastava (2011), to identify the factors affecting quality and productivity in direct-reduced iron (DRI) industry; however, it was limited to small region and it lacked validation.

(4) Available literature indicates that no attempt has yet been made to evolve quality–productivity improvement framework for SI industry anywhere in the world.

(5) There is a need to understand the correlation between CSFs and PMFs for appropriate allocation of resources for better performance of SI industry.

To fulfill the gap, authors have developed the framework for coal-based SI industry in India. The framework is represented in the form of models as given in the section below.

3. Models for SI industry

The data collected from the respondents through a specially prepared instrument was analyzed for factor analysis which resulted into eight CSFs & six PMFs for coal-based SI industry in India as listed in Table 4.

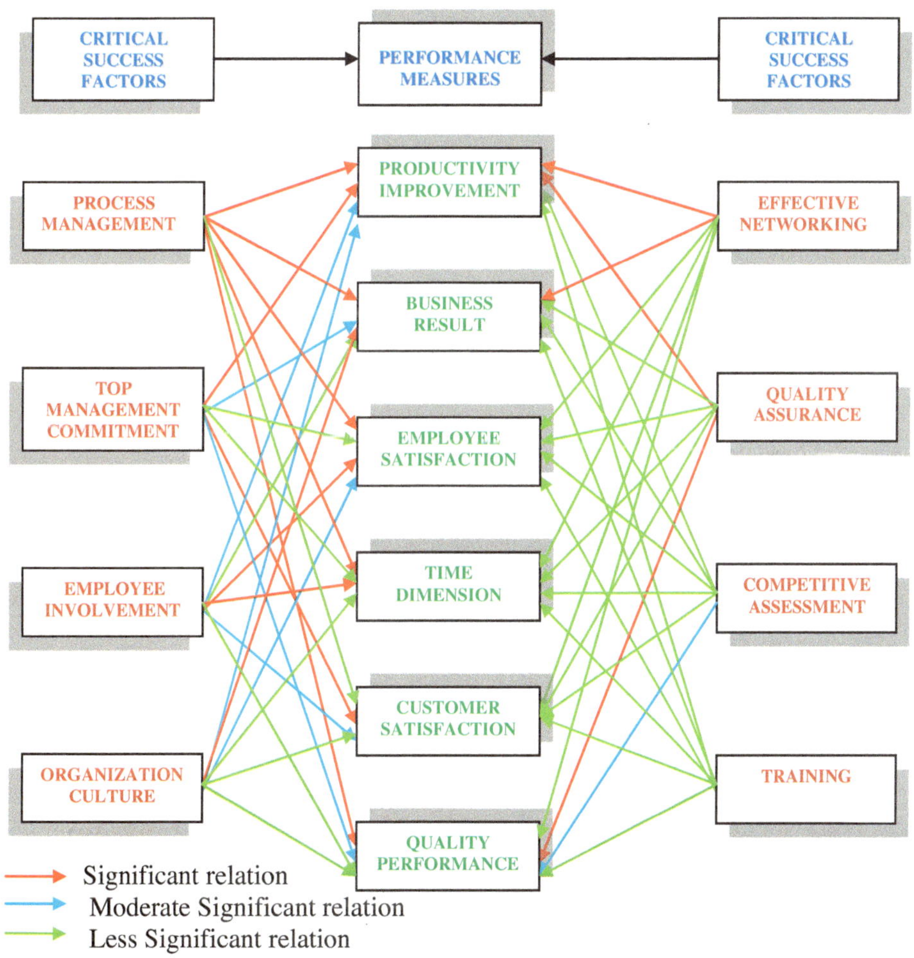

Figure 3. Graphical representation of regression analysis.

The data were regressed to find out the relations between output factors and input factors. Table 5 shows the result of multiple regression analysis. Software SPSS V 20 was used (George & Mallery, 2009; Gupta & Gupta, 2011; IBM SPSS, n.d.; Kerr, Hall, & Kozub, 2002; SPSS online, n.d.). Red color values in the table show significant relations, blue color values show moderate significant relations, and green color values show less significant relations.

The graphical representation of regression analysis is shown in Figure 3.

The regression analysis has resulted into six models for quality–productivity improvement in coal-based SI industry in India as below. They show the relations between output factors and input factors as follows:

(1) *OF 1 (Productivity Improvement)* = 2.210 + .223 (IF1) + .209 (IF2) + .179 (IF3) + .136 (IF4) – .272 (IF5) + .186 (IF6)
(2) *OF 2 (Business Result)* = 2.127 + .472 (IF1) + .147 (IF2) + .254 (IF4) – .137 (IF5)
(3) *OF 3 (Employee Satisfaction)* = 2.392 + .270 (IF1) + .272 (IF3) + .150 (IF4)
(4) *OF 4 (Time Dimension)* = 2.345 + .227 (IF1) + .298 (IF3) + .305 (IF5)
(5) *OF 5 (Customer Satisfaction)* = 3.051 + .283 (IF2) + .185 (IF3)
(6) *OF 6 (Quality Performance)* = 2.632 + .305 (IF1) + .184 (IF2) + .274 (IF6) – .110 (IF7)

Input factors and their contribution in every performance measure can be observed from equations above. It helps managers to allocate the proper share of resources to input factors so as to get desired performance.

4. Validation of models

Model validation is done through model implementation called as case study. According to Zainal (2007), case study is a robust research methodology, when a holistic and in-depth investigation is needed. Case study gives chance to study one aspect of a real-world problem in detail from many different viewpoints. According to Pegram (1999), case study research is a single, in-depth investigation. Yin (1984) has defined the case study method as 'an empirical inquiry that investigates a contemporary phenomenon within its real life context, when the boundaries between phenomenon and context are not clearly evident; and in which multiple sources of evidence are used'. According to Tellis (1997), case studies are designed to bring out the details from the viewpoint of participants by using multiple sources of data. In the opinion of Zainal (2007), case studies are considered useful in research as they enable researchers to examine data at the micro-level. Sroka and Silic (n.d.) found the case studies to be very solid and widely used method by researchers.

Validation of the developed model was done in two different SI manufacturing units. The selection was done in such a way that both the plants differ in many aspects such as production capacity, production method, year of establishment, number of employees, location, ownership, etc.

4.1. Company A (Case study – I)

The plant is equipped with state-of-the-art machineries and the latest technology, making the plant one of the best in India. The usage of most advanced technology helps in keeping emission low and optimize every process in the plants to be energy efficient, thus making the product eco-friendly. The company has pioneered the industry with two induction furnaces of 40 metric tons each along with a common four-strand castor which produces high-quality steel billets. The same is rolled in the most advanced rolling mills to produce TMT bars. With world's largest medium frequency induction melting furnace at its steel plant which is one of the large steel manufacturing units with the capacity to produce .5 million tone steel per annum.

4.1.1. Quality policy

The company has always strived to produce the finest steel in the world. The committed excellence of company in delivering quality products ensures that the integrated steel plant delivers the highest quality steel. The consistent mechanical and chemical properties of steel are attributed to the usage of high-quality virgin steel that is rolled with superior technology. Stringent quality control at every step makes sure that each product comes out with a stamp of consistency. Company adheres to extremely stringent norms to ensure that the higher quality steel reaches the end users. The fully equipped labs with some of the most ultra modern testing and constant upgradation and research make the company best in its class. The company is certified for international quality standard ISO 9001:2008 and ISO 9001:14000 from JNS-ANZ Australia. The company is highly conscious about maintaining the high quality of its product always.

4.2. Company B (Case study – II)

The plant of Company B has three different units in the same premises. SI is produced in 4 kilns of 100 TPD capacities each. The produced SI is supplied as raw material to steel melting shop (SMS) in the same premises for billet making. The waste heat, along with coal is used for power generation. Thus, the three units are:

(1) DRI unit to produce SI
(2) SMS to produce billets
(3) Power plant to produce electricity

4.2.1. Quality policy

The basic principle of the company is, 'there is no survival in long term without quality and services.'

4.3. Data analysis

In order to evaluate the companies' TQM implementation practices and organizational performance, an instrument was prepared as given in Appendix 1. The instrument was administered to evaluate the current situation of TQM implementation and performance. The responses were collected from employees of the plant having different designations and different experiences. The evaluation results are presented in subsequent sections which are translated into score according to scoring methods. It also highlights the extent of relationship between various TQM implementation practices and overall performance.

Table 6. Reliability analysis of instrument.

	Cronbach's alpha
Company A	.916
Company B	.946

4.3.1. *Reliability of research instrument*

The data received from the respondents of Company A as well as from Company B were analyzed for the reliability of research instrument using the software SPSS V 20. The Cronbach's Alpha was found much above .7, indicating the reliability of instrument. Thus, it permits the further analysis of data (Table 6).

4.3.2. *Assessment of company*

The responses in the form of data were analyzed, mean of each item was estimated and items were grouped as respective factors. Further, grand mean of all factors were obtained. The mean of responses were proportionally marked in consultation with the experts. Number '1' indicates that the organization is extremely weak in the practice and number '8' indicates that the organization is extremely strong in the practice. During the process of assessment, the strength and weaknesses of the items of the respective factors were pointed out. Score above and below indicates strong and weak items, respectively. If the company is neither strong nor weak in the particular item, it is indicated as 'Average'. It is relative judgment. Table 7 explains the marking scheme and grading scheme. Based on the total score of factors, the overall grading criterion was decided. The total score is indicated as X.

Table 8 gives the assessment result after detailed analysis of the degree of emphasis of TQM implementation.

Based on above analysis, overall grade of the company for emphasis of TQM implementation and its impact on organizational performance was estimated as follows: (Table 9).

Thus, by using the instrument developed for case study, the degree of emphasis of TQM implementation and its impact on organizational performance was measured. The overall grade achieved as per grading criteria is 'A'. Through the assessment of the companies' quality management practices and overall business performance, most of the established relationship between implementation factors and the performance measures of the companies found to be strong to very strong. Two relations such as Top Management Commitment with Productivity and Quality assurance with Customer Satisfaction

Table 7. Mean and proportionate score and grading criteria.

Mean (response)	Score (strength)	Total score range	Grading
1.00–1.50	1 (Extremely weak)	$X > 90\%$	A^+
1.51–2.00	2 (Very weak)	$80\% < X < 90\%$	A
2.01–2.50	3 (Weak)	$70\% < X < 80\%$	B
2.51–3.00	4 (Average)	$60\% < X < 70\%$	C
3.01–3.50	5 (Above average)	$50\% < X < 60\%$	D
3.51–4.00	6 (Strong)	$X < 50\%$	E
4.01–4.50	7 (Very strong)		
4.51–5.00	8 (Extremely strong)		

Table 8. Assessment results for Company A and Company B.

S. No	Critical success factors	Company – A score (out of 8)	Company – B score (out of 8)
1	Process management	7	7
2	Top management commitment	7	7
3	Employees involvement	6	7
4	Organization culture	7	7
5	Effective networking	7	6
6	Quality assurance	7	6
7	Competitive assessment	7	6
8	Training	7	7

Table 9. Estimation of overall grade of company.

S. No	Particulars	Company A	Company B
1	Maximum numerical weightage assigned to each factor	8	8
2	Total number of factors	8	8
3	Maximum score	$8 \times 8 = 64$	$8 \times 8 = 64$
4	Total score achieved	54	53
5	% of scores achieved	84.375%	82.81%
6	*Overall grade*	*A*	*A*

are found extremely strong. Thus, company is doing well on the front of quality–productivity improvement. This has also validated the instrument used for case study.

4.3.3. *Impact of quality practices on the organizational performance*

As a result of competitiveness and to strive for better performance, the company has implemented various performance improvement approaches. Many steps have been taken by the company during past to enhance overall performance. The aim of the organization is to achieve better result than the past.

4.3.3.1. TQM implementation. Companies have taken many initiatives for the implementation of QMS for the quality–productivity improvement.

(1) *Process management*: every care is taken for the process management at both the companies. All departments are involved in quality improvement; there are robust feedback mechanisms; all employees audit the process regularly; there are faster adoptions of new technology for continuous improvement; automation is promoted in production; employees' suggestions are considered favorably; the process is monitored continuously for improvement; R&D and innovation in the product development are promoted; employees are motivated for QMS implementation.
(2) *Top management commitment*: top management of both the companies are committed in delivering quality products through quality practices. They have clear vision, mission, and strategic direction. Top management is able to take the responsibility for continuous quality improvement. Adequate budgetary allocations are made for quality improvement initiatives. Constructive efforts of

Table 10. Mean and *F*-value of ANOVA for Company – A.

	1-Productivity	2-Business results	3-Employee satisfaction	4-Time dimension	5-Customer satisfaction	6-Quality performance
1-Process management	4.67 (.310)	3.85 (1.308)	3.81 (2.603)	4.15 (1.219)	4.15 (.579)	4.18 (.441)
2-Top management commitment	4.59 (.691)	4.11 (2.639)	3.81 (.463)	4.00 (1.570)	3.70 (.654)	4.04 (.873)
3-Employee involvement	4.26 (1.149)	3.96 (1.503)	4.11 (1.484)	3.89 (4.191)*	3.59 (2.634)	4.00 (1.549)
4-Organization culture	4.26 (3.465)*	4.07 (1.080)	3.85 (.886)	4.07 (2.365)	4.11 (1.020)	4.22 (1.318)
5-Effective networking	4.26 (.221)	4.44 (.024)	3.89 (2.730)	3.74 (.924)	4.04 (2.147)	4.04 (2.648)
6-Quality assurance	4.26 (.993)	3.96 (.242)	4.15 (.388)	4.18 (1.279)	4.55 (1.096)	4.48 (.765)
7-Competitive assessment	4.33 (.135)	4.3 (.298)	3.18 (2.097)	3.67 (2.233)	3.89 (.756)	4.18 (.233)
8-Training	4.41 (5.201)**	4.15 (7.653)**	4.30 (.584)	4.26 (2.109)	4.00 (.265)	4.48 (2.886)

*Indicates one tailed significance.
**Indicates two tailed significance.

Table 11. Mean and *F*-value of ANOVA for Company – B.

	1-Productivity improvement	2-Business results	3-Employee satisfaction	4-Time dimension	5-Customer satisfaction	6-Quality performance
1-Process management	4.06 (.954)	4.18 (1.044)	3.82 (.557)	3.53 (.566)	4.41 (.684)	4.47 (.366)
2-Top management commitment	4.29 (6.898)**	4.06 (.195)	3.94 (.275)	3.65 (2.464)	4.35 (.765)	4.35 (.131)
3-Employee involvement	4.35 (3.524)*	4.06 (2.079)	3.41 (.777)	3.94 (.908)	4.00 (2.569)	4.41 (3.704)*
4-Organization culture	4.23 (1.753)	3.94 (.153)	3.94 (.430)	3.88 (1.058)	3.94 (3.366)	4.12 (2.737)
5-Effective networking	4.29 (1.842)	4.18 (1.488)	3.88 (.698)	3.76 (.801)	3.88 (5.134)*	3.82 (8.125)**
6-Quality assurance	4.00 (.603)	4.06 (.858)	3.70 (.535)	3.29 (.592)	4.06 (.384)	4.29 (.744)
7-Competitive assessment	3.94 (.369)	3.76 (.398)	3.59 (.040)	3.59 (1.034)	3.70 (.986)	3.65 (1.463)
8-Training	4.47 (.566)	4.23 (3.526)*	4.00 (2.208)	3.94 (1.256)	3.70 (3.183)	4.06 (1.824)

*Indicates one tailed significance.
**Indicates two tailed significance.

employees are recognized. Congenial work environment is provided along with fair compensation package to the employees.

(3) *Employee involvement*: both the companies put the efforts for capacity building of employees. Innovation in the work is always encouraged. Good coordination

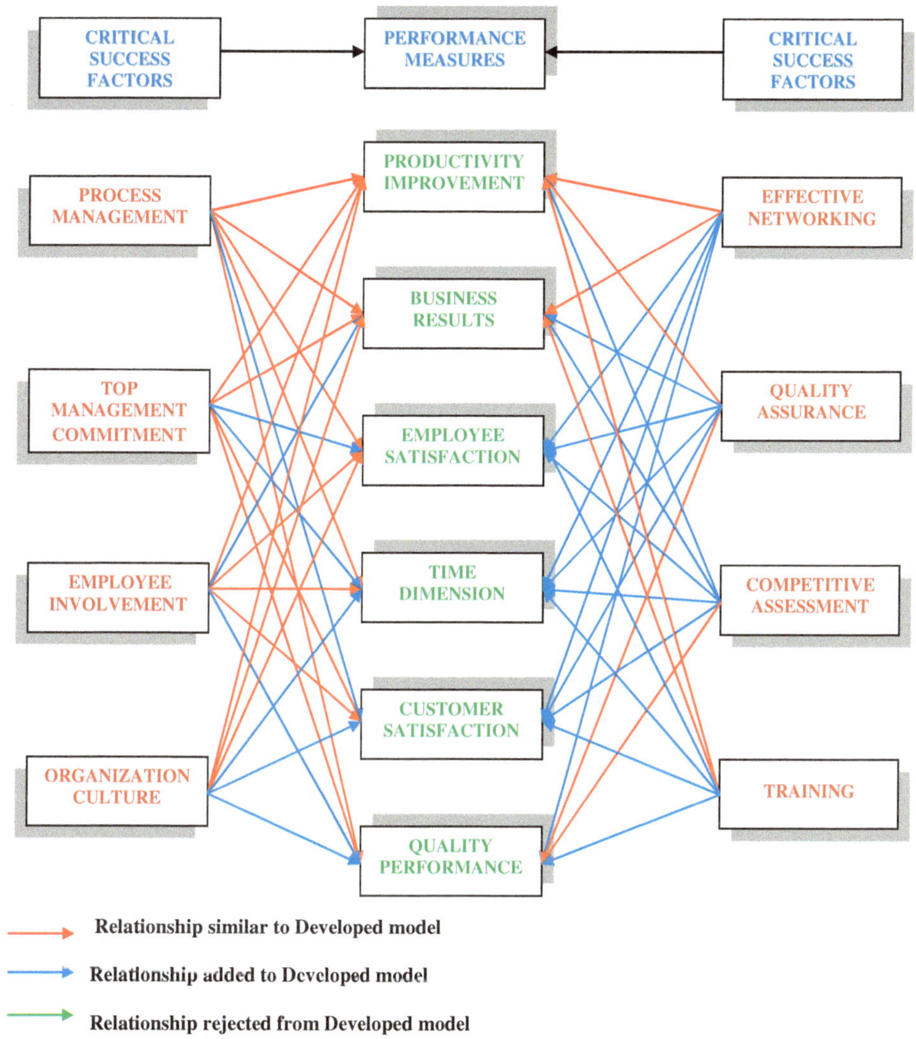

Figure 4. Graphical representation of relations for Company A.

and cooperation is maintained in the plant. All these help in more and more involvement of employees.

(4) *Organization culture*: companies have well-defined organizational structure and governance system. Through the well-defined rules, regulations, and operating procedures and team spirit among the employees, companies get better performance.

(5) *Effective networking*: it is observed that effective networking has been managed by the companies through various ways such as honest and two-way communication between management and employees, regular monitoring of internal and external customer satisfaction; providing technical guidance to suppliers etc.

(6) *Quality assurance*: quality is assured and improved continuously through various ways. Every effort is directed towards the achievement of better and better quality through integration of prevention and correction in production process.

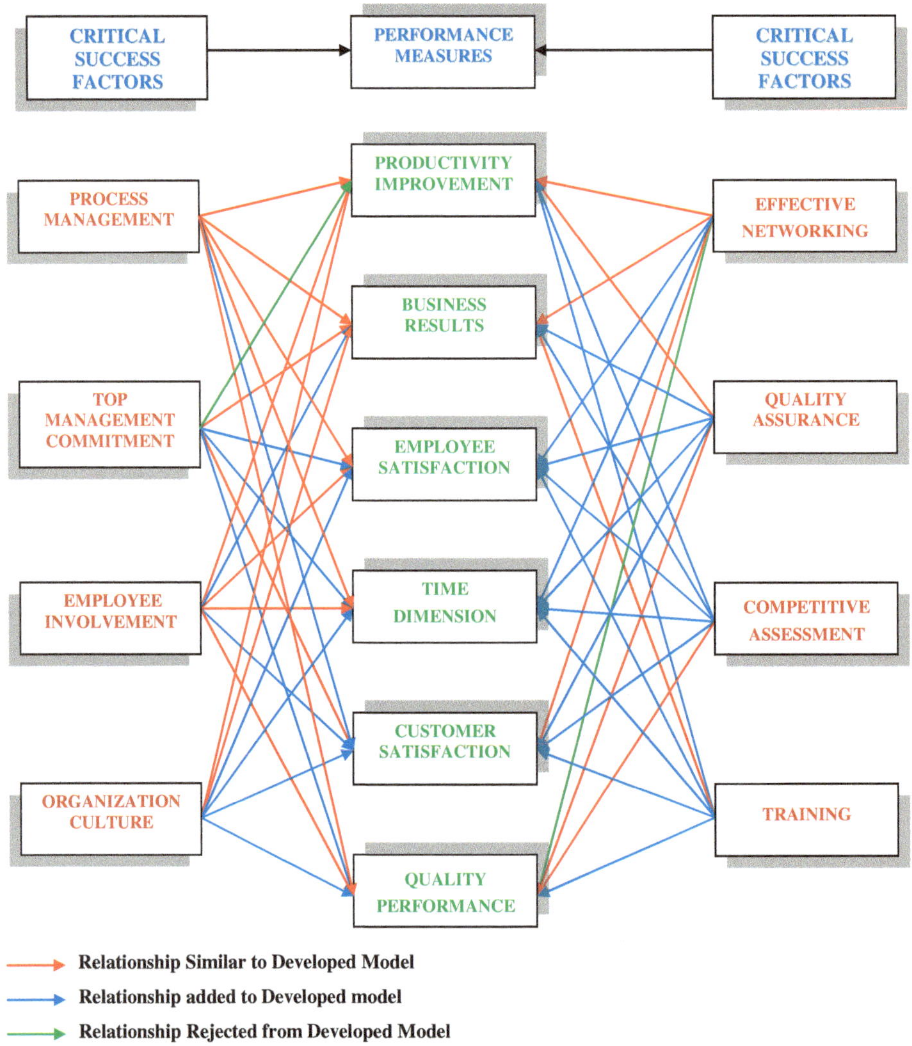

Figure 5. Graphical representation of relations for Company B.

(7) *Competitive assessment*: companies maintain the internal supply chain and manages the relationship with suppliers. Based on the current performance, the goal is set for future.

(8) *Training*: the employees are given on job training to understand the practical aspects of SI manufacturing. Top management of both the companies provides guidance and ensures the effectiveness of training.

Thus, by using the instrument developed for case study, the degree of emphasis of TQM implementation and its impact on organizational performance was measured. The overall grade achieved as per grading criteria is 'A'. Through the assessment of the companies' quality management practices and overall business performance, most of the established relationship between implementation factors and the performance measures of the company found to be strong to very strong. One relation i.e. Training with Time Dimension is found extremely strong, while two relations i.e. Employee Involvement with Employee Satisfaction and Quality Assurance with Time Dimension found above average. Thus, company is doing well on the front of quality–productivity improvement. This has also validated the instrument used for case study.

5. Result

For the validation of the models developed for coal-based SI industry in India as given in section 3 and to find out the significance of developed relations, the collected data for company A and Company B was analyzed using the software SPSS V 20 to carry out ANOVA through mean and F-value. ANOVA is useful to identify the differences between the perceptions of employees based on some parameter. Tables 10 and 11 show the mean and F-value of ANOVA for Company A and Company B, respectively.

The Figure 4 shows similarity of relations with the earlier developed relations for Company A.

The Figure 5 shows similarity of relations with the earlier developed relations for Company B.

6. Conclusion

From the data given in Tables 10 and 11 and subsequently drawn into Figures 4 and 5, it is observed that there are many relations similar to developed model for coal-based SI industry in India, shown in Figure 3.

The relations those are observed similar to the earlier developed relations are:

(1) Process management and productivity improvement
(2) Process management and business result
(3) Process management and employee satisfaction
(4) Process management and time dimension
(5) Process management and quality performance
(6) Top management commitment and business results
(7) Top management commitment and customer satisfaction
(8) Employee involvement and productivity improvement
(9) Employee involvement and employee satisfaction
(10) Organization culture and productivity improvement
(11) Organization culture and business results
(12) Effective networking and productivity improvement
(13) Effective networking and business results
(14) Quality assurance and productivity improvement
(15) Quality assurance and quality performance
(16) Competitive assessment and quality performance
(17) Training and productivity improvement

In both the companies, some relationships have been added to earlier relations as follows –

(18) Process management and customer satisfaction
(19) Top management commitment and time dimension
(20) Organization culture and time dimension
(21) Organization culture and customer satisfaction

(22) Organization culture and quality performance
(23) Effective networking and employee satisfaction
(24) Effective networking and time dimension
(25) Effective networking and quality performance
(26) Quality assurance and business results
(27) Quality assurance and employee satisfaction
(28) Quality assurance and customer satisfaction
(29) Competitive assessment and productivity improvement
(30) Competitive assessment and business results
(31) Competitive assessment and employee satisfaction
(32) Competitive assessment and customer satisfaction
(33) Training and employee satisfaction
(34) Training and time dimension
(35) Training and customer satisfaction
(36) Training and quality performance

Company A does not reject any relation, while only two relations are rejected by Company B. Since almost all the relations indicated by model developed for coal-based SI industry are endorsed by both case studies, *both the case studies validate the model developed for coal-based SI industry in India.*

7. Discussion

Validation of the model, developed for coal-based SI industry in India has been done through the two case studies. Both the plants have different capacity, location, manufacturing process, and number of employees. This has enabled to get the quality–productivity status from the SI manufacturers of different nature. It is found that significant relationships exist between implementation factors and performance measures of companies. It was noticed that various performance indicators have remarkable improvement over the years. The companies have achieved both tangible and intangible benefits by practicing quality management.

8. Future scope

The validated model may be implemented to all SI plants in India for quality–productivity improvement. Ultimately, it will help improve the bottom line of the company – the very essence of any business. The concept can also be expanded to other fields.

Disclosure statement

No potential conflict of interest was reported by the author(s).

References

Battacharjee, S. (2007). *Sponge iron industry: Current scenario* [online]. Retrieved November 20, 2012, from http://www.steelworld.com/report0707.pdf

Borwankar, R. S. (2011, January). Gas based sponge iron industry – Outlook. *DRI Update, A Journal of Sponge Iron Manufacturing Association,* 14–19.

Chatterjee, A. (2010). *Sponge iron industry in India – Scope for clean technology.* Retrieved January 10, 2013, from http://ebookbrowse.com/amit-chatterjee-sponge-iron-industry-in-india-scope-for-clean-technolog

Dhopatkar, V. (2012, February). Coal sector in India. *Steelworld*, pp. 16–20.
Equitymaster.com. (2009). *Steel*. Retrieved November 4, 2012, from www.equitymaster.com/research-it/sector-info/steel/index-jul09.html
George, D., & Mallery, P. (2009). *SPSS for windows step by step* (8th ed.). Pearson Education.
Gorantiwar, V. S., Kshirsagar, V. P., & Shrivastava, R. L. (2011). Quality & productivity in direct reduced iron industry and factors affecting it. In *Proceedings of the international conference on sustainable manufacturing: issues, trends and practices (ICSM – 2011)* (pp. 303–312). November 10–12, 2011, BITS, Pilani, India.
Gupta, S. L., & Gupta, H. (2011). *SPSS 17.0 for researchers* (1st ed.). International Book House.
IBM SPSS. (n.d.). *Statistics 20 brief guide* [online]. Retrieved October 10, 2012, from ftp://public.dhe.ibm.com/software/analytics/spss/documentation/statistics/20.0/en/client/Manuals/IBM_SPSS_Statistics_Brief_Guide.pdf
Infocus. (2011). *Sponge iron industry: An analysis* [online]. Retrieved November 10, 2011, from http://www.dhanbank.com
Invest India. (2012). *A joint venture of Ministry of Commerce & Industry, government of India and Federation of Indian Chambers of Commerce & Industry, 'steel – Investment opportunity'*. Retrieved December 30, 2012, from www.investindia.gov.in/?q=steel-sector
Kalluraya, A. S., Ramchandra, G. K., & Sriniwasan, T. M. (2006, March). Iron ore pellets as feed to sponge iron plants. *Steelworld*, pp. 17–21.
Kashiva, D. (2012, March). Sponge iron industry faces raw material crisis. *Steelworld*, pp. 18–20.
Kerr, A. W., Hall, H. K., & Kozub, S. A. (2002). *Doing statistics with SPSS* (1st ed.). New Delhi: Sage.
Khattoi, S. C. (2010, April). Changing scenario of coal based sponge iron. *DRI Updates, Journal of Sponge Iron Manufacturer's Association of India,* 16–18.
Ministry of Steel, Government of India. (2011–2012). *Annual report*. Retrieved January 5, 2013, from www.steel.gov.in
Ministry of Steel, Government of India. (2012–2013). *Annual report*. Retrieved June 10, 2013, from www.steel.gov.in
Ore Team Research. (2011, January). Sponge iron industry – Indian outlook. *DRI Update, Journal of Sponge Iron Manufacturing Association,* 39–40.
Pandit, A., Sarangi, B. M., Babu, K., & Sheshadri, M. K. (2003, December). Coal based sponge iron industry – A prime movers to enhance steel making capacities in India. *Steelworld*.
Pegram, A. (1999). What is case study research? *Nurse Researcher, 7*, 5–16.
Raja, B. V. R., & Pal, N. (2006, April). India sponge iron industry – Status, potential & prospects. *IIM Metal News, 9*, 11–13.
Raju, P. R. K. (2005, July). Sponge iron industry – An overview of problems & solutions. *Steelworld*, p. 20.
Reddy, J. M. (2010, August). Role of raw materials for production of sponge iron. *Steelworld*, pp. 40–42.
Scribe.com. (2012). *Sponge iron report*. Retrieved October 22, 2012, from http://www.scribd.com/doc/7278475/Sponge-Iron-Report
Sengupta, S. (2000, September). Indian sponge iron industry: Growth despite constraints. *Steelworld*.
Sengupta, S. (2001, October). Indian sponge iron industry in global perspective. *Steelworld*.
Sengupta, S. (2005, July). Gearing up to meet growing metallics requirement. *Steelworld*, pp. 11–18.
Sengupta, S. (2009, June). An outline of Indian ferro alloy industry. *Steelworld*, pp. 14–21.
Sengupta, S. (2011, April). Indian steel today & tomorrow. *Steelworld*, pp. 16–24.
Sengupta, S. (2012a, April). Indian steel: Big challenge ahead. *Steelworld*, pp. 26–42.
Sengupta, S. (2012b, June). Indian steel: Performance in FY'12 and its future prospects. *Steelworld*, pp. 34–46.
SPSS. (n.d.) [online]. Retrieved October 10, 2012, from http://en.wikipedia.org/wiki/SPSS
Sroka, A., & Silic, M. (n.d.). *Which requirements do case studies have to meet and how can they be analyzed?* Report on Ph.D. Seminar. St. Gallen: University of St. Gallen.
Steel Sector Analysis Report. (2011). Retrieved from www.equitymaster.com/research-it/sector-info/steel/Steel-Sector-Analysis-Report.asp
Tellis, W. (1997). Application of a case study methodology. *The Qualitative Report, 3*(3), 1–18.
Yin, R. K. (1984). *Case study research: Design and methods* (1st ed.). Beverly Hills, CA: Sage.
Zainal, Z. (2007). Case study as a research method. *Jurnal Kemanusiaan bil,* (9), 1–6.

Appendix 1

Questionnaire for case study validation

Identification of critical success factors and performance measures for sponge iron industry – model validation (The information provided here will be kept strictly confidential).

Name of the respondent:
Qualification: Designation:
E-mail: Experience: _______ yrs.
Mobile No.:
Industry name & address:

Critical success factors and performance measurement factors

Based on the analysis of responses, eight CSFs and six PMFs were derived. More focus on CSFs (also called as input factors, TQM factors or significant factors) will lead to the improvement in PMFs (also called as output factors). Detailed descriptions of factors are –

(A) Critical success factors

(1) *Process management* includes all those attributes that suggest the need of process management in sponge iron industry, such as involvement of all departments in quality improvement processes; regular measurement and monitoring of quality cost; consideration of employee's suggestions/feedback; robust feedback mechanism; regular process auditing by all employees; cross-functional team participation; collective responsibility; training programs for quality techniques; advanced statistical techniques training programs; continuous process monitoring and improvement; data-driven process improvement; faster adoption of new technology for continuous improvement; promoting R&D and innovation for product development; promoting automation in production; conducive environment for implementation of QMS; and motivation to the employees for implementing QMS etc.

(2) *Top management commitment* signifies an important role of top management's clarity of vision, mission, and strategic direction; top management's ability to take responsibility for continuous quality improvement; adequate budgetary allocation for quality improvement initiatives; recognition of constructive efforts of employees by management; management's commitment for providing good work environment; fair compensation package to employees; and congenial work environment

(3) *Employee involvement* has loading of regular mapping and monitoring of employee's skills and competencies; capacity building of employees; use of bench marking for process improvement; encouraging innovation; good coordination and cooperation; and measurement of employee commitment levels.

(4) *Organization culture* accounts for team spirit; well-defined organizational structure and governance system; well-defined rules, regulations and operating procedures.

(5) *Effective networking* is based on the important role of honest and two-way communication between management and employees; fulfillment of legal, ethical, and social responsibilities by the company; establishment of management information system; providing technical guidance to suppliers; selection of suppliers on the basis of raw material quality; regular monitoring of customer satisfaction; standardization of customer care systems; customer relationship management.

(6) *Quality assurance* is an indicator of use of statistical process control for process management; integration of prevention and correction in production process.

(7) *Competitive assessment* is based on the attributes such as goal setting on the basis of current performance; finalizing strategy after SWOT analysis; supplier relationship management; internal supply chain.
(8) *Training* includes communication skill improvement training programs; leadership development training programs.

(B) Performance measurement factors

(1) *Productivity improvement*: has loading of high capital productivity; high productivity of each employee; high productivity of assets; equipment availability for longer duration; optimum human resource utilization; lower product development time; low machine line downtime; low supplier lead time; etc.
(2) *Business result*: includes high return on investment; top-line growth; sustainable increase in market share; bottom line growth; high processes efficiency; less variation in process; higher process accuracy rate.
(3) *Employee satisfaction*: accounts for higher value addition per employee; high job satisfaction of employees; employee involvement; better environmental compliance; better safety of employees; positive change in corporate culture; better quality of work life.
(4) *Time dimension*: signify an important role of reduced material procurement time; low-order process time; low process through time; high inventory turnover (material productivity); high flexibility in operations / system; etc.
(5) *Customer's satisfaction*: is based on customer retention; low customer complaints; higher customer satisfaction (value for the money spent); etc.
(6) *Quality performance*: accounts for less number of defects in the finished product; less rework and waste; optimum inventory utilization; low cost of quality; etc.

Relationship between critical success factors and performance measures

Based on your experience, please rate the importance of following elements in achieving successful implementation of TQM model in sponge iron industry.

S. No.	Attribute/variables	Very low IMP 1	Low IMP 2	Med IMP 3	High IMP 4	Very high IMP 5
1.	Process management leads to improvement in productivity	☐	☐	☐	☐	☐
2.	Process management leads to improvement in business results	☐	☐	☐	☐	☐
3.	Process management leads to improvement in employee satisfaction	☐	☐	☐	☐	☐
4.	Process management leads to improvement in time dimension	☐	☐	☐	☐	☐
5.	Process management leads to improvement in customer's satisfaction	☐	☐	☐	☐	☐
6.	Process management leads to improvement in quality performance	☐	☐	☐	☐	☐
7.	Top management commitment leads to improvement in productivity	☐	☐	☐	☐	☐

(*Continued*)

Table (*Continued*).

S. No.	Attribute/variables	Very low IMP 1	Low IMP 2	Med IMP 3	High IMP 4	Very high IMP 5
8.	Top management commitment leads to improvement in business results	☐	☐	☐	☐	☐
9.	Top management commitment leads to improvement in employee satisfaction	☐	☐	☐	☐	☐
10.	Top management commitment leads to improvement in time dimension	☐	☐	☐	☐	☐
11.	Top Management Commitment leads to improvement in Customer's Satisfaction	☐	☐	☐	☐	☐
12.	Top management commitment leads to improvement in quality performance	☐	☐	☐	☐	☐
13.	Employees involvement leads to improvement in productivity	☐	☐	☐	☐	☐
14.	Employee involvement leads to improvement in business results	☐	☐	☐	☐	☐
15.	Employee involvement leads to improvement in employee satisfaction	☐	☐	☐	☐	☐
16.	Employee involvement leads to improvement in time dimension	☐	☐	☐	☐	☐
17.	Employee involvement leads to improvement in customer's satisfaction	☐	☐	☐	☐	☐
18.	Employee involvement leads to improvement in quality performance	☐	☐	☐	☐	☐
19.	Organization culture leads to improvement in productivity	☐	☐	☐	☐	☐
20.	Organization culture leads to improvement in business results	☐	☐	☐	☐	☐
21.	Organization culture leads to improvement in employee satisfaction	☐	☐	☐	☐	☐
22.	Organization culture leads to improvement in time dimension	☐	☐	☐	☐	☐
23.	Organization culture leads to improvement in customer's satisfaction	☐	☐	☐	☐	☐
24.	Organization culture leads to improvement in quality performance	☐	☐	☐	☐	☐
25.	Effective networking leads to improvement in productivity	☐	☐	☐	☐	☐
26.	Effective networking leads to improvement in business results	☐	☐	☐	☐	☐
27.	Effective networking leads to improvement in employee satisfaction	☐	☐	☐	☐	☐
28.	Effective networking leads to improvement in time dimension	☐	☐	☐	☐	☐
29.	Effective networking leads to improvement in customer's satisfaction	☐	☐	☐	☐	☐
30.	Effective networking leads to improvement in quality performance	☐	☐	☐	☐	☐
31.	Quality assurance leads to improvement in productivity	☐	☐	☐	☐	☐
32.	Quality assurance leads to improvement in business results	☐	☐	☐	☐	☐

(*Continued*)

Table (*Continued*).

S. No.	Attribute/variables	Very low IMP 1	Low IMP 2	Med IMP 3	High IMP 4	Very high IMP 5
33.	Quality assurance leads to improvement in employee satisfaction	☐	☐	☐	☐	☐
34.	Quality assurance leads to improvement in time dimension	☐	☐	☐	☐	☐
35.	Quality assurance leads to improvement in customer's satisfaction	☐	☐	☐	☐	☐
36.	Quality assurance leads to improvement in quality performance	☐	☐	☐	☐	☐
37.	Competitive assessment leads to improvement in productivity	☐	☐	☐	☐	☐
38.	Competitive assessment leads to improvement in business results	☐	☐	☐	☐	☐
39.	Competitive assessment leads to improvement in employee satisfaction	☐	☐	☐	☐	☐
40.	Competitive assessment leads to improvement in time dimension	☐	☐	☐	☐	☐
41.	Competitive assessment leads to improvement in customer's satisfaction	☐	☐	☐	☐	☐
42.	Competitive assessment leads to improvement in quality performance	☐	☐	☐	☐	☐
43.	Training leads to improvement in productivity	☐	☐	☐	☐	☐
44.	Training leads to improvement in business results	☐	☐	☐	☐	☐
45.	Training leads to improvement in employee satisfaction	☐	☐	☐	☐	☐
46.	Training leads to improvement in time dimension	☐	☐	☐	☐	☐
47.	Training leads to improvement in customer's satisfaction	☐	☐	☐	☐	☐
48.	Training leads to improvement in quality performance	☐	☐	☐	☐	☐

1. In your opinion, what are the areas in Sponge Iron Industry which require immediate attention for quality & productivity improvement?

Ans:- ____

____

2. What solutions you suggest for improvement in quality & productivity in Sponge Iron Industry ?

Ans:- ____

____

3. Kindly suggest, Critical Factors, that you feel are important, but not covered in my study.

Ans:-____

____

4. How important is Industry – Institute partnership for growth of both ?

Ans :-____

____

5. Any other aspects / issues, you would like to mention for improvement / excellence of Sponge Iron Industry.

Ans:- ____

____

Permissions

All chapters in this book were first published in PMR, by Taylor & Francis Online; hereby published with permission under the Creative Commons Attribution License or equivalent. Every chapter published in this book has been scrutinized by our experts. Their significance has been extensively debated. The topics covered herein carry significant findings which will fuel the growth of the discipline. They may even be implemented as practical applications or may be referred to as a beginning point for another development.

The contributors of this book come from diverse backgrounds, making this book a truly international effort. This book will bring forth new frontiers with its revolutionizing research information and detailed analysis of the nascent developments around the world.

We would like to thank all the contributing authors for lending their expertise to make the book truly unique. They have played a crucial role in the development of this book. Without their invaluable contributions this book wouldn't have been possible. They have made vital efforts to compile up to date information on the varied aspects of this subject to make this book a valuable addition to the collection of many professionals and students.

This book was conceptualized with the vision of imparting up-to-date information and advanced data in this field. To ensure the same, a matchless editorial board was set up. Every individual on the board went through rigorous rounds of assessment to prove their worth. After which they invested a large part of their time researching and compiling the most relevant data for our readers.

The editorial board has been involved in producing this book since its inception. They have spent rigorous hours researching and exploring the diverse topics which have resulted in the successful publishing of this book. They have passed on their knowledge of decades through this book. To expedite this challenging task, the publisher supported the team at every step. A small team of assistant editors was also appointed to further simplify the editing procedure and attain best results for the readers.

Apart from the editorial board, the designing team has also invested a significant amount of their time in understanding the subject and creating the most relevant covers. They scrutinized every image to scout for the most suitable representation of the subject and create an appropriate cover for the book.

The publishing team has been an ardent support to the editorial, designing and production team. Their endless efforts to recruit the best for this project, has resulted in the accomplishment of this book. They are a veteran in the field of academics and their pool of knowledge is as vast as their experience in printing. Their expertise and guidance has proved useful at every step. Their uncompromising quality standards have made this book an exceptional effort. Their encouragement from time to time has been an inspiration for everyone.

The publisher and the editorial board hope that this book will prove to be a valuable piece of knowledge for researchers, students, practitioners and scholars across the globe.

List of Contributors

M. Tucci, F. De Carlo, O. Borgia and N. Fanciullacci
IBIS Lab, Department of Industrial Engineering, University of Florence, Viale Morgagni 40, Florence 50134, Italy

Rastee D. Koyee, Uwe Heisel and Rocco Eisseler
aInstitute for Machine Tools, University of Stuttgart, Holzgartenstr. 17, D-70174 Stuttgart, Germany

Siegfried Schmauder
IMWF, University of Stuttgart, Pfaffenwaldring 32, D-70569 Stuttgart, Germany

Morteza Zangeneh, Asadolah Akram, and Alireza Keyhani
Department of Agricultural Machinery Engineering, Faculty of Agricultural Engineering and Technology, School of Agriculture & Natural Resources, University of Tehran, Karaj, Iran

Peter Nielsen
Department of Mechanical and Manufacturing Engineering, Aalborg University, Aalborg, Denmark

Joseph Achebo
Department of Production Engineering, University of Benin, Benin City, Nigeria

Joakim Wikner and Eva Johansson
Department of Industrial Engineering and Management, School of Engineering, Jönköping University, Jönköping, Sweden

Fabian G. Beck, Eric H. Grosse and Ruben Teßmann
Carlo and Karin Giersch Endowed Chair "Business Management: Industrial Management", Technische Universität Darmstadt, Hochschulstr. 1, 64289 Darmstadt, Germany

Joakim Wikner
Department of Management and Engineering, Linköping University, Linköping, Sweden
School of Engineering, Jönköping University, Jönköping, Sweden

Amjad Hussain and Aftab Ahmad
Department of Industrial and Manufacturing Engineering, University of Engineering and Technology, Lahore 54890, Pakistan

Keith Case
Mechanical and Manufacturing Engineering, Loughborough University, Loughborough, Leicestershire LE11 3TU, UK

Vinod S. Gorantiwar and R.L. Shrivastava
Department of Mechanical Engineering, Yeshwantrao Chavan College of Engineering, Nagpur 441 110, India

Index

A

Accelerated Life Test, 1, 15-16

Agricultural Advisers, 67

Agricultural Mechanization Service, 66

Agricultural Service Centers, 65, 67

Although Infrared, 21

Aluminum Alloy 60-61, 90-92

Analytic Hierarchy Process, 69, 88

Analytical Hierarchy Process, 65

Artificial Neural Network, 38

Assemble-to-order, 113-115, 123, 163, 188

C

Capacity Buffers, 141, 150

Chip-breaker, 49

Competitive Market, 1

Compounded Decision Domains, 143, 156-158, 162, 175-176, 184

Corporate Social Responsibility, 189, 193-194, 202

Cost Issues, 106

Cost Reduction, 1

Critical Success Factors, 210, 213, 219, 226-227

Customer Order Decoupling Point, 110, 122-123, 142, 166-167, 186-188

Customer Order Triggers, 117

Customer Requirements, 106, 143, 159, 171

Customer Satisfaction, 1, 7, 189, 195-196, 198, 205, 213, 215-216, 218, 221-224, 226-227

Customer-order-based Management, 140, 142, 174-175, 177, 180-181, 184

Cutting Forces, 17, 19-20, 32-34, 44, 49-50, 52-53, 57

D

D:s-relation, 163, 166-167

Decision Categories, 140, 142-144, 147-148, 154, 156, 158, 161-162, 164-165, 174-175, 177-178, 180, 183-185

Decision-making Techniques, 68

Decoupling Points, 106-108, 112, 116-119, 122-123, 140-145, 149-151, 153, 157-158, 161, 164, 166, 174-177, 179-186, 188

Decoupling Zones, 123, 140, 143, 151-152, 156-158, 161, 164, 170, 174-186

Delphi Survey Process, 71

Demand-based Differentiation, 171

Dependent Supply, 109-110, 118

Drivers And Barriers, 189-193, 196-197, 200, 202

Ductility, 90-91, 102, 104

Duplex Stainless Steel, 17-18, 62-63

Dynamic Lot-sizing, 124-127, 129, 136, 138

E

Economic Order Quantity, 124

Economic Profits, 66

Elastic-viscoplastic, 17

Electrode Diameter, 95, 99-100, 103

Environmental Protection, 66

F

Fahp Model Development, 77

Final Inspection, 1, 4

Financial Resources, 67

Firefly Algorithm (FA), 40

Firefly Algorithm Neural Network System, 17, 19, 38, 43

Flow Driving, 140, 143, 148-149, 165-166, 168, 171, 175-176

G

Gas Metal Arc Welding, 91, 105

H

Heat Affected Zone, 101

Heuristics, 69, 124-127, 133, 136-138

I

Inclusive Design, 189-209

Input Supply Service, 66

Inventory Classification, 106-107, 116, 122-123, 188

Inventory Classifications, 107, 118-119

Inventory Costs, 124

J
Jmatpro, 17-18, 27-29, 61

L
Least Unit Cost, 125, 128
Linear Holding Costs, 125
Logistics System, 106, 109-110, 113, 115-116

M
Machinery Service Availability, 65
Machining Analytical, 17
Machining Performance Index, 17, 57
Make-to-order, 110, 113-115, 163, 166
Make-to-stock, 110, 113, 163, 166, 187
Manufacturer, 7-8, 12-14, 225
Manufacturing Industry, 17, 124
Market Responsiveness, 140
Markets Conditions, 65
Material Requirements Planning, 108, 123, 125, 137-138, 141
Materials Management, 106, 108, 111, 120-123, 140-141, 188
Mechanical Properties, 20, 29-30, 52, 61, 90-91, 98, 102-105
Mechanization Services, 65-67, 86
Metal Cutting Operations, 18
Multi-choice Goal Programming, 65, 69, 88
Multi-layer Perceptron, 38-39

N
Numerical Machining Performance Measure, 19, 57

O
Objectives, 2, 65-66, 68-71, 73-75, 77-78, 84-87, 143, 197, 199, 208
Operations Strategy, 140, 186, 188
Oscillating System, 7, 9

P
Performance Measurement Factors, 210, 213, 226-227
Perishable Product, 74
Planning And Control, 106-107, 114, 119-123, 145, 173-175, 182
Postponement, 106, 140, 142, 144-145, 159, 167, 183, 186-188
Process Parameters, 90
Product Development, 1, 6, 15-16, 219, 226-227
Product Launch, 1
Product Quality, 1
Product Reliability, 1-3, 6-7
Product Transportation, 74
Production Lead Time, 124
Production Stage, 3-4, 14
Purchase Order Decoupling Point, 111, 171-172

R
Raw Material, 121, 124, 213-214, 217, 225-226
Regression Analysis, 90-92, 94-95, 104-105, 214-216
Reliability Performance, 1-5, 7, 11-13, 15

S
Serrated Chip Geometries, 23
Signal-to-noise, 31
Sponge Iron, 210-212, 224-227, 230
Statistical Quality Control, 4
Suppliers Changes, 7
Supply Chains, 65, 88, 123, 145, 187-188
Supply-based Decoupling Point, 111
Supply-based Differentiation, 170-171

T
Tactical Process, 118
Tensile Properties, 90-92, 94-95, 97-104
Thermal Processes, 18
Thermo-graphic Camera, 21
Tool Orientation, 17, 19, 50, 53

U
Ultimate Tensile Strength, 90, 104-105

V
Vikor Method, 17, 19, 35, 38, 60

W
Wagner-whitin, 125, 128, 138
Warranty Analysis, 7
Welding Process, 91-92, 99, 101-105